INTERNETWORKING LANs AND WANs
Concepts, Techniques and Methods

Internetworking is one of the fastest growing markets in the field of computer communications. However, the interconnection of LANs and WANs tends to cause significant technological and administrative difficulties. This book provides valuable guidance, enabling the reader to avoid the pitfalls and achieve successful connection.
1993 0 471 93568 9

THE MULTIPLEXER REFERENCE MANUAL

Designed to provide the reader with a detailed insight into the operation, utilization and networking of six distinct types of multiplexers, this book will appeal to practising electrical, electronic and communications engineers, students in electronics, network analysts and designers.
1993 0 471 93484 4

PRACTICAL NETWORK DESIGN TECHNIQUES

Many network design problems are addressed and solved in this informative volume. Gil Held confronts a range of issues including through-put problems, line facilities, economic trade-offs and multiplexers. Readers are also shown how to determine the numbers of ports, dial-in lines and channels to install on communications equipment in order to provide a defined level of service
1991 0 471 93007 5 (Book)
**   0 471 92942 5 (Disk)**
**   0 471 92938 7 (Set)**

NETWORK MANAGEMENT
Techniques, Tools and Systems

Techniques, tools and systems form the basis of network management. Exploring and evaluating these three key areas, this book shows the reader how to operate an effective network.
1992 0 471 92781 3

Please refer to the inside-back cover for further details

DATA AND IMAGE
COMPRESSION

Fourth Edition

DATA AND IMAGE COMPRESSION

Tools and Techniques

Fourth Edition

Gilbert Held

4-Degree Consulting,
Macon, Georgia, USA

and

Thomas R. Marshall

(Software Author, Second
and Third Editions)

JOHN WILEY & SONS LTD
Chichester · New York · Brisbane · Toronto · Singapore

Copyright © 1996 by John Wiley & Sons Ltd.
Baffins Lane, Chichester
West Sussex PO19 1UD, England

Other Wiley Editorial Offices

John Wiley & Sons, Inc., 605 Third Avenue,
New York, NY 10158–0012, USA

Jacaranda Wiley Ltd, G.P.O. Box 859, Brisbane,
Queensland 4001, Australia

John Wiley & Sons (Canada) Ltd, 22 Worcester Road,
Rexdale, Ontario M9W 1L1, Canada

John Wiley & Sons (SEA) Pte Ltd, 2 Clementi Loop #02-01,
Jin Xing Distripark, Singapore 0512

Library of Congress Cataloging-in-Publication Data:

Held, Gilbert, 1943–
    Data and image compression : tools and techniques / Gilbert Held
    and Thomas R. Marshall — 4th ed.
        p.   cm.
    Previous eds. pub. under title: Data compression.
    Includes bibliographical references and index.
    ISBN 0 471 952478 (alk. paper)
    1. Data compression (Computer science)   2. Image processing-
Digital techniques.   I. Marshall, Thomas (Thomas R.)   II. Held,
Gilbert, 1943– Data compression.   III. Title.
    QA76.9.D33H473   1996
    005.74′6—dc20
                                                    95-41961
                                                    CIP

British Library Cataloguing in Publication Data

A catalogue record for this book is available from the British Library
ISBN 0 471 95247 8

Typeset by Photo·graphics, Honiton, Devon
Printed in Great Britain by Bookcraft (Bath) Ltd

CONTENTS

PREFACE

Since the publication of the third edition of this book the field of compression has considerably expanded. Today it is difficult, if not impossible, to work without encountering the effect of compression. If you use a CCITT compatible V.42 bis modem, operate a fax, transmit data via a remote bridge or router, view a graphics image or load a newly purchased software program onto your personal computer you will directly or indirectly encounter the effect of compression. From a niche market just a few years ago, compression-performing products now represent a large and rapidly expanding product category. Accompanying this growth in compression-performing products was the development of new compression-performing techniques, several of which represent modifications to algorithms developed during the late 1970s and early 1980s which considerably enhance the efficiency and effect of compression.

Based upon the increased use of images in computer applications, this new edition was expanded to cover image compression techniques. Other significant additions to this new edition include the recoding of previously covered compression algorithms as well as new algorithms in the C++ programming language, a considerable expansion of coverage of Lempel–Ziv string compression and its derivatives, and an expansion of information concerning file archiving programs to include archive header formats.

Although the goal of compression is to reduce the quantity of data necessary for storage or transmission, this new edition may appear as a paradox to readers familiar with prior editions. Instead of getting smaller, each edition of this book has grown in size and this fourth edition is no exception to this trend. With an expansion of coverage to images this book is now better oriented to the requirements of personal computer users, who commonly work with both text and images.

A second significant change you will find in this fourth edition reflects the new subtitle of this book. In addition to describing and discussing compression-performing techniques, this book provides you with a large number of compression tools. Some tools are represented by shareware programs, while other tools are represented by programs that perform different compression algorithms. Instead of a convenience diskette obtainable by mail order, this new edition includes a diskette of compression performing tools.

As a professional author I welcome readers' comments. Please feel free to write me at my publisher's address in this book. Your thoughts concerning the current contents of this new edition as well as any suggestions you may wish to provide concerning a future fifth edition are most welcome.

Gilbert Held
Macon, Georgia

ACKNOWLEDGEMENTS

The author wishes to thank *Telecommunications* for permission received to reprint the tables and figures from the article 'Data Compression in High-Speed Digital Facsimile' which appeared in their June 1977 issue. The author is indebted to Mr Thomas R. Marshall who rightfully must be considered to be the software author of the second and third editions of this book for his cooperation and assistance in developing the encoding and decoding programming examples as well as the BASIC and FORTRAN versions of the DATANALYSIS computer program listed in Appendix B.

In developing a series of updated new editions to this book a large number of persons require recognition. The first three editions of this book resulted in Mrs Carol Ferrell performing magic by converting this author's notes into a professional manuscript. Continuing this magic, I am indebted to Mrs Linda Hayes for converting this author's fourth edition effort into a professional manuscript that resulted in the book you are reading. I would also like to take the opportunity to thank Christine Washburn of Microcom for providing information on the different levels of MNP data compression, Kimberly Bromley of InfoTel for providing information on that firm's INFOPAK Data Compression software products, Irene M. Henderson of System Enhancement Associates for providing information on that firm's ARC archive utility that compresses data as well as permitting the program to be added to the disk accompanying this book, and a similar 'thank you' to Mr Phil Katz of PKWARE for supplying information on his firm's file compression program as well as granting permission for its distribution.

Other persons who warrant a loud 'thank you' include Nico de Vries for permission to include UltraCompressor II revision 2 created and maintained by AIP-NL on the enclosed diskette, Robert K. Jung for permission to distribute ARJ on the enclosed diskette, Phillip G. Gage for permission to describe his BPE code and include his SIXPACK2.C program on the enclosed diskette, and the *C/C++ Users Journal* for permission to excerpt text and figures of Mr Gage's article on BPE that appeared in that journal.

In addition to the previously mentioned, there are a number of persons who indirectly contributed to this book by developing compression-performing programs that they graciously placed in the public domain. At appropriate locations in this book the efforts of those persons are acknowledged.

1

RATIONALE AND UTILIZATION

In the chronology of computer development, large-scale information transfer by remote computing and the development of massive information storage and retrieval systems have witnessed a tremendous growth. Concurrent with this growth, several problem areas have developed which can result in major, but unnecessary, economic expenditures.

One problem is the so-called 'run-away database'. Here the size of the database used by an organization for its information storage and retrieval programs becomes larger and larger, requiring additional disk drives for online systems and reels of magnetic tapes for those systems that can be processed in a batch environment.

Another problem from a data storage perspective is the growing use of images in computer applications. A screen of text consisting of 80 columns by 25 rows can contain a maximum of 2000 characters. In comparison, consider a VGA 640 by 480 pixel 16-color image displayed on a personal computer screen. The image would require 640×480 bits of storage without considering the fact that four bits are required to represent the color of each pixel. Thus, the image would require 153 600 bytes of storage, or approximately 77 times the storage required to represent a full screen of text. While the old adage 'one picture is worth a thousand words' is usually true, another newer adage is 'images eat storage'.

Accompanying the growth in the size of databases and the use of images has been a large increase in the number of users and duration of usage by personnel at remote locations. These factors result in tremendous amounts of data being transferred between computers and remote terminals. To provide transmission facilities for the required data transfers, communications lines and auxiliary devices, such as modems and multiplexers, have been continuously upgraded by many organizations to permit higher data transfer capability.

Although the obvious solutions to these problems of data storage and information transfer are to install additional storage devices and expand existing communications facilities, to do so requires an additional increase in an organization's equipment and operating costs. One method that can be employed to alleviate a portion of data storage and information transfer problems is through the representation of data by more efficient codes. If you examine an organization's database or monitor a transmission line, there is an excellent chance that the individual characters that both make up the database and the transmission sequence could be encoded more effectively. Two techniques that can result in a more effective encoded data representation are logical and physical data compression.

1.1 LOGICAL COMPRESSION

When a database is designed, one of the first steps of the analyst is to obtain as much data reduction as possible. This data reduction results from the elimination of redundant fields of information while representing the data elements in the remaining fields with as few logical indicators as is feasible.

Although logical compression is data dependent and the method employed can vary based upon the analyst's foresight, the following two examples will illustrate the ease of implementation and benefits of this compression technique.

One simple example of logical data compression is the occupational field on a personnel database. Suppose 30 alphanumeric positions are allocated to this field. If the field is fixed, occupations such as the 10-character occupational description 'DISHWASHER' have 20 blanks inserted into the remainder of the field. Then, 30 million characters of storage would be required for the occupational field of 1 million workers. Suppose at most there were 32 768 distinct occupations. Instead of indicating the occupational title, you could encode the equivalent 5-digit data code, eliminating 25 character positions per field. The size of the field could be reduced further by allocating the binary value of 1 or more characters to the occupational code. As an example, an 8-bit character could represent $2^8 - 1$ or 255 distinct values or occupational codes. Linking two 8-bit characters through appropriate software would provide $2^{16} - 1$ or 65 535 distinct codes. This would reduce the field size from 30 to 2 characters, saving 28 million characters of storage. If our counting begins at zero instead of a conventional starting place of 1, an 8-bit character could represent 256 codes, while a 16-bit character could be employed to represent 65 536 distinct values.

A second example of logical compression is a date field. This type of field frequently occurs in databases. Normally, the numeric equivalents of the subfields representing day, month and year are used in place of longhand notation. Thus, 01 04 81 would represent 1 April 1981. While this logical compression results in six numeric characters of storage, additional data reduction can result from storing the date as a binary value. Since the day will never exceed 31, five bits would suffice to represent the data field. Similarly, four bits could be used to represent the month value, while seven bits could represent 127 years, permitting a relative year ranging from 1900 to 2027.

Logical compression using numerical and binary representation is illustrated in Figure 1.1 for the preceding date field example. It is interesting to note that employing binary representation reduces the date field to 16 binary digits or two 8-bit concatenated characters of storage. As discussed, many logical compression methods can be considered by an analyst during the database design process. Each method may result in a distinct degree of data storage reduction. Correspondingly, when logically compressed databases or portions of such databases are transmitted between locations, transmission time is reduced since fewer data characters are transmitted.

While logical compression can be an effective tool in minimizing the size of a database, it only reduces transmission time when logically compressed data is transmitted. Thus, the transmission of inquiry and response data, which are typically encoded as separate and distinct entities in the appropriate bit representation of the code for each character, is not normally affected. Similarly, the occurrence of repeating patterns and groups of characters, which are normally contained in reports transmitted between computers or from computer systems to terminal devices, would not be affected. For such situations, a reduction in data transmission time depends upon the physical compression of the data as it is encountered.

1.2 PHYSICAL COMPRESSION

Physical compression can be viewed as the process of reducing the quantity of data prior to its entering a transmission medium and the

Longhand	DAY	MONTH	YEAR
Example	1	APRIL	1981
Logical compression using numerical representation			
Example	01	04	81
Logical compression using binary representation			
Example	00001	0100	1010001

Figure 1.1 Logical compression methods. Logical compression can result from alphanumeric, numeric or binary representation of data in a shorthand notation

expansion of such data into its original format upon receipt at a distant location. Although both physical and logical compression can result in reduced transmission time, distinct application differences exist between the two techniques. Logical compression is normally used to represent databases more efficiently and does not consider the frequency of occurrence of characters or groups of characters.

Physical compression takes advantage of the fact that when data are encoded as separate and distinct entities, the probabilities of occurrence of characters and groups of characters differ. Since frequently occurring characters are encoded into as many bits as those characters that only rarely occur, data reduction becomes possible by encoding frequently occurring characters into short bit codes while representing infrequently occurring characters by longer bit codes. Like logical compression, many physical compression techniques exist. Some techniques replace repeating strings of characters by a special compression-indicator character and a quantity-count character. Other techniques replace frequently occurring characters or strings of characters with a short binary code, while infrequently encountered characters and strings are replaced by longer binary codes. In Chapter 3, a number of distinct physical compression methods are covered in detail. For the remainder of this book we will focus our attention upon physical data compression.

1.3 COMPRESSION BENEFITS

When data compression is used to reduce storage requirements, overall program execution time may be reduced. This is because the reduction in storage will result in a reduction of disk-access attempts, while the encoding and decoding required by the compression technique employed will result in additional program instructions being executed. Since the execution time of a group of program instructions is normally significantly less than the time required to access and transfer data to a peripheral device, overall program execution time may be reduced.

With respect to the transmission of data, compression provides the network planner with several benefits in addition to the potential cost savings associated with sending less data over the switched telephone network where the cost of the call is usually based upon its duration. First, compression can reduce the probability of transmission errors occurring, since fewer characters are transmitted when data is compressed while the probability of an error occurring remains constant. Second, since compression increases efficiency, it may reduce or even eliminate extra workshifts. Finally, by converting text that is represented by a conventional code such as standard ASCII into a different code, compression algorithms may provide a level of security against illicit monitoring.

For data communications, the transfer of compressed data over a medium results in an increase in the effective rate of information transfer, even though the actual data transfer rate expressed in bits per second remains the same. Data compression can be implemented on most existing hardware by software or through the use of a special hardware device that incorporates one or more compression techniques.

In Figure 1.2, a basic data-compression block diagram is illustrated. Shown as a black box, compression and decompression may occur within a processor built into a personal computer, an intelligent terminal or a device foreign to the processor, such as a specialized communications component. Foremost among these components were data concentrators and statistical multiplexers during the 1970s and early 1980s. From approximately the mid-1980s, a revolution in the utilization of data-compression-performing products has occurred in the areas of switched network modems, remote bridges, routers, personal computers and mainframe disk storage. Literally hundreds of switched network modem, bridge and router manufacturers have incorporated a variety of data-compression algorithms into their products, while numerous software developers marketed programs which increase the storage capacity of disks by compressing data prior to storage.

To examine in some detail a portion of the benefits that may result from the employment of one or more compression techniques requires a review of some fundamental compression terminology.

1.4 TERMINOLOGY

A large vocabulary of terminology was developed and continues to be developed to reference compression-related technology. In this introductory chapter, we will limit our examination of compression-related technology to two general areas that are relevant to the remainder of this book—compression efficiency and compression methods. Thereafter, throughout this book we will describe and discuss compression-related terminology as it relates to technology covered at appropriate points in this book.

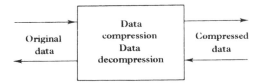

Figure 1.2 Basic data-compression block diagram. An original data stream operated upon according to one or more compression algorithms results in the generation of a compressed data stream

Compression efficiency

As illustrated in Figure 1.2, an original data stream is operated upon according to a particular algorithm to produce a compressed data stream. This compression of the original data stream is sometimes referred to as an encoding process with the result that the compressed data stream is also called an encoded data stream. Reversing the process, the compressed data stream is decompressed to reproduce the original data stream. Since this decompression process results in the decoding of the compressed data stream, the result is sometimes referred to as the decoded data stream. We will use the terms original data stream and decoded data stream synonymously, as well as the terms compressed data stream and encoded data stream.

The degree of data reduction obtained as a result of the compression process is known as the compression ratio. This ratio measures the quantity of compressed data in comparison with the quantity of original data, such that (Ruth and Kreutzer, 1972):

$$\text{Compression ratio} = \frac{\text{Length of original data string}}{\text{Length of compressed data string}}.$$

From the above equation, it is obvious that the higher the compression ratio the more effective the compression technique employed. Another term used when talking about compression is the figure of merit, where:

$$\text{Figure of merit} = \frac{\text{Length of compressed data string}}{\text{Length of original data string}}.$$

The figure of merit is the reciprocal of the compression ratio and must always be less than unity for the compression process to be effective. The fraction of data reduction is one minus the figure of merit. Thus, a compression technique that results in one character of compressed data for every three characters in the original data stream would have a compression ratio of 3, a figure of merit of 0.33 and a fraction of data reduction of 0.66.

Pure random data, by definition, should not have any redundancy. While the compression ratio for this type of data should be unity, in many instances the improper design of a compression algorithm or its improper application may result in a degree of data expansion, resulting in a compression ratio falling below unity.

When data is compressed, the compression ratio will vary in proportion to the susceptibility of the data to the algorithm or algorithms used. Thus, you should focus your attention upon an average compression ratio and not a ratio achieved at a particular time. In general, good algorithms operating on text can be expected to achieve an average compression ratio of 2.0, while excellent algorithms based upon sophisticated processing techniques will achieve an average compression ratio exceeding 3.0.

Compression methods

Compression techniques can be classified into one of two general categories or methods—'lossless' or 'lossy'.

Lossless compression techniques are fully reproducible and are primarily restricted to data operations. After all, since data can consist of accounts receivable, payroll and health insurance claims, it is extremely important that once such data is compressed, its decompression should result in the exact reconstruction of the original data. Other common terms used to reference lossless compression include 'reversible' and 'non-destructive' compression.

Lossy compression techniques may or may not be fully reproducible and are primarily restricted to operations on images, video and audio. Although the result of decompression may not provide an exact duplication of the original data, the differences between the original and reconstructed data may be so minor as to be difficult to see or hear. Since the compression ratio obtainable by the use of lossy compression can significantly exceed the compression ratio obtainable from lossless compression, the primary tradeoff concerns your need for reproducibility versus your requirements for storage and transmission.

To illustrate the effect of lossy compression the well-known painting 'The Doni Madonna' by Michelangelo Buonarroti, which reflects his view of his statues, will be used. Figure 1.3 illustrates the printed version of a GIF file obtained by the author, with permission, from the CompuServe graphics library. The Graphics Interchange Format (GIF) is the copyright property of CompuServe Incorporated and GIF is a Service Mark property of CompuServe Incorporated. A GIF file is an image file stored using a lossless string compression technique. Thus, the decompression and subsequent printing of the image results in its exact reproduction. Unfortunately, its printing on a black-and-white laser printer does not do justice to the talents of the artist. The storage requirements for the GIF file illustrated in Figure 1.3 was 200 739 bytes. Later in this book we will examine GIF files in detail.

To illustrate the effect of lossy compression, the Independent JPEG Group's software program CJPEG was used to create a series of compressed files by varying the scale quantization tables maintained by the program to adjust image quality. JPEG is an acronym for Joint Photographic Experts Group, whose members work with various standards committees to develop a standard for color image com-

Figure 1.3 The Doni Madonna by Michelangelo printed from a GIF formats file

pression. Later in this book we will discuss the JPEG standard. Compressed JPEG files were created using quality levels of 75, 50, 25, 10 and 5, where 100 would represent the best quality and 0 the worst. As quality decreases, the size of the resulting file decreases as we will soon note. However, the reproducibility of the image also decreases as the quality level decreases.

The first use of the CJPEG program was with a quality level of 75. The resulting file required 48 958 bytes of storage and its resulting printout on a laser printer is illustrated in Figure 1.4. The display of Figure 1.4 as well as all GIF and JPEG files in this book were performed using PRINTGF/D from Ravitz Software, Inc. of Lexington, KY. In comparing the images of Figure 1.3 to Figure 1.4, the quantization process which reduced data storage requirements by approximately 150 000 bytes both lightened the image and removed a small degree of its original detail.

A second use of CJPEG was performed using a quality level of 50. The resulting file required 31 935 bytes of storage and its printout is

Figure 1.4 The Doni Madonna by Michelangelo printed from a JPEG file created using a quality scale of 75

illustrated in Figure 1.5. If you focus your attention upon the faces of the statues in Figures 1.4 and 1.5, you can notice a small degree of loss of clarity due to the decrease in the quality level of Figure 1.5 from Figure 1.4.

Additional operations involving the use of CJPEG were performed using quality levels of 25, 10 and 5. Figures 1.6 through 1.8 illustrate the resulting JPEG printed images. The use of a 25 quality level resulted in a file of 20 964 bytes, while a quality level of 10 reduced the size of the resulting file to 12 439 bytes. With a quality level of 25,

Figure 1.5 The Doni Madonna by Michelangelo printed from a JPEG file created using a quality scale of 50

Figure 1.6 The Doni Madonna by Michelangelo printed from a JPEG file created using a quality scale of 25

the reduction in image clarity becomes more noticeable, while at a quality level of 10 you can begin to see blocks of pixels which form one mechanism by which JPEG compresses images.

In Figure 1.8 the use of a quality level of 5, which reduced file storage requirements to 8876 bytes, expanded the size of blocks of pixels so they are fully noticeable throughout the image, resulting in a very large degree of image distortion. Table 1.1 summarizes the quality level and data storage requirements of the series of JPEG images illustrated in Figures 1.3 through 1.8.

Figure 1.7   The Doni Madonna by Michelangelo printed from a JPEG file created using a quality scale of 10

1.5 COMMUNICATIONS APPLICATIONS

To obtain an overview of some of the communications benefits available through the incorporation of data compression, we can consider a typical data communications application. As illustrated in the top portion of Figure 1.9, a remote computer or terminal is connected to a central computer with transmission occurring at a 19.2 kbps data rate. Let us assume that the data to be transmitted has not been compressed. If through the programming of one or more compression

Figure 1.8 The Doni Madonna by Michelangelo printed from a JPEG file created using a quality scale of 5

algorithms or the installation of a hardware compression device a compression ratio of 2 is obtained, several alternatives may be available with respect to your data communications methodology. First, your data transmission time is reduced, since the effective information transfer rate has increased to approximately 38.4 kbps as shown in the middle portion of Figure 1.9. Ignoring communications software overhead, the data transmission time is halved. Thus, you may now consider using the remote computer or terminal for other remote processing applications or perhaps an expensive after-hours shift or portion of such a shift can be alleviated.

Table 1.1 Quantization vs file storage (Figures 1.3 through 1.8)

Quality level	File storage
100	200 739
75	48 958
50	31 935
25	20 964
10	12 439
5	8 876

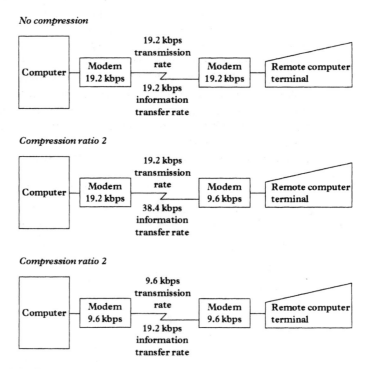

Figure 1.9 Data compression affects the information transfer ratio (ITR); through the use of data compression, the methodology and structure of a data communications facility may be changed

In the lower portion of Figure 1.9 another user option is illustrated. Here the transmission rate may be reduced to 9600 bps. With a compression ratio of 2, this is equivalent to an information transfer rate of 19 200 bps. By lowering the data transmission rate, more expensive 19 200 bps modems may be replaced by 9600 bps modems and line conditioning which is normally required when transmitting data at 19 200 bps may not be required, resulting in an additional cost reduction.

A second type of communications application that can benefit from the utilization of data compression is illustrated in Figure 1.10. A typical multidrop network is illustrated in the top portion of Figure 1.10, connecting terminals at diverse geographical locations via a common leased line to a computer site. Typically, the transmission activity of the terminals is the governing factor that limits the multidrop line to a maximum number of drops. In the bottom portion of Figure 1.10, it is assumed that compression performing modems were substituted for the conventional modems used in the original multidrop configuration. Since data compression on a multidrop line reduces the flow of data on the line, its utilization will normally enable additional drops to be added to the line prior to the occurrence of throughput delays that affect the response time of the terminals attached to each drop. In this particular example, it is assumed that the use of compression performing modems permitted an increase in the number of line-drops from 4 to 6.

For switched network modems, data compression provides the ability to obtain an information transfer capability at a fraction of the cost of higher-speed modems. This is due to the complex modulation schemes used by modems operating at 19 200 bps in comparison with less complex schemes used by 9600 bps modems. By incorporating a compression algorithm into a lower operating rate modem, it becomes

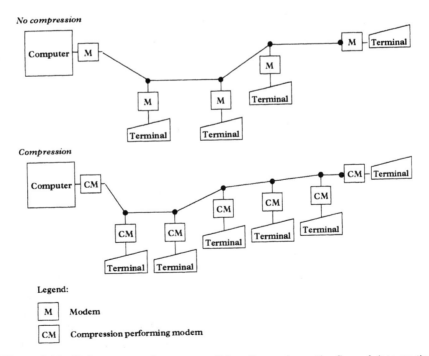

Figure 1.10 Data compression on a multidrop line reduces the flow of data on the line, permitting additional terminals to be serviced

possible to achieve a data throughput between 9600 and 19 200 bps for the cost of a read only memory (ROM) chip that might have a wholesale cost of $5. In comparison, the analog circuitry required by a modem to operate at 19 200 bps may have a wholesale cost of $50 or more over the circuitry required for a 9600 bps modem. Since most manufacturers must sell their products at two to three times their cost, the use of compression in lower-speed modems can translate into significant retail savings in comparison with the cost of modems operating at higher data rates. In addition, the incorporation of data compression into high-speed switched network modems provides users with an information transfer capability beyond the capacity afforded by current modem technology.

1.6 DATA STORAGE APPLICATIONS

To illustrate the applicability of data compression to data storage applications, let us examine how data is stored in a diskette. During the formatting process, a computer divides the recording space into tracks, which are concentric circles. Diskettes used with the original IBM PC and PC XT personal computers have 40 tracks, numbered from 0 at the outer edge to 39 at the inner edge. Each track is subdivided into sectors, with the number of sectors per track a function of the storage capacity of the diskette. More modern $3\frac{1}{2}$-inch diskettes introduced with the Apple Macintosh and IBM PS/2 series as well as DOS clones have 80 tracks, numbered from 0 at the outer edge to 79 at the inner edge.

Figure 1.11 illustrates the track and sector relationship for $5\frac{1}{4}$-inch diskettes used with the IBM PC and PC XT personal computers. Diskettes used with those computers have 40 tracks and eight or nine sectors per track, with the number of sectors per track based upon

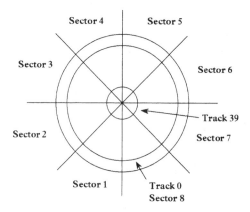

Figure 1.11 Track and sector relationship (note: diskettes can be formatted to hold either eight or nine sectors per track)

the FORMAT command parameter specified. Each sector can store up to 512 characters and represents the minimum amount of data that can be stored or retrieved with one disk read/write operation. If you write one character to a file you will use one sector of 512 characters, while saving a file consisting of 513 characters would require the use of two sectors.

The total storage capacity of a 5¼-inch double-sided diskette formatted to hold eight sectors per track is:

2 sides × 40 tracks/side × 8 sectors/track × 512 characters/sector = 327 680 characters.

Suppose you wish to download a file that contains 350 000 characters. Without a compression program you could not store this file on one diskette. In addition, if the file is not capable of being subdivided, such as a binary file, you might have to consider upgrading your computer's storage capacity by adding a hard disk or a higher-capacity diskette drive. As an alternative, you may be able to download a compressed version of the file. Today, many bulletin boards use public domain programs to archive files into a compressed format both to reduce downloading time as well as storage requirements. If we assume that a 2:1 compression ratio is achieved, the 350 000 character file would be compressed into 175 000 characters. Not only would the file fit on one diskette, but, in addition, it could be stored using 342 sectors. This would leave 298 sectors available for other purposes, such as storing a decompression program. Once downloaded you may be able to expand the program in memory or, if you have a hard disk, expand the program onto that storage facility. If you want to distribute the program to other personal computer users you could consider distributing the compressed version of the program. This would allow you to distribute the program on one diskette. In comparison, the distribution of the normal version of the file would require two diskettes if the file could be logically separated.

Detailed information concerning the operation and utilization of two programs developed to compress IBM PC and compatible personal computer files are presented later in this book. In addition, many of the programs discussed in this book are included on the diskette bundled with this book. Those programs can assist you in compressing and transmitting or distributing information on magnetic media.

1.7 OTHER APPLICATIONS

As previously mentioned in this book, the use of compression is pervasive. Other applications for compression include extending voice recordings stored on CD-ROM, fax transmission, video, and the relatively new term of multimedia used to represent the merging of text, images, audio and video.

Examples of audio compression include communications carriers that use a technique referred to as Adaptive Pulse Code Modulation (ADPCM) instead of PCM to encode digitized voice conversations on international circuits. Doing so reduces the data rate of PCM which requires 64 kbps to 32 kbps under ADPCM. Other audio compression techniques reduce the data rate required to transmit a digitized voice conversation to as low as 2.4 kbps; however, reproducible voice quality diminishes as the data rate decreases.

Although the most popular type of audio storage is the Compact Disk (CD), audio recordings on CD are not compressed and are limited to approximately one hour. The reason for this is the fact that the development of the CD player dates to the 1970s, a time when compression required expensive circuitry. As a result of this, the CD player was designed without compression. In comparison, multimedia applications stored on Compact Disk Read Only Memory (CD-ROM) are designed to work with compression-performing hardware or software that operates on personal computers. Doing so enables the storage of images, video and audio in compressed form.

Concerning video, movies as well as a teleconferencing session can be considered as a sequence of still images repeated so many frames per second. By noting redundancy between frames and eliminating that redundancy you can reduce the storage and transmission of video.

Today there are literally hundreds of compression techniques. Some techniques, such as run length compression, are applicable to voice, data, images and video. Other techniques, such as JPEG, are only applicable to images. Unfortunately, book size constraints as well as a requirement to prepare this book revision in a reasonable length of time requires our primary focus to be upon data and image compression techniques and tools.

1.8 DATA COMPRESSION AND INFORMATION TRANSFER

When data is transmitted between terminals, a terminal and a computer or two computers, several delay factors may be encountered which cumulatively affect the information transfer rate. Data transmitted over a transmission medium must be converted into an acceptable format for that medium. When digital data is transmitted over analog telephone lines, modems must be employed to convert the digital pulses of the business machine into a modulated signal acceptable for transmission on the analog telephone circuit. The time between the first bit entering the modem and the first modulated signal produced by the device is known as the modem's internal delay time. Since two such devices are required for a point-to-point circuit, the total internal delay time encountered during a transmission sequence equals twice the modem's internal delay time. Such times can range from a few to 10 or more milliseconds (ms). The second delay encountered on a circuit is a function of the distance between points and is known as the circuit or propagation delay time. This is the time required for the signal to be propagated or transferred down the line to the distant end. Propagation delay time can be approximated by equating 1 millisecond for every 150 circuit miles and adding 12 milliseconds to the total.

Once data is received at the distant end it must be acted upon, resulting in a processing delay which is a function of the computer or terminal employed as well as the quantity of transmitted data which must be acted upon. Processing delay time can range from a few milliseconds where a simple error check is performed to determine if the transmitted data was received correctly to many seconds where a search of a database must occur in response to a transmitted query.

Each time the direction of transmission changes in a typical half duplex protocol, control signals change at the associated modem to computer and modem to terminal interfaces. The time required to switch control signals to change the direction of transmission is known as line turnaround time and can result in delays up to 250 or more milliseconds, depending upon the transmission protocol employed. We can denote the effect of data compression by examining the transmission protocol commonly known as BISYNC communications and a few of its derivations.

BISYNC communications

One of the earliest commonly employed transmission protocols is the Binary Synchronous Communications (BISYNC) communications control structure. This line control structure was introduced in 1966 by International Business Machines Corporation and is used for transmission by many medium-speed and high-speed devices to include terminal and computer systems. BISYNC provides a set of rules which govern the synchronous transmission of binary-coded data. While this protocol can be used with a variety of transmission codes, it is limited to the half duplex transmission mode and requires the acknowledgement of the receipt of every block of transmitted data. In an evolutionary process, a number of synchronous protocols have been developed to supplement or serve as a replacement to BISYNC, the most prominent being the high-level data link control (HDLC) protocol defined by the International Standards Organization (ISO).

The key difference between BISYNC and HDLC protocols is that BISYNC is a half duplex, character-oriented transmission control structure while HDLC is a bit-oriented, full duplex transmission con-

trol structure. We can investigate the efficiency of these basic transmission control structures and the effect of data compression upon
their information transfer efficiency. To do so, an examination of some
typical error control procedures is first required.

Error control

The most commonly employed error-control procedure is known as
automatic request for repeat (ARQ). In this type of control procedure,
upon detection of an error, a request is made by the receiving station
to the sending station to retransmit the message. Two types of ARQ
procedures have been developed: 'stop and wait ARQ' and 'go back n
ARQ', which is sometimes called continuous ARQ.

'Stop and wait ARQ' is a simple type of error-control procedure.
Here the transmitting station stops at the end of each block and waits
for a reply from the receiving terminal pertaining to the block's accu-
racy (ACK) or error (NAK) prior to transmitting the next block. This
type of error-control procedure is illustrated in Figure 1.12. Here the
time between transmitted blocks is referred to as dead time which
acts to reduce the effective data rate on a transmission medium.
When the transmission mode is half duplex, a circuit must be turned
around twice for each block transmitted, once to receive the reply
(ACK or NAK) and once again to resume transmitting. These line turnarounds, as well as such factors as the propagation delay time,
station message processing time and the modem internal delay time,
all contribute to what is shown as the cumulative delay factors.

The ARQ model illustrated in Figure 1.12 is also applicable to local

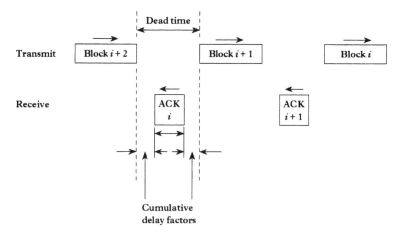

Figure 1.12 Stop and wait ARQ. In this type of error control procedure, the receiver
transmits an acknowledgement after each block. This can result in a significant
amount of cumulative delay time between data blocks

area networks. However, the internal delay time associated with modems is replaced by the delay time of LAN adapter cards. Transmission from a workstation or file server onto a LAN is not instantaneous. Instead, the adapter requires time to format data into the protocol format used to transport data on the LAN media. Similarly, a degree of delay results when frames are read from the LAN by the network adapter card and processed into a format the workstation or file server understands.

Unlike modems that have a near-uniform delay time, the delays associated with network adapters are not as symmetrical between input and output data flow. Normally, the delay associated with placing information on a LAN via a network adapter exceeds the delay in retrieving data from the network. Due to this, LAN models are slightly more difficult to develop; however, the principles illustrated in the remainder of this chapter are applicable to both WAN and LAN transmission as well as LAN to WAN transmission via a wide area network.

When the 'go back n ARQ' type of error control procedure is employed, the dead time can be substantially reduced to the point where it may be insignificant. One way to implement this type of error control procedure is by the utilization of a simultaneous reverse channel for acknowledgement signalling as illustrated in Figure 1.13. In this type of operating mode, the receiving station sends back the ACK or NAK response on the reverse channel for each transmitted block. If the primary channel operates at a much higher data rate than the reverse channel, many blocks may have been received prior to the transmitting station receiving the NAK in response to a block at the receiving station being found in error. The number of blocks one may go back to request a transmission, n, is a function of the block size and buffer area available in the business machines and terminals at the transmitting and receiving stations, the ratio of the data transfer rates of the primary and reverse channels and the processing time

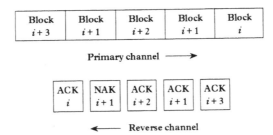

Figure 1.13 Go back n ARQ. In a 'go back n ARQ' error-control procedure, the transmitter continuously sends messages until the receiver detects an error. The receiver then transmits a negative acknowledgement (NAK) on the reverse channel and the transmitter retransmits the block received in error. Some versions of this technique require blocks sent before the error indication was encountered to be retransmitted in addition to the block received in error

required to compute the block check character and transmit an acknowledgement. For the latter, this time is shown as small gaps between the ACK and NAK blocks in Figure 1.13.

Half duplex throughout model

When a message block is transmitted in the BISYNC control structure, a number of control characters are contained in that block in addition to the message text. If the variable C is assigned to represent the number of control characters per block and the variable D is used to represent the number of data characters, then the total block length is $C + D$. If the data transfer rate expressed in bps is denoted as T_R and the number of bits per character is denoted as B_C, then the transmission time for one character is equal to B_C/T_R which can be denoted as T_C. Since $D + C$ characters are contained in a message block, the time required to transmit the block will become $T_C*(D + C)$. Once the block is received, it must be acknowledged. To do so, the receiving station is required to first compute a block check character (BCC) and compare it with the transmitted BCC character appended to the end of the transmitted block. Although the BCC character is computed as the data is received, a comparison is performed after the entire block is received and only then can an acknowledgement be transmitted. The time to check the transmitted and computed BCC characters and form and transmit the acknowledgement is known as the processing and acknowledgement time (T_{PA}).

When transmission is half duplex, the line turnaround time (T_L) required to reverse the transmission direction of the line must be added. Normally, this time includes the request-to-send/clear-to-send (RTS/CTS) modem delay time as well as each of the modems' internal delay time. For the acknowledgement to reach its destination, it must propagate down the circuit and this propagation delay time, denoted as T_p, must also be considered. If the acknowledgement message contains A characters then, when transmitted on the primary channel, $A*B_C/T_R$ seconds are required to send the acknowledgement.

Once the original transmitting station receives the acknowledgement it must determine if it is required to retransmit the previously sent message block. This time is similar to the processing and acknowledgement time previously discussed. To transmit either a new message block or repeat the previously sent message block, the line must be turned around again and the message block will require time to propagate down the line to the receiving station. Thus, the total time to transmit a message block and receive an acknowledgement, denoted as T_B, becomes:

$$T_B = T_C*(D + C) + 2* (T_{PA} + T_L + T_p) + (A*B_C/T_R). \qquad (1.1)$$

Since efficiency is the data transfer rate divided by the theoretical data transfer rate, the transmission control structure efficiency (E_{TCS}) becomes:

$$E_{TCS} = \frac{B_C * D * (1 - P)}{T_R * T_B}.$$

(1.2)

In equation (1.2) the numerator represents the number of actual data bits that are received correctly. Here B_C is the number of bits per character and D is the number of data characters in the block. Thus, $B_C * D$ represents the number of bits contained in the data characters which form a block. Since there is a probability, P, that one or more bits in the block will be in error, multiplying $B_C * D * (1 - P)$ results in the average number of bits per block not in error.

Focusing our attention upon the denominator, T_R represents the data transfer rate, while T_B represents the total time to transmit a message block and receive an acknowledgement. This results in the denominator containing the theoretical number of bits that could be transferred and received. Dividing the number of bits received correctly (numerator) by the theoretical number of bits that could be transferred (denominator) provides us with the efficiency of the transmission control structure.

Although the preceding is a measurement of the transmission control structure efficiency, it does not consider the data code efficiency which is the ratio of information bits to total bits per character. When the data code efficiency is included, we obtain a measurement of the information transfer efficiency. We can call this ratio the information transfer ratio (ITR) which will provide us with a measurement of the protocol's information transfer efficiency. This results in:

$$ITR = \frac{B_{IC} * E_{TCS}}{B_C}$$

(1.3)

where:

ITR $=$ Information transfer ratio
B_{IC} $=$ Information bits per character
B_C $=$ Total bits per character
D $=$ Data characters per message block
A $=$ Characters in the acknowledgement message
C $=$ Control characters per message block
T_R $=$ Data transfer rate (bps)
T_C $=$ Transmission time per character (B_C / T_R)
T_{PA} $=$ Processing and acknowledgment time
T_L $=$ Line turnaround time
T_p $=$ Propagation delay time

P = Probability of one or more errors in block.

From the preceding, the information transfer ratio provides us with a measurement of the efficiency of the transmission control structure without considering the effect of compression. When compression is considered we obtain a new term which we will denote as the effective information transfer ratio (EITR).

When data is compressed, the original data stream will be reduced in size prior to transmission, the actual reduction being dependent upon the compression algorithms employed as well as the composition of the data acted upon. In general, we can assume that the compression ratio considers the number of characters in the compressed data stream to include special control characters, if required, to indicate one or more compression algorithms. This reasonable assumption simplifies the effect of considering data compression when examining a particular protocol. As an example, consider a 160-character data block compressed into 78 data characters plus 2 compression indicator characters. Here the compression ratio would be $160/(78+2)$ or 2. The effect upon the previously developed equation to compute the information transfer ratio would be to change D in the numerator to the non-compressed string length of 160 characters, while D in the denominator would be the actual 78 compressed data characters plus the two additional special characters required to indicate data compression, resulting in a total of 80 characters. If the total number of control characters framing the data block is relatively small, the effective information transfer ratio can be approximated by multiplying the information transfer ratio by the compression ratio.

Computation examples

For illustrative purposes let us assume that our data transmission rate is 4800 bps and we will transmit information using a BISYNC transmission control structure employing a 'stop and wait ARQ' error control procedure. Furthermore, let us assume the following parameters:

A = 4 characters per acknowledgement
B_{IC} = 8 bits per character
B_C = 8 bits per character
D = 80 data characters per block
C = 10 control characters per block
T_R = 4800 bps
T_C = $8/4800$ = 0.001 66 seconds (s) per character
T_{PA} = 20 ms = 0.02 s
T_L = 100 ms = 0.10 s
T_p = 30 ms = 0.03 s

P $= 0.01.$

Then:

$$ITR = \frac{8*80*(1 - 0.01)}{4800*[0.001\ 66*(80 + 10) + 2*(0.02 + 0.03 + 0.1) + 4*8/4800]}$$

$= 0.2894.$

Since the transfer rate of information in bits (TRIB) is equal to the product of the data transfer rate and the information transfer ratio, we obtain:

$$TRIB = ITR*T_R = 0.2894*4800 = 1389 \text{ bps.}$$

For the preceding example, approximately 29% of the data transfer rate is effectively used.

Let us now examine the effect of doubling the text size to 160 characters, while the remaining parameters except P continue as before. Since the block size has doubled, P approximately doubles, resulting in the ITR becoming:

$$ITR = \frac{8*160*(1 - 0.02)}{4800*[0.001\ 66*(1600 + 10) + 2*(0.02 + 0.03 + 0.1) + 4*8/4800]}$$

$= 0.4438.$

With an ITR of 0.4438 the TRIB now becomes:

$$TRIB = ITR*T_R = 0.4438*4800 = 2130 \text{ bps.}$$

Here, doubling the block size raises the percentage of the data transfer rate effectively used to 44.38%.

Compression effect

Suppose one or more data-compression algorithms are employed which result in an average compression ratio of 2. What effect would this have upon the effective information transfer ratio?

The effective information transfer ratio (EITR) can be obtained by modifying equation (1.2) as follows:

$$EITR = \frac{B_{IC}*D_1*(1 - P)}{T_R*[T_C*(D_2 + C) + 2*(T_{PA} + T_L + T_P) + (A*B_C/T_R)]}$$

(1.4)

where:

D_1 = original data block size in characters prior to compression
D_2 = compressed data block size in characters to include special compression indication characters

If on the average 160 data characters are transmitted in a compressed format of 80 characters we obtain:

$$EITR = \frac{8*160*(1-0.01)}{4800*[0.001\,66*(80+10)+2*(0.02+0.03+0.1)+(4*8/4800)]}$$

$$= 0.5788.$$

As previously discussed, the effective information transfer ratio can be approximated as follows:

$$EITR \simeq ITR*CR. \qquad (1.5)$$

Substituting, we obtain:

$$EITR \simeq 0.2861*2 \simeq 0.5722.$$

From the preceding you will note that the difference between the computed EITR (0.5788) and the approximated EITR (0.5722) is essentially insignificant. Thus, if you know the ITR you can simply multiply it by the compression ratio to obtain a reasonable approximation of the effective information transfer ratio.

Since the transfer rate of information in bits (TRIB) is the product of the effective information transfer ratio and the operating data rate, we obtain:

$$TRIB = 0.5788*4800 = 2778 \text{ bps.}$$

In Table 1.2, the reader will find a comparison of the variations in the ITR, EITR and TRIB when non-compressed and compressed data are transmitted for two different block sizes.

From Table 1.2, it is apparent that two methods can be employed to increase transmission efficiency. First, you may alter the protocol

Table 1.2 Compression effect comparison

	Non-compressed data		Compressed data	
Block size (characters)	80	160	80	160
ITR (dimensionless)	0.2894	0.4438	N/A	N/A
EITR (dimensionless)	N/A	N/A	0.5788	0.8678
TRIB (bps)	1389	2130	2778	4165

or transmission control sequence by varying the size of the data blocks transmitted. Alternatively, you can compress data prior to transmitting a block of information. Both methods can result in more information passing over a transmission line per unit time.

To facilitate the computation of the ITR based upon varying block sizes, a BASIC language program was developed. This program, which is listed in Table 1.3, computes the ITR for block sizes varying from 40 to 4000 characters in increments of 40 characters. In examining the program listing in Table 1.3, note that the probability of a block being in error is 0.005. This is one-half of 0.01 since we are now varying block sizes by 40 characters instead of 80 characters. Also note that as each block increases by 40 characters the probability that the block contains one or more bits in error has increased by 0.005 since the probability of a block having one or more bits in error is pro-portion to its size.

In Table 1.4, you will find a tabulation of the execution of the pro-gram which calculated the ITR as the block size varied from 40 to 4000 characters in increments of 40. In examining this table note that the maximum ITR of 0.7358 is obtained when the block size is 1080 characters. This indicates that as the block size increases with a constant error rate, a certain point is reached where the time to retransmit a long block every so often negates the enlargement of the block size. For the parameters considered, the optimum block size is 1080 characters. Only for the ideal situation, where $P = 0$ would a continuous increase in block size produce additional efficiencies.

Table 1.3 BASIC program listing to compute ITR as a function of block size

```
A = 4
BIC = 8
BC = 8
D = 80
C = 10
TR = 4800
TC = 8 / 4800
TPA = 0.02
TL = 0.1
TP = 0.03
P = 0.005
FOR BLOCK = 40 TO 4000 STEP 40
NUMERATOR = BIC * BLOCK * (1 − P)
DENOM = (TR * (0.00166 * (BLOCK + 10) + 2 * (TPA + TL + TP) + A * BIC /
TR))
ITR = NUMERATOR / DENOM
PRINT USING " #.#### ####"; ITR; BLOCK
P = P + 0.005
NEXT BLOCK
```

Table 1.4 Information transfer ratio and block size. Probability of block error = 0.01 per 80 characters

ITR	Block size	ITR	Block size
0.1702	40	0.6828	2040
0.2894	80	0.6794	2080
0.3771	120	0.6759	2120
0.4438	160	0.6723	2160
0.4960	200	0.6687	2200
0.5376	240	0.6651	2240
0.5714	280	0.6614	2280
0.5992	320	0.6576	2320
0.6222	360	0.6539	2360
0.6415	400	0.6501	2400
0.6577	440	0.6462	2440
0.6714	480	0.6423	2480
0.6830	520	0.6384	2520
0.6928	560	0.6345	2560
0.7011	600	0.6305	2600
0.7082	640	0.6265	2640
0.7142	680	0.6224	2680
0.7191	720	0.6184	2720
0.7233	760	0.6143	2760
0.7267	800	0.6102	2800
0.7295	840	0.6060	2840
0.7317	880	0.6019	2880
0.7333	920	0.5977	2920
0.7345	960	0.5935	2960
		0.5893	3000
0.7353	1000	0.5850	3040
0.7357	1040	0.5808	3080
0.7358	1080	0.5765	3120
0.7356	1120	0.5722	3160
0.7350	1160	0.5679	3200
0.7343	1200	0.5635	3240
0.7332	1240	0.5592	3280
0.7320	1280	0.5548	3320
0.7306	1320	0.5504	3360
0.7290	1360	0.5460	3400
0.7272	1400	0.5416	3440
0.7252	1440	0.5372	3480
0.7231	1480	0.5328	3520
0.7209	1520	0.5283	3560
0.7185	1560	0.5239	3600
0.7161	1600	0.5194	3640
0.7135	1640	0.5149	3680
0.7108	1680	0.5104	3720
0.7080	1720		

Table 1.4 (continued)

ITR	Block size	ITR	Block size
0.7051	1760	0.5059	3760
0.7021	1800	0.5014	3800
0.6991	1840	0.4969	3840
0.6960	1880	0.4924	3880
0.6928	1920	0.4878	3920
0.6895	1960	0.4833	3960
0.6862	2000	0.4787	4000

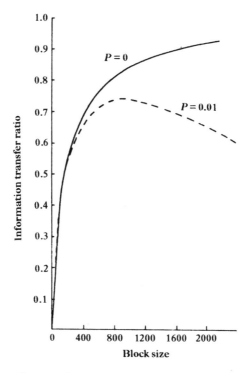

Figure 1.14 ITR and error rate

In Figure 1.14 the ITR is plotted as a function of block size for the error-free condition and 0.01 probability of error conditions. The 0.01 probability of error condition per 80-character block was held constant by incrementing the error rate in proportion to the increase in the block size. Since an error-free line is not something a transmission engineer can reasonably expect, a maximum block size will exist beyond which our line efficiency will decrease. At this point, only data compression will result in additional transmission efficiencies. In addition, from a physical standpoint, the buffer area of some devices may prohibit block sizes exceeding a certain number of characters. Once again, data compression can become an effective mechanism for increasing transmission efficiency while keeping data buffer requirements within an acceptable level.

In a LAN environment the error rate is several orders of magnitude less than the error rate on a WAN. While this might indicate that you can simply continuously increase the block size to increase the information transfer rate, other constraints preclude this from occurring. The primary constraint is the type of LAN you are using. For example, an Ethernet LAN has a maximum information field length of 1500 bytes per frame. Thus, the only way to increase the efficiency of the transportation of information is to make each byte in the information field represent more data. This, as you might expect, can be accomplished by transporting compressed data in each LAN frame.

Return channel model

Consider a 'stop and wait ARQ' error control procedure where a return channel is available for the transmission of acknowledgements. The use of this return channel eliminates the necessity of line turnarounds; however, transmission is still half duplex since an acknowledgement is only transmitted after each received message block is processed.

When the message block is sent to the receiving station, both propagation delay and processing delay are encountered. When the acknowledgement is returned, one additional propagation delay and processing delay result. In addition to these delays, you must also consider the time required to transmit the acknowledgement message. If A denotes the length in characters of the acknowledgement message and T_S is the reverse channel data rate in bps, then the transmission time for the acknowledgement becomes $(A*B_C)/T_S$.

The total delay time due to the propagation and processing as well as the acknowledgement transmission time becomes:

$$2*(T_{PA} + T_p) + \frac{A*B_C}{T_S}$$

Thus, the information transfer ratio becomes:

$$\text{ITR} = \frac{B_{IC}*D_1(1 - P)}{T_R*[T_C*(D_2 + C) + 2*(T_{PA} + T_p) + A*B_C/T_S]}. \tag{1.6}$$

Let us examine the effect of this modified transmission procedure on the previous example where data was packed 80 characters per block. Let us assume that a 75 bps reverse channel is available and our acknowledgement message is comprised of four 8-bit characters. Then:

$$ITR = \frac{8*80*(1 - 0.01)}{4800*[0.001\ 66*(80 + 10) + 2*(0.02 + 0.03) + 4*8/75]}$$

$$= 0.1953.$$

Note that the ITR actually decreased. This was caused by the slowness of the reverse channel where it took 0.4266 (4*8/75) seconds to transmit an acknowledgement. In comparison, the two-line turnarounds that were eliminated only required 0.2 s when the acknowledgement was sent at 4800 bps on the primary channel. This modified procedure is basically effectively when the line turnaround time exceeds the transmission time of the acknowledgement on the return channel. This situation normally occurs when the primary data transfer rate is 2400 bps or less. If the data is compressed prior to transmission and a compression ratio of 2 results in 160 data characters transmitted as a block of 80 compressed characters, the EITR can be computed as follows:

$$EITR = \frac{8*160*(1 - 0.01)}{4800*[0.001\ 66*(80 + 10) + 2*(0.02 + 0.03) + 4*8/75]} = 0.39.$$

In comparing the effect of compression, note that the transfer rate of information in bits (TRIB) rises from 0.1953*4800 or 937 to 0.39* 4800 or 1872 bps. Thus, a compression ratio of 2 can be expected to approximately double throughout.

Full duplex model

A much greater throughput efficiency with the 'stop and wait ARQ' error control procedure can be obtained when a full duplex mode of transmission is employed. Although this requires the use of a four-wire leased line or modems that either splits a two-wire circuit into independent frequency bands or uses echo cancellation technology, the modems and line do not have to be reversed. This permits an acknowledgement to be transmitted at the same data rate as the message block but in the reverse direction without the line turnaround. Thus, the information transfer ratio becomes:

$$ITR = \frac{B_{IC}*D_1*(1 - P)}{T_R*[T_C*(D_2 + C) + 2*(T_{PA + Tp})]}. \tag{1.7}$$

Again, returning to the original 80-character block example, we obtain:

$$\text{ITR} = \frac{8*80*(1 - 0.01)}{4800*[0.001\ 66*(80 + 10) + 2*(0.02 + 0.03)]} = 0.5293.$$

When data compression results in a compression ratio of 2, we obtain:

$$\text{EITR} = \frac{8*160*(1 - 0.01)}{4800*[0.001\ 66*(80 + 10) + 2*(0.02 + 0.03)]} = 1.06.$$

With an EITR greater than unity this means that the bits of information per unit time (in compressed format) exceed the data transmission rate of the equipment connected to the line. This illustrates the value of data compression, permitting you to obtain very high effective data transfer rates without requiring additional communications facilities.

A second variation of the full duplex model results if a 'go back n ARQ' error control procedure is employed. In this situation, only the block received in error is retransmitted. Here, the information transfer ratio becomes:

$$\text{ITR} = \frac{B_{\text{IC}}*D_1*(1 - P)}{T_\text{R}*[T_\text{C}*(D_2 + C)]}. \tag{1.8}$$

Again, substituting values from the original example we obtain:

$$\text{ITR} = \frac{8*80*(1 - 0.01)}{4800*[0.001\ 66*(80 + 10)]} = 0.8835.$$

This is obviously the most efficient technique since the line turnaround is eliminated and the processing and acknowledgment time (T_{PA}) and propagation delay time (T_p) in each direction are nullified due to simultaneous message block transmission and acknowledgement response. If we consider the effect of a compression ratio of 2, the effective information transfer ratio can be computed as follows:

$$\text{EITR} = \frac{8*160\ (1 - 0.01)}{4800*[0.001\ 66*(80 + 10)]} = 1.767.$$

As an alternative to the previous computation you could estimate the EITR by multiplying the ITR by the compression ratio. Doing so in this situation would result in exactly the same result. For the preceding example, the TRIB becomes 1.767*4800 bps or 8482 bps. Here, compression and protocol structure permit an effective information transfer of 8482 bps on a 4800 bps data path. In effect, the selection of an appropriate protocol coupled with effective data compression algorithms can result in a very effective data transfer. This will result in data transfers normally associated with high speed facilities occurring over conventional voice data transmission facilities.

In a LAN networking environment, both bridges and routers now commonly employ data compression on WAN facilities. Here the use of data compression minimizes network bottlenecks that typically occur when a frame transmitted at a Mbps operating rate on a LAN is transferred onto an analog or digital WAN circuit operating at a fraction of the LAN rate. For example, a digital circuit operating at 512 kbps to interconnect geographically separated LANs through the use of bridges or routers would provide an information transfer rate of over 1.5 Mbps if the bridges or routers used compression and obtained an average compression ratio of 3:1.

The commonly used T1 and E1 circuits operating at 1.544 Mbps in North America and 2.048 Mbps in Europe that just a few years ago were considered to represent an operating rate organizations may never fully utilize now represent bottlenecks to the flow of data in some organizations. Recognizing that the use of parallel T1 or E1 circuits can represent a considerable expense, manufacturers now offer compression-performing products that can be used with those high speed digital circuits to obtain a 3 to 4 Mbps information transfer rate over those circuits. The preceding examples discussed in this chapter represent but a small fraction of the use of compression-performing products. The use of compression represents one of the fastest growing areas of applied mathematical technology and applications are only limited by one's imagination.

2

DATA CODES AND COMPRESSION- INDICATING CHARACTERS

Prior to discussing data-compression techniques, it is important to become familiar with several data codes and the binary representation of characters of such codes. This familiarity is required since character-oriented compression techniques are based upon the employment of one or more characters in a particular character set. Such characters are used to indicate the occurrence of a compression algorithm and compression and decompression could not occur without their use. In addition, other compression-indicating characters are required to indicate the occurrence of physical conditions upon which the compression routine must initiate a predefined operation. Examples of physical conditions include codes to indicate the end of a file, the filling of a buffer that must be flushed, or a requirement to change the number of bits used to represent the occurrence of a string. We can refer to characters that indicate the occurrence of a compression algorithm as logical compression-indicating characters, while characters that indicate the occurrence of physical conditions can be referred to as physical compression-indicating characters.

In this chapter we will first examine the composition of four popular data codes. Once this is accomplished we will use the preceding knowledge to discuss several methods that can be used to develop compression-indicating characters as well as the advantages and disadvantages associated with the use of each method. Initially this discussion will be primarily focused upon the creation of logical compression-indicating characters. This will be followed by a limited discussion of the creation of physical compression-indicating characters since a detailed explanation of their use depends upon knowledge of statistical and string compression techniques for both data and images which are covered later in this book. Thus, we will defer a detailed discussion of the creation and utilization of physical compression-indicating characters to appropriate locations in this book in which compression techniques use those types of characters.

2.1 DATA CODES

Baudot code

The 5-level Baudot code was devised to permit teletypewriters to operate faster and more accurately than relays used to transmit information via telegraph. Since the number of different characters which can be derived from a code having two different (binary) states is 2^m, where m is the number of positions in the code, the 5-level Baudot code permits 32 unique character bit combinations. Although 32 characters could be represented normally with such a code, the necessity to transmit digits, letters of the alphabet and punctuation marks made it necessary to devise a mechanism to extend the capacity of the code to include additional character representations. The extension mechanism was accomplished by the use of two 'shift' characters—'letters shift' and 'figures shift'.

The transmission of a shift character informs the receiver that the characters which follow the shift character should be interpreted as characters from a symbol and numeric set or from the alphabetic set of characters. The 5-level Baudot code is illustrated in Figure 2.1 for one particular terminal pallet arrangement. A transmission of all 1s in bit positions 1 to 5 indicates a 'letters shift' and the characters following the transmission of that character are interpreted as letters. Similarly, the transmission of 1s in bit positions 1, 2, 4 and 5 would indicate a 'figures shift' and the following characters would be interpreted as numerics or symbols based upon their code structure.

BCD and EBCDIC codes

The development of computer systems required the implementation of coding systems to convert alphanumeric characters into binary notation. One of the earliest codes used to convert data to a computer-acceptable form was the binary coded decimal (BCD) system. This coding technique permits numeric information to be represented by four binary bits and permits an alphanumeric character set to be represented through the use of six bits of information. This code is illustrated in Figure 2.2. One advantage of this code is that two decimal digits can be stored in an 8-bit computer word and manipulated with appropriate computer instructions. Although only 36 characters are shown, a BCD code is capable of containing a set of 2^6 or 64 different characters.

Characters

Letters	Figures	Bit selection				
		1	2	3	4	5
	–	1				
A	?	1				1
B	:	.			1	
C	$	1	1	1		
D	3	1			1	
E	!	1	1		1	
F	¢			1		1
G					1	1
H	8			1	1	
I	'	1	1	1		
J	(	1				
K	)		1	1	1	1
L	.		1		1	1
M	,		1	1		
N	9		1	1	1	1
O	0				1	1
P	1	1		1	1	1
Q	4			1		
R		1	1			
S	5		1	1	1	
T	7	1	1			
U	:			1	1	1
V	2	1		1		1
W	/	1	1		1	1
X	6	1	1	1		1
Z	"	1	1	1	1	1

Functions

		1	2	3	4	5
Carriage return						
Line feed			1		1	
Space						
Letters shift	<	1	1	1	1	1
Figures shift	=	1	1	1	1	1

Figure 2.1   Baudot code data representation

In addition to transmitting letters, numerics and punctuation marks, a considerable number of control characters may be required to promote line discipline. These control characters may be used to switch on and off devices which are connected to the communications line, control the actual transmission of data, manipulate message formats and perform additional functions. Thus, an extended character set is usually required for data communications. One such character set is the extended binary-coded decimal interchange code (EBCDIC) shown in Figure 2.3. This code is an extension of the binary-coded

Bit position						
b_6	b_5	b_4	b_3	b_2	b_1	Character
0	0	0	0	0	1	A
0	0	0	0	1	0	B
0	0	0	0	1	1	C
0	0	0	1	0	0	D
0	0	0	1	0	1	E
0	0	0	1	1	0	F
0	0	0	1	1	1	G
0	0	1	0	0	0	H
0	0	1	0	0	1	I
0	1	0	0	0	1	J
0	1	0	0	1	0	K
0	1	0	0	1	1	L
0	1	0	1	0	0	M
0	1	0	1	0	1	N
0	1	0	1	1	0	O
0	1	0	1	1	1	P
0	1	1	0	0	0	Q
0	1	1	0	0	1	R
1	0	0	0	1	0	S
1	0	0	0	1	1	T
1	0	0	1	0	0	U
1	0	0	1	0	1	V
1	0	0	1	1	0	W
1	0	0	1	1	1	X
1	0	1	0	0	0	Y
1	0	1	0	0	1	Z
1	1	0	0	0	0	0
1	1	0	0	0	1	1
1	1	0	0	1	0	2
1	1	0	0	1	1	3
1	1	0	1	0	0	4
1	1	0	1	0	1	5
1	1	0	1	1	0	6
1	1	0	1	1	1	7
1	1	1	0	0	0	8
1	1	1	0	0	1	9

Figure 2.2 Binary-coded decimal (BCD) data representation

decimal system and uses eight bits for character representation. This code permits 2^8 or 256 unique characters to be represented although, currently, a lesser number are assigned meanings. This code is primarily used for transmission by byte-oriented computers. The use of EBCDIC by computers may alleviate the necessity of having the computer perform code conversion if the connected terminals transmit data in that code. If not, the computer will perform code conversion prior to operating upon the data. In Figure 2.3, the EBCDIC character set is illustrated. Note that the undefined characters in this code, or any code for that matter, can be used to signify predefined special

	00				01				10				11			
	00	01	10	11	00	01	10	11	00	01	10	11	00	01	10	11
0000	NUL	DEL	DS		SP	&										0
0001	SOH	DCI	SOS			/			a	j			A	J		1
0010	STX	DC2	FS	SYN					b	k	s		B	K	S	2
0011	ETX	TM							c	l	t		C	L	T	3
0010	PF	RES	BYP	PN					d	m	u		D	M	U	4
0101	HT	NL	LF	RS					e	n	v		E	N	V	5
0110	LC	BS	ETB	UC					f	o	w		F	O	W	6
0111	DEL	IL	ESC	EOT					g	p	x		G	P	X	7
1000		CAN							h	q	y		H	Q	Y	8
1001		EM							i	r	z		I	R	Z	9
1010	SMM	CC	SM		¢	!		:								
1001	VT	CU1	CU2	CU3	.	$	,	*								
1100	FF	IFS		DC4	<	*	%	®								
1101	CR	IGS	ENQ	NAK	{	}	–	'								
1110	SP	IPS	ACK		+	;	>	=								
1111	SI	IUS	BEL	SUB				?	"							

ACK	Acknowledge	EOT	End of transmission	PN	Punch on
BEL	Bell	ESC	Escape	RES	Restore
BS	Backspace	ETB	End of transmission block	RS	Reader stop
BYP	By pass	ETX	End of text	SI	Shift in
CAN	Cancel	FF	Form feed	SM	Set mode
CC	Cursor control	FS	Field separator	SMM	Start of manual message
CR	Carriage return	HT	Horizontal tab	SO	Shift out
CU1	Customer use 1	IFS	Interchange file separator	SOH	Start of heading
CU2	Customer use 2	IGS	Interchange group separator	SOS	Start of significance
CU3	Customer use 3	IL	Idle	SP	Space
DC1	Device control 1	IRS	Interchange record separator	STX	Start of text
DC2	Device control 2	IUS	Interchange unit separator	SUB	Substitute
DC4	Device control 4	LC	Lower case	SYN	Synchronous idle
DEL	Delete	LF	Line feed	TM	Tape mark
DLE	Data link escape	NAK	Negative acknowledge	UC	Upper case
DS	Digit select	NL	New line	VT	Vertical tab
EDM	End of medium	NUL	Null		
ENQ	Enquiry	PF	Punch off		

Special graphic characters

¢	Cent sign	–	Minus sign, hyphen
.	Period, decimal point	/	Slash
<	Less-than sign	,	Comma
(	Left parenthesis	%	Per-cent
+	Plus sign	—	Underscore
|	Logical OR	>	Greater-than sign
&	Ampersand	?	Question mark
!	Exclamation mark	:	Colon
$	Dollar sign	#	Number sign
*	Asterisk	@	At sign
)	Right parenthesis	'	Prime, apostrophe
;	Semicolon	=	Equal sign
¬	Logical NOT	"	Quotation mark

Figure 2.3 EBCDIC character set

operations or data representations. Thus, one or more data-compression techniques could be indicated by an 8-bit sequence that represents a character, defined or undefined, from the EBCDIC character set. Normally, it is best to employ an undefined character since the delineated characters, in addition to being numeric, graphic and alphabetic characters, include control characters which govern the operation of the terminal and the communications sequence.

When data is compressed according to a certain algorithm, an 8-bit sequence is typically inserted into a data stream to signify both the fact that compression has occurred as well as the type of algorithm employed. In this book, we refer to the use of such characters as special logical and physical compression-indicator characters or special characters for brevity. When the compressed data is received, the same 8-bit sequence would indicate the presence of compressed data. In addition, it would inform the receiving station of the data compression algorithm employed which would enable the receiver to decompress the compressed data stream.

The ASCII code

Due to the proliferation of the number of data transmission codes, attempts to standardize such codes resulted. Out of this quest for standardization was the development of the American Standard Code for Information Interchange (ASCII). This 7-level code is illustrated in Figure 2.4 and is based upon a 7-bit code developed by the International Standards Organization (ISO).

ASCII characters are encoded in seven bits while an eighth bit is available for use as a parity bit. Eight-level ASCII is commonly used for transmission by interactive asynchronous teletype devices where the parity bit can represent odd or even parity.

Another popular version of ASCII is represented by the character set used by the IBM PC and compatible personal computers. Known as extended ASCII, this character set uses eight bits instead of the seven used by conventional ASCII. The first 128 ASCII characters whose codes are 0 to 127 are exactly the same as those characters listed in Figure 2.4. The next 128 characters are formed by the setting of the eighth bit and define block graphics and other special characters.

2.2 SELECTING COMPRESSION-INDICATING CHARACTERS

In this section we turn our attention to methods you can use to select compression-indicating characters. First we will focus our attention upon developing logical compression-indicating characters. Once this

b7 →					0	0	0	0	1	1	1	1
b6 →					0	0	1	1	0	0	1	1
b5 →					0	1	0	1	0	1	0	1
b4	b3	b2	b1	→	0	1	2	3	4	5	6	7
0	0	0	0	0	NUL	DLE	SP	0	@	P		P
0	0	0	1	1	SOH	DC1	!	1	A	Q	a	q
0	0	1	0	2	STX	DC2	"	2	B	R	b	f
0	0	1	1	3	ETX	DC3	*	3	C	S	c	s
0	1	0	0	4	EOT	DC4	$	4	D	T	d	t
0	1	0	1	5	ENQ	NAK	%	5	E	U	e	u
0	1	1	0	6	ACK	SYN	&	6	F	V	f	v
0	1	1	1	7	BEL	ETB	'	7	G	W	g	w
1	0	0	0	8	BS	CAN	(	8	H	X	h	x
1	0	0	1	9	HT	EM	)	9	I	Y	i	y
1	0	1	0	10	LF	SUB	*	:	J	Z	j	z
1	0	1	1	11	VT	ESC	+	;	K	[	k	{
1	1	0	0	12	FF	FS	,	<	L	\	l	|
1	1	0	1	13	CR	GS	-	=	M	]	m	}
1	1	1	0	14	SO	RS	.	>	N	`	n	~
1	1	1	1	15	SI	US	/	?	£	—	o	DEL

Control characters

NUL	Null/idle	DLE	Data link escape (CC)
SOH	Start of heading (CC)	DC1	
STX	Start of text (CC)	DC2	Device controls
ETX	End of text (CC)	DC3	
EOT	End of transmission (CC)	DC4	Device control (stop)
ENQ	Enquiry (CC)	NAK	Negative acknowledge (CC)
ACK	Acknowledge (CC)	SYN	Synchronous idle (CC)
BEL	Audible or attention signal	ETB	End of transmission block (CC)
BS	Backspace (FE)	CAN	Cancel
HT	Horizontal tabulation (punch card skip) (FE)	EM	End of medium
		ESC	Escape
LF	Line feed	FS	File separator (IS)
VT	Vertical tabulation (FE)	GS	Group separator (IS)
FF	Form feed (FE)	RS	Record separator (IS)
CR	Carriage return (FE)	US	Unit separator (IS)
SO	Shift out	DEL	Delete
SI	Shift in		

Note: (CC) Communication control
(FE) Formal effector
(IS) Information separator

Special graphic characters

SP	Space	(	Opening parenthesis
!	Exclamation mark	)	Closing parenthesis
	Logical OR	*	Asterisk
"	Quotation marks	+	Plus
*	Number signs	,	Comma
$	Dollar sign	-	Hyphen
%	Per cent	.	Period
&	Ampersand	/	Slant
;	Semicolon	:	Colon
<	Less than	]	Closing bracket
=	Equals	—	Underline
>	Greater than	`	Grave accent
?	Question mark	{	Opening brace
@	Commercial at	|	Vertical line
[	Opening bracket	}	Closing brace
\	Reverse slant	~	Tilde

Figure 2.4 The ASCII data code

is accomplished, we will examine why physical compression-indicating characters are required and how they are normally formed.

Logical compression-indicating characters

The key to character-oriented compression is the selection and utilization of one or more compression-indicating characters. The logical compression-indicating character will serve as a flag, informing the receiving device that compression has occurred as well as the type of compression that was performed. The receiving device can then apply an appropriate decompression algorithm to restore the data to its original form.

To illustrate the use of a logical compression-indicating character, assume a string of repeating characters occurred, such as XXXXXX. One possible method to compress this string would be to encode a special compression-indicating character, followed by a count and the actual character that was compressed. This encoding technique is known as run length encoding and will be discussed in more detail in Chapter 3 when we examine a variety of character-oriented compression techniques. Under this encoding technique any repeating string of characters in excess of three can be encoded as illustrated in Figure 2.5.

The key to the selection of appropriate logical compression-indicating characters depends upon several factors, including the data code employed, the number of character-oriented compression schemes you will use, and, upon occasion, prior knowledge of the data to be transmitted or stored. Of these three factors, by far the most important is the data code employed which will have a major bearing upon the other factors.

Some data codes, such as EBCDIC, have unassigned characters in the character set. In such situations, the selection of special compression-indicating characters involves selecting unassigned characters from the character set you are using.

As you incorporate a mixture of character-oriented compression techniques you may run out of unassigned characters. In addition, if you use Baudot, 7-level ASCII, or other codes that do not have unassigned characters in the character set you must use a different tech-

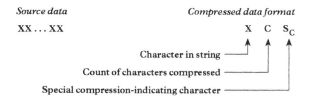

Figure 2.5 Run length encoding

nique to obtain logical compression-indicating characters. In such situations you can consider using the extended code capability of certain data codes or the insert and delete technique.

Extended code utilization

Some character sets, including ASCII and EBCDIC, have the built-in capability to redefine characters through the utilization of Shift Out (SO) and Shift In (SI) characters. The Shift Out character, in effect, denotes that all characters following that character until a Shift In (SI) character is encountered have new meanings represented by a new character set you have defined. Figure 2.6 illustrates the use of an extended code to implement run length compression. Note that the compressed data format now requires five characters. In comparison, the use of an existing character in the character set for the special compression-indicating character only required three characters. Thus, an extended code for run length compression would only be beneficial when a string of repeating characters exceeded five.

Insert and delete technique

The insert and delete technique can be considered as similar to techniques used in many data transmission schemes to effect data transparency and prevent sequences of runs of set bits from being misinterpreted as a flag in the HDLC protocol. Under the insert and delete technique you should select logical compression-indicating characters that are not frequently encountered in the data to be compressed. In the English language you might select such characters as Q and Z, since their frequency of occurrence is very low in comparison to other characters.

Once your logical compression-indicating characters are selected, your compression algorithm must also examine data for the natural occurrence of compression-indicating characters. If those characters naturally appear in data, the encoding process will add a second

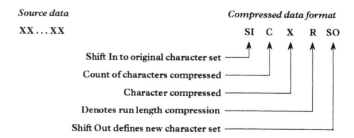

Figure 2.6 Run length encoding using an extended code

character, while the decoding process must then delete the previously inserted character, hence the term insertion and deletion.

Figure 2.7 illustrates the use of the character Q as a special compression-indicating character when two strings are operated upon using the insert and delete technique to form the compression-indicating character. Note that in the first example the compressed data stream actually represents data expansion since run length compression could not be used to compress the source data and the character Q naturally occurred in the data string. In the second example, the string of six Xs was compressed with Q serving as the compression-indicating character.

Physical compression-indicating characters

Physical compression-indicating characters are normally used to denote a compression-related event other than the occurrence of a specific type of compression. As indicated earlier in this section, examples of events noted by physical compression-indicating characters can include the initiation of a buffer flush and a change in the number of bits used to represent strings. Since the creation of physical compression-indicating characters is highly related to the method of compression used, we will examine a 'generic' dictionary compression scheme to obtain an appreciation for how such characters are normally formed. However, we will defer a discussion of the formation of specific physical compression-indicating characters until we examine specific dictionary compression techniques, due to the close correlation between the formation of physical compression-indicating characters and different dictionary-based compression methods.

One common type of dictionary-based compression technique is created by initially assigning each character in an 8-bit character code to the first 256 locations in a dictionary. Each character read from the input data stream becomes a suffix used to form a new string which will be matched against entries in the dictionary. The prefix of the string is the null character for the first read operation and the suffix of the prior string for all subsequent read operations. Once the string is formed it is compared against entries in the dictionary.

If the string matches an entry in the dictionary, the prefix is assigned the value of the suffix, the next character is read as the suffix, and a new string is formed which is matched against the entries in the dictionary. If the newly formed string is not encountered

Source data	Compressed data
JOHN Q. PUBLIC	JOHN QQ. PUBLIC
$XXXXXX 5.15	$QX65.15

Figure 2.7 Run length encoding using the insert and delete technique to form a special compression-indicating character

in the dictionary, the position of the prefix of the string in the dictionary is output, the new string is added to the dictionary, the prefix for the next string search is assigned the value of the suffix of the current string, and the next character in the input data string is read as the suffix for a new string search of entries in the dictionary.

The previously described dictionary-based compression method is more commonly referred to as Lempel–Ziv compression after the two mathematicians who developed two 'core' dictionary-based compression methods. As we will note later in this book, the previously described compression method is formally referred to as LZ78 since the technique was published by the authors in 1978.

To illustrate the operation of the previously described dictionary-based compression technique let's assume the input data string consists of the sequence 'abadabada'. Table 2.1 lists the values of the prefix and suffix substrings used to form each string for dictionary searching, the resulting string, whether or not a dictionary match occurred, and if not, the resulting placement of the string in the dictionary as well as the output for each character processed from the input string.

In examining Table 2.1, note that the nine-character input string was compressed to an output sequence of seven codes. Until now we have not discussed the codes used to represent dictionary entries, so let's do so. Those codes can be considered to represent the physical location of strings in the dictionary, enabling you to compress strings by replacing a string with its code or position in the dictionary. If the 8-bit ASCII code is used to form the initial dictionary, each position in the table represents the ASCII code of the character in the string. For example, the space character has an ASCII code of 32, uppercase

Table 2.1 Dictionary-based string compression operation (input string: abadabada)

Prefix	Suffix	String	Dictionary match	
			Dictionary modification	Output
null	a	a	yes	none
a	b	ab	no, add as 256	code(a)
b	a	ba	no, add as 257	code(b)
a	d	ad	no, add as 2558	code(a)
d	a	da	no, add as 259	code(d)
a	b	ab	yes	none
ab	a	aba	no, add as 260	code(ab)
a	d	ad	yes	none
ad	a	ada	no, add as 261	code(ad)
a	—	a	yes	code(a)
Output codes:	97	98	97 100 260 261 97	

A through Z have the ASCII codes 65 through 90, and lowercase a through z have the ASCII codes 97 through 122. The first 127 characters in 8-bit ASCII are exactly the same characters in conventional 7-bit ASCII, while characters whose codes range from 128 to 255 are more commonly referred to as extended ASCII since they must be represented by using 8-bit codes. The column labeled 'Output' in Table 2.1 indicates the position of the relevant strings in the dictionary that are output. If the string represents a character in the first 256 positions in the dictionary, the output code can be the 8-bit ASCII code of the character. However, as the dictionary expands beyond its initial 256 entries, additional bits are required to represent each character. Thus, if the dictionary holds 512 entries, each code will be nine bits. Then, the seven output codes listed at the bottom of Table 2.1 would require 63 bits. Since the input string was nine 8-bit characters, the compression ratio for this small example of dictionary-based compression becomes:

$$\text{Compression ratio} = \frac{9 \text{ bytes} \times 8 \text{ bits/byte}}{7 \text{ bytes} \times 9 \text{ bits/byte}} = 1.14$$

Although the preceding example did not result in a high compression ratio, it indicates that a dictionary-based compression technique offers a potential to replace relatively long strings by short codes. For example, common English words to include 'the', 'what', 'when', 'where', and 'who' could each be replaced by a 10-, 11- or 12-bit code from a dictionary with 512, 1024, 2048 or 4096 entries.

The efficiency of a dictionary-based compression technique depends upon the size of the dictionary used to store strings as well as the composition of the data to be compressed. In a communications environment there are several physical events that the transmitter must convey to the receiver. First, the transmitter and receiver must agree on the size of the dictionary which will govern the number of bits used to represent each string. Next, as data is compressed and transmitted, the dictionary buffer will rapidly fill. Some dictionary-based compression techniques are implemented to examine the frequency of strings that are not compressible. When a certain threshold is reached, the transmitter will flush or empty its dictionary of strings beyond the first 256 entries. Doing so enables the compression technique to reconstruct its dictionary with newly created entries that may result in a higher compression ratio being obtained. To indicate the fact that the buffer was flushed, a special code must be sent to the receiver which will perform a similar operation upon its receipt.

Other implementations of dictionary-based compression techniques may compute statistics concerning the use of a buffer of n characters and determine that a buffer of n^2 would be more efficient. To communicate the fact that the size of the dictionary has doubled and an extra bit is now used to represent each code would require the use of another physical compression-indicating character.

Most software developers begin their use of physical compression-indicating characters at code position 256, since codes 0 through 255 are used for each character in an 8-bit character set. Placing physical compression-indicating characters beginning at code position 256 facilitates the development of programming code to check for the occurrence of physical events that must be acted upon. After one or more physical compression-indicating character codes, the remaining positions in the dictionary are then used to store strings developed based upon the composition of the input data. When we examine the implementation of specific dictionary-based compression techniques later in this book we will also examine specific examples of the use of physical compression-indicating character codes.

3

CHARACTER-ORIENTED COMPRESSION TECHNIQUES

The tremendous growth in remote computing during the last decade has focused the interest of communications personnel upon data compression techniques. Originally brought to data-processing user attention during the 1960s as a mechanism for increasing the capacity of mass storage devices, compression is now commonly applied to the data communications field. Here, compression results in the transfer of data in shorter time periods than if such data was transmitted without the employment of a compression technique.

In this chapter, eight distinct methods that can be employed to compress data are covered. Each of the methods described in this chapter is implemented based upon the use of a special compression-indicating character and can be classified as a character-oriented data-compression technique. Although each of the techniques can be implemented on an individual basis, from a practical standpoint it is very common to implement two or more character-oriented compression techniques since the data to be compressed will normally be more susceptible to a mixture of compression methods.

Character-oriented compression, while offering the potential to significantly reduce data storage or transmission requirements, should not be viewed as an end to itself. In many instances, character-oriented compression may be the first level of a multi-level compression scheme, with other levels employing a statistical-based compression method which encodes data based upon the frequency of occurrence of all characters, including special compression-indicating characters. The reader is referred to Chapter 4 for specific information concerning statistical-based methods of data compression.

Another common mechanism by which character-oriented compression can be used is with dictionary-based string compression.

The reader is referred to Chapter 5 for specific information concerning dictionary-based string compression techniques.

In addition to examining character-oriented compression methods, various combinations of techniques are discussed with emphasis placed upon their utilization and efficiency. Some of the techniques covered in this chapter require a careful analysis of current or projected data traffic to be effective. None of the techniques presented requires more than a moderate level of difficulty in developing software to conduct the encoding and decoding algorithms. In fact, most of the techniques in this chapter should be easy for end-users to implement and their implementation may result in a high degree of data reduction for a minimal amount of effort.

By the application of one or more character-oriented compression techniques, operational efficiencies may be increased or transmission costs reduced. For the former, data compression will permit an increase in information transferred over a data link per unit time interval. Concerning the latter, reducing the amount of data to be physically transferred may make the employment of a lower-speed data link permissible, resulting in a reduction in cost in comparison with the expense of a data link operated at a higher data rate.

3.1 NULL SUPPRESSION

Null or blank suppression was one of the earliest data compression techniques developed. Today, this simplistic technique is employed in the IBM 3780 BISYNC transmission protocol. In addition, null suppression is commonly used with a mixture of other character-oriented compression techniques to reduce data storage and data transmission.

Technique overview

As the name implies, null suppression is a data-compression technique that scans a data stream for repeated blanks or nulls. Upon encountering such a sequence, the blank or null characters are replaced by a special ordered pair of characters whose format is illustrated in Figure 3.1A. First, a compression-indicator character is employed to denote that null suppression has occurred. The second character is used to indicate the quantity of null characters that were encountered and replaced by the two-character sequence (Aronson, 1977; Ruth and Kreutzer, 1972).

When the two-character sequence is transmitted within a data stream, the receiving device performs a search for the special character used to indicate null suppression. Upon detection of that character, the receiver knows that the next character contains the count of

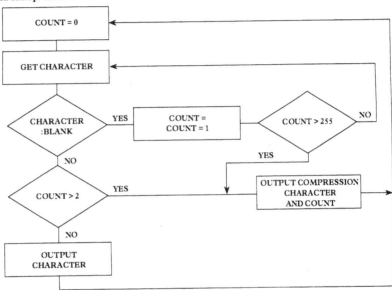

Compression format

NULL COUNT	COMPRESSION INDICATOR CHARACTER

Data compression example

Original data stream   XYZ ƀ ƀ ƀ ƀ QRX

Compressed data stream   XYZS$_C$5QRX

where S$_C$ = special compression indicator character

Data scan process

Figure 3.1   Null suppression

the number of nulls that were compressed. From this information, the original data stream can be reconstructed.

In the middle portion of Figure 3.1 is an example of the application of null suppression upon a data stream. Here, the character S$_C$ indicates a special compression-indicating character, denoting that null suppression has transpired.

In the lower portion of Figure 3.1, a flow chart of the null suppression scanning process is illustrated. If we assume an 8-bit format for data characters, then the character counter can store values for up to 255 sequentially encountered nulls prior to overflowing if we start our numbering at 1, or 256 if our numbering commences by assuming a zero counter represents a value of 1.

Programming examples

As a fourth edition this book builds upon the prior three versions. The first edition included a FORTRAN program listing whose execution could be used to analyze the susceptibility of a data file to different compression methods. Recognizing the expanding role of the personal computer, the second and third editions of this book were expanded to include program listings for compressing and decompressing data files based upon a series of independent character-oriented compression techniques and a convenience diskette could be obtained from the publisher to alleviate the necessity to enter program listings from the keyboard.

Programs developed for the second and third editions of this book were based upon the BASICA version of the BASIC programming language. BASICA was bundled with the IBM PC and most compatible computers as it was included on the PC-DOS operating system diskette. Due to its wide acceptance as a programming language and the ability of many readers to read BASIC code, BASICA was selected as a learning tool for coding programming examples.

Since the publication of prior editions of this book, the C language and its derivatives, C+ and C++, have gained widespread acceptance. In addition, Compiled C code will run considerably faster than interpretative BASIC code generated by the use of BASICA. Recognizing the preceding, this author recoded all BASICA examples from the third edition using the Borland Turbo C language and also wrote several new programs for this fourth edition in that language. In addition, selected public domain compression-performing programs as well as contributed programs were selected to provide readers with a 'toolbox' of programs available for use or for experimentation. Each of those programs was developed using the C language. Although many readers will prefer the C language examples, I have retained prior BASICA examples since many people continue to prefer to use those examples as a learning tool.

File naming conventions

To provide readers with an easy mechanism to note the function of programs and contents of data files, a file naming convention was developed. This convention builds upon the conventions established in previous editions of this book and facilitates reference to most of the files included on the convenience diskette which is included with this book. The exceptions to this naming convention are shareware and contributed programs whose names were retained.

Table 3.1 lists the file naming conventions used throughout this book and on the accompanying diskette. The letters 'XXXXXXX' rep-

Table 3.1 File naming conventions

Filename	File type
XXXXXXXC.BAS	Program written in BASIC which performs compression
XXXXXXXD.BAS	Program written in BASIC which performs decompression
XXXXXXXC.C	Program written in C which performs compression
XXXXXXXD.C	Program written in C which performs decompression
XXXXXXX.DAT	Non-compressed data file
XXXXXXXC.DAT	Compressed data file
XXXXXXXD.DAT	Decompressed data file
XXXXXXXC.EXE	Executable program written to perform compression
XXXXXXXD.EXE	Executable program written to perform decompression

resent a mnemonic of up to seven characters that will be used to further describe the function of a file. For example, the seven Xs would be replaced by 'NULL', resulting in the filename NULLC.C to represent the file that contains a program written in C to demonstrate null compression. Similarly, the file NULL.DAT would represent the non-compressed data file used by the file NULLC.C or NULL.BAS to demonstrate the effect of compression. In examining the entries in Table 3.1, you might be confused by the difference between the files XXXXXXX.DAT and XXXXXXXD.DAT. In actuality, the contents of both files should be exactly the same, as the decompressed file should be an exact copy of the original file. However, as noted in Chapter 1, when working with lossy compression this may not be the case. In addition, the programming examples created by this author were developed to retain the original non-compressed data files for comparison purposes. Thus, the filename convention XXXXXXXD.DAT was used to denote the resulting decompressed file created by the execution of the program stored on the file named XXXXXXXD.EXE.

The files XXXXXXXC.EXE and XXXXXXXD.EXE represent executable programs to perform compression and decompression, respectively. As .EXE programs you can execute those programs by simply entering their name without a file extension. The executable programs included on the accompanying diskette represent compiled versions of C language programs which provide readers with an ability to use the C language programs without having to have a C language compiler.

Compression program

To facilitate an examination of the operation of C language programs developed to illustrate different types of compression and decompression, a series of common or near-common functions were developed for inclusion in each program developed by the author. Those common or near-common functions include file operation statements, compression and decompression statements, statements for tallying the compression count or number of inserted characters, and statements which illustrate the contents of the file operated upon before and after the program. The statements in the last function also provide a summary of the characters eliminated or inserted based upon the type of operation performed.

Figure 3.2 lists the C language code for the program NULLC.C which is contained on the diskette accompanying this book. Although we will examine the entire program contained in Figure 3.2, for subsequent programs we will primarily focus our attention upon the major functions of each program and simply reference the operation of several previously described functions.

In examining the coding prior to the 'main' function, note that the define statements are used to associate the suggested names of I/O files to statements in the 'filein' and 'summary' functions which generate file-related messages. For example, the statement 'printf("Enter non-compressed ASCII Filename. Eg, %s :",infile);' results in the display of the message 'Enter non-compressed ASCII Filename. Eg NULL.DAT' since the DEFINE statement associates the string 'NULL.DAT' to the variable 'infile'. The use of the DEFINE statements permits the functions 'filein' and 'summary' to be used either 'as is', or with slight modification throughout this chapter.

The arrays 0 and buffer were both set to 256 as that array size can store the longest line length supported by DOS. The array 0 will be used as the output buffer, while the array buffer will be used as the input buffer.

After the use of the DEFINE statements and allocation of arrays, each program in this chapter will initialize appropriate counters. Next, the 'main' function will call the 'filein' function, which will prompt the user to enter the appropriate ASCII file. The 'while' loop cycles through each line in the file, setting n to the length of the recently read line. This is followed by the execution of the functions 'nullit' which performs the actual compression and 'tally' which keeps a running total of characters eliminated due to null compression. Once the input file is processed, the function 'summary' displays the 'before and after' contents of the file just operated upon and a summary of the characters eliminated.

In developing each compression-performing program, the author included an abundance of comments to document the actions of each program. Most decompression versions of the compression program will contain fewer comments, mostly used to document the actual decompression operations. Although most coding is self-explanatory for persons with a background in C programming, a few statements require a degree of elaboration. First, the IF statement in the NULLIT function tests for the current and next position in the input buffer equaling the space character which has an ASCII value of 32. Thus, the program actually suppresses space characters instead of nulls which have an ASCII value of 127. This was done for illustrative purposes since it is easier to generate spaces from the keyboard. Secondly, the statement o[j]=125 results in the use of the right brace } as the null suppression-indicating character.

In examining the program listing for NULLC.C and NULLD.C readers should note that the programs do not check for the natural occurrence of a right brace in the datastream. Thus, readers may wish to consider modifying the code to insert a right brace if this character is encountered in the data stream when performing null compression. When performing null decompression, you would modify the program NULLD.C to check for the occurrence of two right braces in sequence which would indicate that a brace naturally occurred in the data. Similarly, since the ASCII value of the right brace is 125, you would not compress the occurrence of exactly 125 nulls as doing so would result in the replacement of that sequence by two right braces. Instead, you would break the occurrence of a string of 125 nulls into two strings so you could continue to use the insert/delete technique to preclude the adverse effect of a compression-indicating character naturally occurring in a data file.

Program execution

The execution of the program NULLC.C is illustrated in Figure 3.3. At the top of that illustration the program lists the contents of the test file NULL.DAT developed for illustrative purposes. The lower portion of Figure 3.3 lists the contents of the compressed file. Since ASCII values less than 32 are non-printable, you cannot note the count character displayed after the right brace in line 4. However, you will note that line 3 has 49 spaces. Since ASCII 49 is the digit 1, line 3 in the compressed file printout verifies that 49 spaces were replaced by the right brace followed by the numeric 1.

Decompression program

In following our naming conventions the program NULLD.C represents the C language program developed to decompress files previously compressed using the NULLC.C program. The complete C language program is included on the diskette accompanying this book

```
/* NULLC.C - Program to demonstrate null surpression (actually ASCII 32    */

#include <stdio.h>
#define infile "NULL.DAT"      /* define name of I/O files suggested for  */
#define outfile "NULLC.DAT"    /* use in this program                     */
      char  o[256],            /* output buffer                           */
      filename[13],            /* store filename                          */
      buffer[256];             /* input buffer - max string length        */

int   t=0,                     /*                                         */
      t1=0,
      n=0,
      n1=0,

FILE  *input,
      *output;

main()
{
   filein();
   while( fgets(buffer,256,input) != NULL )
   {
      n = strlen(buffer);              /* get length of string             */
      tally( nullit() );               /* check for nulls, & tally         */
   }
   summary();
   return 0;
}

nullit()
{
   int   i=0                           /* set indices                      */
         j=0,                          /* buffer position                  */
         k=1;                          /* null count                       */

   unsigned char   a,
                   b;

   for(i=0;i < n; i++)                 /*step through record               */
   {
      a = buffer[i];                   /* extract a character              */
      if ((a ==' ') && (a == buffer[i+1]))  /* two nulls in a row          */
      {
         b=a;                          /* set next = current               */
         k++;                          /* increment count                  */
         continue;
      }
      if ( k> 2)                       /* have null suppression            */
      {
         o[j] = 125;                   /* set flag                         */
         o[j+1] = k;                   /* insert null count                */
         j+=2;                         /* increment buffer position        */
         k = 1;                        /* reset count                      */
      }
       else
        {
         if ( k== 2)                   /* only two nulls in sequence       */
         {
            o[j]= b;
            o[j+1] = b;
            j+=2;
            k=1;
         }
         else
         {
          o[j] = a;
          j++;
```

Figure 3.2 NULLC.C program listing

```
            }
          }
        }
    return(j);
}

tally(j)                                    /* tally the compression count   */
int    j;
{
    int    i=0;

    n1=n1+n;
    t=n-j;
    t1=t1+t;
    for(i=0; i<j; i++)
    {
        fputc(o[i],output);
    }
    return 0;
}

filein()
{
    printf("Enter non-compressed ASCII Filename.Eg, %s  :", infile);
    scanf("%13s",filename);

    if ( (input  = fopen(filename,"rb")) == NULL )
    {
        printf("ERROR: Cannot open the file: %s \n",filename);
        exit(-1);
    }
    if ( (output = fopen(outfile,"wb")) == NULL )
    {
        printf("ERROR: Cannot open the file %s for output\n",outfile);
        exit(-2);
    }
    return 0;
}

summary()
{
    fclose(input);
    fclose(output);
    if ( (input  = fopen(filename,"RB")) == NULL )
    {
        printf("ERROR: Cannot open the file %s for input\n",filename);
        exit(-1);
    }
    printf("File  %s Before Compression:\n",filename);

    while( fgets(buffer,256,input) != NULL )
    {
        printf("%s",buffer);
    }
    fclose(input);
    if ( (input = fopen(outfile,"rb")) == NULL )
    {
        printf("ERROR: Cannot open the file %s for output\n",outfile);
        exit(-2);
    }
    printf("\fFile %s After Compression :\n", outfile);

    while( fgets(buffer,256,input) != NULL )
    {
        printf("%s",buffer);
    }
    fclose(input);
    printf("\n%d Total Characters Eliminated From %d Or %d%%\n\n",
    t1,n1,(int)((float)((float)t1/(float)n1)*100));
    return 0;
}
```

Figure 3.2 Continued

```
C:\TC>nullc
Enter non-compressed ASCII Filename.Eg,      NULL.DAT :null.dat
File  null.dat Before Compression:
1)this is a test
2)012345678901234567890123456789012345678901234567890
3)*                                                 *
4)* of null suppression                  only      *

File NULLC.DAT After Compression :
1)this is a test
2)012345678901234567890123456789012345678901234567890
3)*}1*
4)* of null suppression}only}*

68 Total Characters Eliminated From 183 Or 37%


C:\TC>
```

Figure 3.3   Execution of the program NULLC.C

on the file NULLD.C. In this section we will focus our attention upon the decompression function and point out some minor changes between this program and the previously described NULLC.C program in the program header and some functions.

Since the input file for NULLD.C is now NULLC.DAT while the output file becomes NULLD.DAT, the two DEFINE statements were changed to reflect those files. That is, #define infile "NULLC.DAT" and #define outfile "NULLD.DAT" are now used to assign the strings NULLC.DAT and NULLD.DAT to the variables infile and outfile for use in the program.

Figure 3.4 lists the coding for the functions 'main', 'nullit' and 'tally' for the program NULLD.C. Since NULLC.C was developed to compress sequences of spaces, NULLD.C was developed to restore spaces once the compression-indicating character is encountered. Thus, the statement IF (BUFFER[i]==125) checks for the compression-indicating right brace. When encountered, the statement o[j++]=" "; sets the *j*th position of the output buffer to a space and increments the *j* index by one.

Program execution

The execution of NULLD.C reads the previously compressed file, creates a decompressed file and displays the contents of the compressed and decompressed files. Figure 3.5 illustrates the execution of the executable program file NULLD which was created using the Borland Turbo C compiler.

In examining the decompressed file shown in the lower portion of Figure 3.5 you will note that it matches the top portion of Figure 3.3. In addition, note that the characters inserted equal the characters previously eliminated.

```
main()
{
    filein();
    while( fgets(buffer,256,input) != NULL )
    {
        n = strlen(buffer);              /* get length of string         */
        tally( nullit() );               /* check for nulls, & tally     */
    }
    summary();
    return 0;
}

nullit()
{
    int   i=0,                           /* set indices                  */
          k=0,                           /* null count                   */
          l=0,                           /* loop for output buffer       */
          j=0;                           /* buffer position              */

    for(i=0;8 <n; i++)                   /* step through record          */
    {
        if (buffer[i] == 125)            /* compression indicator ?      */
        {
            k = buffer(i+1);             /* get null count               */

            for (1=0; 1<k; 1++)          /* set nulls in output buffer   */
            {
                o[j++]=' ';              /* set to null                  */
            }
            i++;                         /* skip over null count         */
        }
        else
        {
            o[j++] = buffer[i];          /* just a non null character    */
        }
    }
    return(j);
}
tally(j)                                 /* tally the compression count  */
int   j;
{
    int    i=0;

    n1=n1+n;
    t-n-j;
    t1=t1-t;
    for(i=0; i<j; i++)
    {
        fputc(o[i],output);
    }
    return 0;
}
```

Figure 3.4 Major functions of the program NULLD.C

Limitations

Since a two-character compression sequence always results from the compression of up to 255 sequentially encountered nulls, no savings are possible unless three or more sequential nulls are found. Thus a

```
C:\TC>nulld
Enter compressed ASCII Filename.Eg,  NULLC.DAT  :nullc.dat
Contents of compressed file nullc.dat :
1)this is a test
2)012345678901234567890123456789012345678901234567890
3)*}1*
4)* of null suppression}only}*

Decompressed file contents :

1)this is a test
2)012345678901234567890123456789012345678901234567890
3)*                                                   *
4)* of null suppression                     only      *

68 Total Characters Inserted


C:\TC>
```

Figure 3.5 Execution of the program NULLD.C

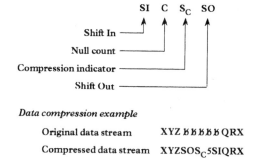

Figure 3.6 Null suppression using an extended code

sequence of two nulls should not be placed into the null suppression compressed format. This is because no savings would result, while the compression and decompression process requires a portion of processor time. In addition, if one is employing several datacompression techniques, we will see that two sequentially encountered nulls can be effectively compressed by the Diatomic encoding process. This compression technique results in a 100% data reduction for the two-null sequence situation where the null suppression technique is ineffective.

One key limitation associated with the use of null suppression concerns the selection of the special compression-indicating character. If an undefined character is available from the character set you are using, then savings begin to accrue when three or more sequential nulls are encountered. If you must use an extended character set first defined by the shift out (SO) character, you will require a four-character sequence to encode a run of nulls. Figure 3.6 illustrates the compression format and an example of null suppression using an extended code.

As noted from the example in Figure 3.6, null suppression is only viable for a sequence of five or more nulls when an extended code must be used to form the special compression-indicating character.

To consider the insert and delete technique for the use of the compression-indicating character you should have knowledge of the frequency of occurrence of the selected character for insertion and deletion in comparison to the potential saving that may accrue from null suppression. If the frequency of occurrence exceeds the potential percentage of data reduction, your algorithm would result in data expansion even though it would actually suppress all occurrences of three or more nulls.

While null suppression is viewed as an elementary data-compression technique, it is very easy to implement and its payoff can be substantial. Throughput gains of between 30 and 50% have been reported for a number of computer installations that switched from the 2780 bisynchronous transmission control sequence that does not compress data to the 3780 sequence that performs null suppression.

Technique variations

Two variations of null suppression can be used to compress portions of documents containing predefined or variable indentations. In one situation, it might be beneficial to reserve a group of characters from the character set to represent several predefined numbers of spaces or nulls. Thus, one character might then represent the indentation in a letter of five spaces, while a second character could be used to represent 20 spaces required to tab over to the beginning of a column within a document.

Since predefined indentations represent tab stop positions, a second variation of null suppression is obtained from the employment of the tab character. If tab stops are predefined, you have only to replace a sequence of spaces or nulls by the tab character to signify that the next character begins in a particular column on the line, and all columns between the last character and the location where the next character begins are spaces or null characters. To illustrate this concept, let us assume that a portion of the document we wish to transmit is as follows:

Now is the time to examine the relationship of defence expenditures upon the economy. For the years 1980 to 1984 our analysis shows:

Year	Guns	Butter
xxxx	yyyy	zzzz

Note that there are four distinct tab stop locations in this document—the indentation of a paragraph and the three column positions. Thus, a tab stop followed by the character 'N' could be used to position the beginning of the paragraph into its appropriate location. Since the indentation occurs prior to the first column position, to position the 'Y' in year would require two tab stops to be issued. Similarly, the 'G' in guns would have to be preceded by three tab stop characters and so on.

As the number of unique indentation and column location positions increases in a document or between different documents, the number of tab stop characters that may have to be issued to represent a predefined location could result in the expansion of data instead of its compression. To prevent such situations from occurring, as well as to eliminate the requirement of having prior knowledge about indentation and column locations, a variable tab stop procedure can be employed. In using variable tab stops you simply substitute a tab stop character and the column position to tab to in place of the spacing between columns. Returning to the previous example, if 'year' began in column 15 while 'guns' and 'butter' began in columns 30 and 45 respectively, the line column heading labels could be replaced by the sequence Ts15 Year Ts30 Guns Ts45 Butter, where Ts represents the tab stop character.

One of the key applications for the replacement of strings of nulls by tab stops is IBM mainframe CICS transmissions. Several third party software developers market products which intercept outbound CICS transmissions to terminals and, among many functions, replace strings of nulls by tabbing sequences. By transmitting a lesser amount of data both the efficiency of the transmission system as well as user response times are improved.

3.2 BIT MAPPING

This compression technique is effective when the data to be operated upon consists of a high proportion of specific data types, such as numerics, or a large proportion of a specific character, such as blanks. As the name implies, a bit map is employed to indicate the presence or absence of data characters or the fact that certain data characters have been operated upon previously and must be operated upon again to return the data into its original format.

Encoding process

To examine the bit mapping technique and its applications, we will first see how it can be employed to implement a version of null suppression. In the left-hand portion of Figure 3.7, a portion of a data stream consisting of three data characters and five nulls is illustrated. Here, the five nulls represent 62½% of the content of the string and are spread throughout the data stream in a random sequence. Since null suppression is only effective when three or more sequential blanks are encountered, its use would only reduce the string from eight to seven characters in length.

Through the use of a bit map appended in front of the string, we can indicate the presence or absence of nulls and thereby reduce the size of the data string. In the lower portion of Figure 3.7, the employment of a bit map character is illustrated where all nulls are dropped from the data string and the bit which corresponds to the null position is set to zero while the bit position in the map which corresponds to a non-null or data character is set to one.

In comparing the compressed data string with the original data string, the eight characters of data to include nulls have been reduced to four characters, three data characters and the bit map character. This results in a compression ratio of 2:1 for this particular application.

Hardware considerations

The bit map character illustrated in Figure 3.7 denotes non-null data character positions by location, from left to right. By reversing the bit map order, the data element positions can be indicated from right to left. Figure 3.8 indicates the two different methods of forming the bit map to represent the compressed data string. Using the bit map data element positioning technique illustrated in the lower portion of Figure 3.8, the bit map character resulting from the original data stream as illustrated in Figure 3.7 would become 10001001. The instruction set of the hardware device under consideration for performing the bit

Original data string

Data character 1	Null character	Null character	Data character 4	Null character	Null character	Null character	Data character 8

Compressed data string

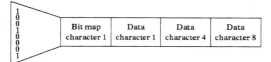

Figure 3.7 The bit mapping process. In a typical data stream, there is a high probability that one or more characters are repeated. Using one character to serve as a bit map can serve to eliminate the high frequency of occurrence of characters from a data stream

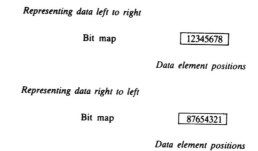

Figure 3.8 Bit map element positioning. Two methods can be employed to represent the compressed data string in the bit map—data represented left to right and right to left

map suppression technique will govern the method of bit map element positioning to be employed. This can be easily explained by first examining a flow chart of the functions that have to be performed on the original data string in order to construct the bit map and the compressed data string.

The bit map suppression process is illustrated functionally in Figure 3.9. The software routine to compress data must first initialize the bit map position counter (1), the bit map (2) and a character counter (3). After a character is obtained (4), the character counter is compared with eight (5). If a match occurs, eight incoming characters have been processed and we can exit from the routine (10). If no match occurs, the character counter is incremented (6) and the character under examination is compared with a null character (7). If the character under examination is not a null, the bit map position is set equal to a binary one (8). If the character is a null, this function (8) is bypassed. Next, the bit map position is either incremented or decremented (9) so that the bit map is prepared to be set to a zero in the following bit location if the next character examined is a null. Finally, after eight characters have been processed, the count equals eight (5) and the routine exit branch is taken (10).

From a hardware standpoint, the method used to perform the functions indicated in blocks (8) and (9) of Figure 3.9 depends upon the shift and logical instructions available for programmer utilization. This interrelationship can be viewed by denoting the effect on the bit map character as succeeding data characters are examined. In Figure 3.10, the effect on the bit map and 'mask' as a progression of data characters is examined is illustrated. Here, the mask is simply a binary one that is shifted through the 8-bit map positions and logically 'OR'd' with the bit map when the data character is not a null. In examining the mask, we can note that a logical or arithmetic left shift operation is required if we wish to position our bit map so that the right-hand bit indicates the presence or absence of a null character in the first element of the original data string. Thus, from a hard-

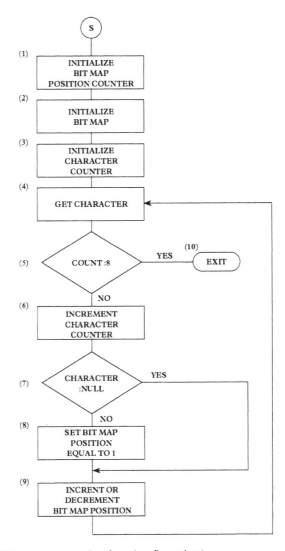

Figure 3.9 Bit map suppression function flow chart

ware viewpoint, the shift instruction available will be a governing factor with respect to how the bit map elements are positioned. Although
most microprocessors, minicomputers and certainly all large
computers have both left and right shift functions, a few microprocessors may have limited shifting capability. Such capability
should be examined prior to attempting to implement this technique.

Suppression efficiency

In the previous example, the bit map character contained eight bits.
While the example showed a 50% reduction in characters from the

Data character	Initial bit map	Mask	Bit map ⊖ mask (or bit map if null)
Data	00000000	00000001	00000001
Null	00000001	00000010	00000001
Data	00000001	00000100	00000101
Data	00000101	00001000	00001101
Null	00001101	00010000	00001101
Null	00001101	00100000	00001101
Null	00001101	01000000	00001101
Data	00001101	10000000	10001101

Figure 3.10 The bit map masking process. The mask character is a binary one shifted through all bit positions and logically OR'd with the bit map when the data character is not a null

original data string to the compressed data string, consider what happens to the compression efficiency when the percentage of nulls in the data string decreases. Table 3.2 shows the compression ratio based upon the percentage of nulls contained in the string for an 8-bit map. When there are no null characters, the resultant data string increases in size by one character as a result of the addition of the bit map character, producing a compression ratio of 0.888. This means that for the worst case situation where there are no null characters to be suppressed, an extra 12.5% of data will result from the employment of this compression technique.

We can develop a mathematical model of suppression efficiency as follows. If p is the probability that any given character is a null, the expected number of nulls in a string of length S characters is Sp. Using null compression this will be encoded as a string of length

$$S*(1 - p) + \left\lceil \frac{S}{8} \right\rceil$$

Table 3.2 Compression efficiency and null percentage

Null percentage	Resultant string size	Compression ratio
0.0	9	0.888
12.5	8	1.000
25.0	7	1.143
37.5	6	1.334
50.0	5	1.600
62.5	4	2.000
75.0	3	2.667
87.5	2	4.000
100.0	1	8.000

and the compression ratio is then

$$\left(\frac{1}{S} \left\{ S(1-p) + \left[\frac{S}{8} \right] \right\} \right)^{-1} \simeq \left((1-p) + \frac{1}{8} \right)^{-1}$$

for large values of S.

Bit map variations

In the previous discussion of the bit map procedure, we have assumed that either a null or another character appearing in large proportion to the remainder of the data is to be suppressed. For some applications, there is no particular character that is encountered more frequently than other characters; however, in certain cases one may encounter a situation where a specific type of data, such as numerics, frequently appears. One application where such a situation could exist is the process control area where numeric readings of various equipment are transmitted to a central site for processing and control signals are returned to the devices based upon certain predefined criteria. Depending upon the transmission code employed, certain economies may be obtained by the use of the bit map technique. If the data to be transmitted is in the extended binary-coded decimal interchange code (EBCDIC), then the first four bit positions of each numeric character are all ones. Thus, the bit map character could be employed to denote the number of packed characters in the compressed string, each character containing two digits with the leading four bit positions stripped. This technique is illustrated in Figure 3.11 and is quite similar to the half-byte packing technique that is covered later in this chapter.

Original data string

1
8
4
6
2
7
9
8
3
2

Compressed data string

00001010 BIT MAP	
8	1
6	4
7	2
8	9
2	3

Figure 3.11 Half-digit suppression. In the half-digit suppression technique, the contents of the bit map specify the number of digits that follow, packed two per character

Technique constraints

One key limitation of the bit map technique is that it is applicable to data having fixed size units, such as characters, bytes or words. When used to suppress a particular character, such as a null, the compression ratio of this technique is directly proportional to the percentage of occurrence of that character in the original data stream. Thus, if one character in a data string occurs 30% of the time while the second most frequently encountered character occurs, say, 25% of the time, this technique ignores the high percentage of occurrence of the second character or any other characters. As we shall see, a technique known as run-length encoding can be employed to take advantage of the adjacent redundancy of occurrence of all characters in a data stream.

Two other limitations associated with the bit map technique include the requirement to operate upon only one type of character, such as nulls or digits, and developing a method to recognize that a character represents a bit map instead of data. Let us examine how we can overcome the second limitation when using 7-level ASCII data and how we can develop a method to overcome that limitation when using other character sets.

If we are using 7-level ASCII we can strip the parity bit and use that eighth bit position as a bit map indicator. Although this technique reduces the maximum compression ratio that can be achieved through the use of the bit map technique, it also eliminates the possibility of data expansion. To illustrate how we can avoid data expansion let us examine the operation of this revised technique.

Figure 3.12 illustrates the formation of a bit map indicator through the setting of bit position 8 which in 7-level ASCII is used as a parity bit. In the lower portion of Figure 3.12 the resulting compressed data stream is shown based upon an original data stream of seven characters. Now let us assume that the composition of the next seven characters changes so that no nulls are encountered. Under this revised technique the characters are simply placed into the compressed data stream without the necessity of adding a bit map which would result in the expansion of data. Here the data is recognized as data since the parity bit position is not set. Another advantage associated with this revised compression technique is the fact that the setting of the former parity bit position results in the automatic recognition of the character as a 7-position bit map. This allows you to eliminate coding to keep track of bit maps versus data.

If you are using each bit position within a character, such as with 8-level EBCDIC, you must then keep track of the relationship between bit map characters and data characters. To do so, you can simply count the number of set bits in the first bit map which then tells you the number of characters to skip prior to reading the next bit map. Figure 3.13 illustrates the use of the count of the set bits in the bit map to determine the location of the next bit map in the data stream.

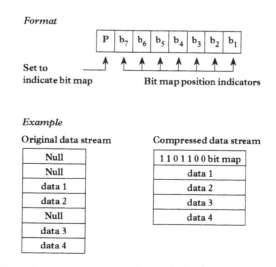

Figure 3.12 Using the parity bit as a bit map indicator

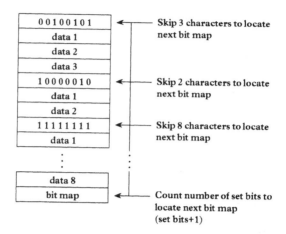

Figure 3.13 Locating bit maps. Succeeding bit maps can be located in the data stream by adding 1 to the count of the number of bit sets in the prior bit map

3.3 RUN LENGTH

Run-length encoding is a data-compression method that will physically reduce any type of repeating character sequence, once the sequence of characters reaches a predefined level of occurrence. For the special situation where the null character is the repeated character, run-length compression can be viewed as a superset of null suppression (Rubin, 1976; Ruth and Kreutzer, 1972).

Operation

In a similar way to the method used to effect null suppression, the employment of run-length encoding normally requires the use of a special character to denote that this type of compression has occurred. One exception to the use of a special compression-indicating character is the Microcom Networking Protocol (MNP) Class 5 method of data compression which is described later in this book. When a compression-indicator character is used, it is normally followed by one of the repeating characters which was in the encountered string of repetitious characters. Finally, a count character signifies the number of times the repeated character occurred in the sequence.

When codes such as ASCII or EBCDIC are employed, a good choice for the special character is one that will not occur in the data string. For each of these codes there are numerous unassigned characters with unique bit representations that can be used. For situations where the character set contains no unused character, such as in the

Baudot 5-level (bit) code, this technique may still be used by selecting a character that may not be used twice in succession, such as a letter shift or figure shift, to indicate that compression has occurred. The reader should refer to Chapter 2 for additional information concerning the selection and utilization of compression indicating characters from different character codes.

Encoding process

The run-length compression process results in a string of repeated characters being converted into a compressed data string as shown in Figure 3.14 (Aronson, 1977). With three characters required to denote compression, run-length encoding is only effective when a data string contains a sequence of four or more repeated data characters.

Three examples of the application of run-length encoding upon repeating character sequences are presented in Table 3.3. Note that S_C represents the special character used to indicate the occurrence of run-length encoding, while the symbol $\emptyset$ is used to indicate the presence of a blank character.

With the null suppression format requiring two characters,

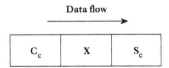

Data flow

| C_c | X | S_c |

Figure 3.14 Run-length encoding, general compression format. In run-length encoding, a special character, repeated data character and character count character are required to indicate the compression parameters

S_c = Special character indicating compression follows.

X = Any repeated data character.

C_c = Character count. This count is the number of times the compressed character is to be repeated

Table 3.3 Applying run-length encoding

Original data string	Encoded data string
$******55.72	$$S_c$*655.72
---------	S_c-9
Guns$\emptyset\emptyset\emptyset\emptyset\emptyset\emptyset\emptyset\emptyset\emptyset\emptyset\emptyset$Butter	Guns$S_c$$\emptyset$10Butter

employing run-length compression to suppress nulls always results in one additional character generated in the compressed data stream. While this is not significant when long strings of nulls are compressed, numerous short strings of nulls could result in an excess quantity of compressed data. This suggests that you should consider the use of a mixture of several algorithms to perform data compression.

The major steps in the run-length encoding process are shown in Figure 3.15 through the use of a systems flow chart. Initially, a character counter (1) and character repetition counter (2) are set to zero. After a character in the original data string is obtained (3), the character counter is incremented (4) by one. The character count is

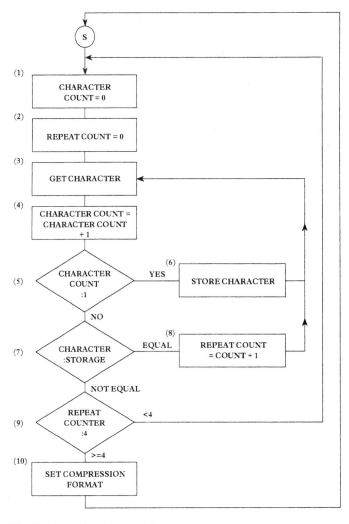

Figure 3.15 Basic run-length encoding process

then compared with one (5). In the first cycle, this comparison always holds true and the character is then placed in a buffer (temporary storage) area (6) for later processing if the original data string is found to contain four or more repetitive data characters. For the second and subsequent cycles, the character obtained from the original, data string (3) is compared with the character placed in storage (7). If the present character is equal to the character in storage, compression may be possible if four or more identical characters are encountered in sequence. Thus, when the character equals the stored character, the repeat counter (8) is incremented by one and another character is obtained from the original data string (3). If the present character under examination does not equal the character stored (7), the repeat counter is compared with four (9). If less than four, no compression is worthwhile since three characters must be used to encode compressed data. When the repeat counter is equal to or greater than four (9), the compression format (10) can now be set.

Special considerations

In the basic encoding flow chart illustrated in Figure 3.15, it was assumed that the repeat counter was capable of having an unlimited range of values. In reality, the maximum value that the repeat counter can contain is a function of the character code level employed. For an 8-level (eight bits per character) character code, a maximum of between 255 and 260 repetitive characters can be represented by the character counter. The exact value will depend upon how the character counter is employed. In most situations, the actual character counter value is used as the number of repetitive characters. In this mode, the counter's maximum value is $2^8 - 1$ or 255. Since the compression format illustrated in Figure 3.14 occurs only when four or more repetitive characters are encountered, the presence of a character count character in itself implies that four or more repetitive characters exist. Thus, a character counter of all bits zero can be used to indicate four repetitive characters, while a character counter of all bits set to 1 would then indicate 260 repetitive characters. Once the method of employing the repeat counter is determined, the flow chart in Figure 3.15 must be modified to add an additional repeat counter comparison to test for the maximum value permitted to be stored in the character counter.

Decoding

The functions necessary to decompress data compressed according to the run-length encoding process are illustrated in Figure 3.16 in flow chart format. At the beginning of the decompression procedure,

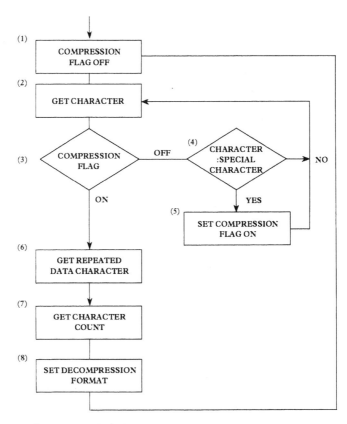

Figure 3.16 Run-length decoding process

a compression flag is turned off (1) and a character is obtained from the compressed data string (2). Next, if the compression flag is off (3), the character is compared with the special run-length compression-indicator character (4) to determine if run-length compression has occurred. If the character is not the special character, the next character is obtained (2). If the character is the run-length compression-indicator character, the compression flag is turned on (5) and the next character is obtained (2). On the next pass, since the compression flag is on (3), the following character obtained (6) is the repeated data character while the next character (7) contains the character count. Once these characters are obtained, the decompression format can be initiated (8).

Utilization

One of the most popular utilizations of run-length encoding is the subset known as null suppression. This compression technique is primarily encountered in the IBM 3780 BISYNC protocol. Space compression is a standard feature of this protocol when the 3780 device is operating in the line mode with non-transparent data. Here, each group of two or more consecutive space characters, up to 63, is replaced by an IGS character if the transmission code is EBCDIC or a GS character if the code is ASCII. Either character is followed by a space-count character that defines the number of spaces removed. For the situation where 64 or more consecutive space characters occur, an additional IGS or GS character and space-count character are inserted.

On Honeywell and some Bull successor computer systems, a version of run-length compression is used in their general remote terminal system (GRTS) software on front-end processors communicating with remote terminals under the remote computer (RC) protocol. In addition, the same type of compression is used on the Honeywell stand alone tape-to-tape system (SATTS) for the transmission of reels of magnetic tape between locations. In this version of run-length encoding, a record is examined for a series of three or more occurrences of the same data character. When such a situation occurs, the series is compressed and a string of repeated characters is converted into a three-character sequence as illustrated in Figure 3.17.

The most popular implementation of run-length encoding actually represents a slight variation of the previously described data-compression technique. The variation of run-length encoding that has achieved an installed base of approximately one million products is the run-length encoding technique employed by Class 5 of the Microcom Networking Protocol (MNP) which is licensed for use by over 50 manufacturers of switched network modems.

MNP Class 5 data compression uses a combination of two data compression techniques—run-length encoding and adaptive-frequency encoding. Since the adaptive-frequency encoding technique can be classified as a statistical-based compression method we will examine that technique in Chapter 4.

Under the MNP Class 5 run-length encoding method, a repetition count is inserted into the data stream to represent the number of

Figure 3.17 Honeywell and Bull version of run-length encoding. Run-length encoding as implemented on Honeywell and Bull computer systems differs slightly from most other computer manufacturers

X = Any repeated data character.

US = The ASCII character (0011111).

C_c = A 6-bit binary count. The BCD character represented by the binary count must be translated to ASCII for transmission to the communications subsystem. This count is the number of times the compressed character is to be repeated (maximum 63)

repeated octets which follow the first three occurrences of a sequence. Since the MNP version of run-length encoding sends the first three repeated octets, those repeating characters serve as an indicator of the beginning of run-length encoding and, in effect, eliminate the necessity of utilizing a special compression-indicating character.

The top portion of Figure 3.18 illustrates the format of MNP Class 5 run-length encoding.

The lower portion of Figure 3.18 contains five examples of the operation of MNP Class 5 run-length encoding upon different strings. Note that under this version of run-length encoding a repeated sequence of precisely three characters results in the use of a repetition count of zero being inserted into the compressed-data stream. In addition, the maximum repetition count supported by MNP is limited to 250 decimal. Thus, this portion of MNP Class 5 data compression can result in data expansion when sequences of three repeating characters are encountered in the original data stream. In addition, the MNP version of run-length encoding requires four characters, whereas the previously described method of run-length encoding uses three characters. Lastly, conventional run-length encoding can be implemented to compress up to 256 repeating characters, whereas MNP Class 5 is limited to 250. These disadvantages must be weighted against the fact that the MNP Class 5 version of run-length encoding eliminates the necessity of using a special compression-indicating character since the three repeating characters both indicate that compression has occurred and identify the character that was compressed.

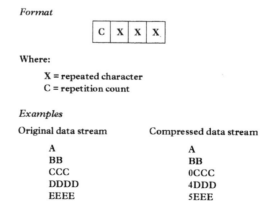

Format

| C | X | X | X |

Where:

X = repeated character
C = repetition count

Examples

Original data stream	Compressed data stream
A	A
BB	BB
CCC	0CCC
DDDD	4DDD
EEEE	5EEE

Figure 3.18 MNP Class 5 run-length encoding. Since three repeating characters indicate run-length encoding the MNP Class 5 run-length compression format does not require the use of a special compression-indicator character

Efficiency

Run-length encoding efficiency depends upon the number of repeated character occurrences in the data to be compressed, the average repeated character length and the technique employed to perform compression. In Table 3.4, the reader will find a listing of the results

Table 3.4   Run-length encoding efficiency based upon original data string of 1000 characters

Number of repeated character occurrences	Average repeated character length	Compression ratio
10	4	1.010
10	5	1.020
10	6	1.031
10	7	1.042
10	8	1.053
10	9	1.064
10	10	1.075
20	4	1.020
20	5	1.042
20	6	1.064
20	7	1.087
20	8	1.111
20	9	1.136
20	10	1.163
30	4	1.031
30	5	1.064
30	6	1.099
30	7	1.136
30	8	1.176
30	9	1.220
30	10	1.266
40	4	1.042
40	5	1.087
40	6	1.136
40	7	1.190
40	8	1.250
40	9	1.316
40	10	1.384
50	4	1.053
50	5	1.111
50	6	1.176
50	7	1.250
50	8	1.333
50	9	1.429
50	10	1.538

of the execution of a computer program written to compute the overall compression ratio based upon a varied number of repeated character occurrences in a string of 1000 data characters. Here, the number of repeated character occurrences was varied from 10 to 50 while the average repeated character length was varied from 4 to 10. It was assumed that three characters were used for the compressed data format. The computed compression ratios listed in Table 3.4 ranged from a low of 1.0101 to a high of 1.5384. Table 3.4 is a synthetic representation due to the wide divergence of actual text. Since this table covers most common compressible occurrences, it provides a handy tabular reference for readers to determine the effect of run-length encoding.

A second factor affecting the efficiency of run-length encoding is the mechanism used to implement this data-compression technique. Table 3.5 indicates the compression ratio for three common methods by which run-length encoding is implemented. The first column indicates the compression ratio when a special compression character is followed by the compressed character and a count, using a sequence of three characters to denote the occurrence of run-length compression. In the second column the compression ratio resulting from the use of an extended code set (ECS) is indicated. In this situation the SO and SI characters must be added to the beginning (SO) and terminate (SI) the previously described three-character sequence,

Table 3.5 Efficiency of run-length encoding methods

Number of repeated characters	Compression ratio		
	Run length	ECS run length	MNP run length
3	1.0000	0.6000	0.7500
4	1.3333	0.8000	1.0000
5	1.6667	1.0000	1.2500
6	2.0000	1.2000	1.5000
7	2.3333	1.4000	1.7500
8	2.6667	1.6000	2.0000
9	3.0000	1.8000	2.2500
10	3.3333	2.0000	2.5000
11	3.6667	2.2000	2.7500
12	4.0000	2.4000	3.0000
13	4.3333	2.6000	3.2500
14	4.6667	2.8000	3.5000
15	5.0000	3.0000	3.7500
16	5.3333	3.2000	4.0000
17	5.6667	3.4000	4.2500
18	6.0000	3.6000	4.5000
19	6.3333	3.8000	4.7500
20	6.6667	4.0000	5.0000

resulting in five characters being required to encode a string of repeating characters. The third column, which is labeled MNP Run Length, indicates the compression ratio when the MNP Class 5 method of run-length encoding is used. Under this technique, strings of three or more repeating characters are encoded as four characters.

As indicated in Table 3.5, the conventional method of run-length encoding provides the greatest compression ratio. This is because it requires the fewest number of characters to implement. MNP Class 5 is the next most efficient as it requires four characters, while the use of an extended character set is the least efficient method for implementing run-length encoding.

Programming examples

To illustrate the programming required to implement run-length encoding and other compression techniques several computer programs were developed. Each of these small program segments was originally written in the BASICA version of the BASIC programming language which operates on the IBM PC and compatible computers. The utilization of BASICA was based upon its wide acceptance as a programming language and the ability to use the language as a learning tool for a maximum number of readers to follow. For optimum usage of the programming examples presented in this book, we suggest that you should employ a BASIC compiler to speed up the execution of the examples. In this new edition, each BASICA program segment was recorded in Borland's C++ programming language. The C++ programs are contained in both source and compiled executable form on the diskette included with this book. The original BASICA program segments are included in their original source form. Throughout this book the original description of the BASICA program segment will be followed by a description of the C language version of the program.

In its internal operation, the IBM PC uses an 8-bit extended ASCII character code. This extended character code results in the assignment of distinct characters to ASCII values 128 through 255. Since every character from ASCII value 0 through 255 is defined and can occur when transmitting data from an IBM PC to another computer system, you might normally employ the ASCII SO (shift out) and SI (shift in) characters in developing a compression module designed to operate on ASCII data whenever there is a probability of occurrence for each character in the character set.

The SO character is used to shift out of the current ASCII character set, resulting in the ability of the user to redefine each character in the character set. Similarly, the SI character is used to shift back into the defined ASCII character set.

By using the SO and SI characters in ASCII you obtain a set of either 128 or 256 new characters, depending upon whether you are using a system that uses the 7-bit or an extended 8-bit ASCII code. This new character set can then be used to represent compression indicating characters.

Figure 3.19 illustrates the utilization of ASCII SO and SI characters to obtain a new character set where the ASCII value 082 (conventional ASCII R) is used to denote run-length encoding. In this example, a string of six Xs was assumed to be followed by a string of seven Ys. Since the ASCII value 082 in a newly defined ASCII code will be used to indicate run-length compression, you must first shift out (SO) to the new code, issue the compression-indicating character (R) and then shift back into (SI) the normal ASCII code to transmit the character that was compressed (X) and the quantity of X characters compressed (6). Due to the requirement to shift out of the character set to issue the compression indicating character and then to shift back to the normal ASCII character set, two additional characters are required to represent a run-length encoded string. In addition to this technique requiring two extra characters, the use of a shift out code of 14 is used to turn on the double-width mode setting of most dot matrix printers while the shift in code of 15 is used to turn on the compressed character mode setting of such printers, making the graphic illustration of this technique tedious at best.

Based upon the preceding information, it was determined that for the examples presented in this book, the use of a single character in the ASCII character set would sufficiently serve as a compression flag in addition to actually saving two characters in representing the compression of data based upon the use of run-length encoding.

Compression program

Figure 3.20 contains the listing of a BASIC program that illustrates the coding required to perform run-length compression.

In the RUNLENC.BAS program, the ASCII value of 125 (right brace, }) was used as the compression-indicating character, which was then

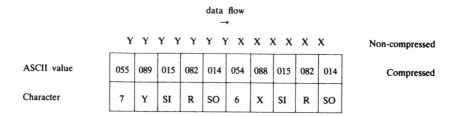

Figure 3.19 Using SO and SI characters. Using SO and SI provides a new set of characters that can be used to indicate different compression techniques, in this example, the ASCII value 082 is used to denote run-length encoding

followed by the ASCII character being compressed and its repetitious count in decimal notation. Thus, any string in excess of three repeating characters would be subject to compression.

Several statements in the program listing contained in Figure 3.20 warrant discussion for those readers unfamiliar with the IBM PC BASICA version of the BASIC programming language. The LINE INPUT statement in line 130 results in an entire line from a sequential file being read and assigned to the string variable X$. In line 140, the length of the string that represents one line in the data file is determined. The length of the string is then used in the FOR-NEXT loop bounded by lines 240 through 310 to process the string for repeating characters. The MID$ functions in lines 250 and 260 extract the Ith and Ith + 1 characters from the string and compare these characters to one another. When they are equal, the repeated character is saved (line 330) and the count of repeating characters is incremented (line 340). When the repeating string of characters is broken, line 260 is FALSE and a comparison of the repeating count occurs (lines 270 and 280). When the count exceeds three (line 270) data is compressed by the coding contained in lines 360 through 400. If the count equals three there is no advantage to be gained from run-length compression and the routine bounded by lines 420 through 450 simply adds the input characters to the output buffer. When the Ith and Ith + 1 characters are not equal, the Ith character in the input buffer is simply placed in the output buffer (line 290). Lines 900 through 9999 are not actually part of the run-length encoding process and are only included to facilitate file operations and comparison of the input and output buffers to obtain a measurement of the efficiency of this technique when applied to a data file containing a variety of repeating data strings.

Figure 3.21 illustrates a sample execution of the RUNLENC.BAS program using an ASCII file named RUNLENC.DAT as input to the program. Note that RUNLENC.BAS was purposely written to first list the contents of the file prior to its compression which is illustrated in lines 1 to 8 at the top of Figure 3.21. Next, the program lists the file after its contents were compressed based upon the application of run-length encoding to the data contained in the file.

It should be noted that string decimal values ranging below ASCII 32 were purposely omitted from inclusion in the test file since they would cause unwanted carriage returns, line feeds and other non-printable characters to be displayed, which would make an illustration of this compression technique difficult to comprehend. They would, however, be quite appropriate in normal string compression and decompression applications.

```
10 REM RUNLENC.BAS PROGRAM
20 DIM O$(132)
30 WIDTH 80:CLS
40 '**********MAIN ROUTINE************************
50 '* THIS ROUTINE READS RECORDS FROM AN ASCII *
60 '* FILE INTO A STRING CALLED X$ WHICH IS     *
70 '* THEN PASSED TO SUBROUTINES FOR COMPRESSION
80 '**********************************************
90 PRINT "ENTER ASCII FILENAME. EG, RUNLEN.DAT"
100 INPUT F$: OPEN F$ FOR INPUT AS #2
105 OPEN "RUNLENC.DAT" FOR OUTPUT AS #3
110 PRINT "PATIENCE - INPUT PROCESSING"
120 IF EOF(2) THEN GOTO 9000
130 LINE INPUT #2, X$
140 N= LEN(X$)
150 GOSUB 180
160 GOSUB 900
170 GOTO 120
180 '*****RUN LENGTH ENCODING SUBROUTINE********
190 '* THIS ROUTINE PROCESSES RECORDS FROM X$  *
200 '* AND COMPRESSES OUT REPETITIVE CHARACTERS*
210 '* USING O$ AS THE OUTPUT BUFFER.          *
220 '*********************************************
230 K=1:J=1                      'RESET INDICES
240 FOR I= 1 TO N                'STEP THRU RECORD
250 A$= MID$(X$,I,1)             'EXTRACT A CHAR
260 IF A$= MID$(X$,I+1,1) THEN 330 'SAME AS NEXT?
270 IF K>3 THEN 360              'COMPRESS
280 IF K=3 THEN 420              'DON'T COMPRESS
290 O$(J)=A$                     'STUFF IN OUTPUT BUFFER
300 J=J+1                        'BUMP BUFFER INDEX
310 NEXT I                       'GO BACK FOR MORE
320 RETURN                       'END OF STRING
330 B$=A$                        'SAVE REPEATED CHAR
340 K=K+1                        'BUMP COUNT
350 GOTO 310                     'KEEP LOOKING
355 '**************************************************************
360 'INSERT COMPRESSION NOTATION IN OUTPUT BUFFER
365 '**************************************************************
370 O$(J)=CHR$(125)             'SET FLAG FOR RUN-LENGTH
380 O$(J+1)=B$                  'INSERT REPEATED CHAR
390 O$(J+2)=CHR$(K)             'INSERT COUNT
400 J=J+3:K=1                   'RESET INDEX
410 GOTO 310
420 O$(J)=B$                    'STUFF 1ST REPEAT CHAR
430 O$(J+1)=B$                  'STUFF 2ND REPEAT CHAR
440 J=J+2:K=1                   'RESET INDEX
450 GOTO 310
```

Figure 3.20 RUNLENC.BAS program listing

C language program

To facilitate the use of programs by readers who prefer to use the C language, RUNLENC.BAS was rewritten using Borland's Turbo C. The resulting program, which is contained on the file RUNLENC.C on the

```
900 '*****TALLY THE COMPRESSION COUNT & WRITE BUFFER*****
910 '* DISPLAY BEFORE & AFTER RESULTS OF COMPRESSION     *
920 '* AND SHOW THE NET RESULTS OBTAINED BY EACH METHOD  *
930 '***************************************************
931 N1=N1+N                              'TALLY INPUT CHAR COUNT
932 T=N-J+1                              'NET DIFFERENCE IN BUFFERS
936 T1=T1+T                              'SAVE COUNT FOR SUMMARY
940 FOR I= 1 TO J-1
950 PRINT #3, O$(I);
960 NEXT I
965 PRINT #3, ""
970 RETURN
9000 CLOSE: OPEN F$ FOR INPUT AS #2
9010 PRINT "FILE ";F$;" BEFORE COMPRESSION:"
9020 LINE INPUT #2,X$
9030 IF EOF(2) THEN 9060
9040 PRINT X$
9050 GOTO 9020
9060 PRINT X$:OPEN "RUNLENC.DAT" FOR INPUT AS #3
9070 PRINT "FILE ";F$;" AFTER COMPRESSION:"
9080 LINE INPUT #3,O$
9090 IF EOF(3) THEN 9998
9100 PRINT O$
9110 GOTO 9080
9998 PRINT O$:PRINT T1;" TOTAL CHARACTERS ELIMINATED FROM ";
9999 PRINT N1;"OR ";INT((T1/N1)*100);"%":CLOSE:END
```

Figure 3.20 *Continued*

```
ENTER ASCII FILENAME. EG. RUNLEN.DAT
? RUNLEN.DAT
PATIENCE - INPUT PROCESSING
FILE RUNLEN.DAT BEFORE COMPRESSION:
 1 BEGIN*********************************
 2 RRRRRRRRRRRRRRRRRRRRRRRRRRRRRRRRRR
 3 EEEEEEEEEEEEEEEEEEEEEEEEEEEEEEEEE
 4 PPPPPPPPPPPPPPPPPPPPPPPPPPPPPPPPP
 5 EEEEEEEEEEEEEEEEEEEEEEEEEEEEEEEEEEE
 6 AAAAAAAAAAAAAAAAAAAAAAAAAAAAAAAAAA
 7 TTTTTTTTTTTTTTTTTTTTTTTTTTTTTTTTTT
 8 ********************************END
FILE RUNLEN.DAT AFTER COMPRESSION:
 1 BEGIN}*#
 2 }R!
 3 }E"
 4 }P#
 5 }E$
 6 }A%
 7 }T&
 8 }*%END
  261  TOTAL CHARACTERS ELIMINATED FROM  309 OR  84 %
Ok
```

Figure 3.21 Sample execution of RUNLENC.BAS

diskette accompanying this book, uses DEFINE statements to assign the string 'RUNLENC.DAT' to the variable 'infile' and the string 'RUN-LENC.DAT' to the variable 'outfile' to provide consistency with previously developed C language programs.

Figure 3.22 lists the statements in the program RUNLEN.C. Note that when possible variable labels similar to the variable labels used in the equivalent BASIC program were used. In addition, the C language program makes liberal use of comments to facilitate understanding of its operation. The execution of RUNLENC.C will produce similar results to the execution of RUNLENC.BAS illustrated in Figure 3.21. The difference between the execution of RUNLENC.BAS and RUN-LENC.C is limited to the count of characters in the file. The C language program counts the line feed and carriage return characters on each line as well as the end-of-file character placed by the operating system at the end of the file. Thus, the eight-line file illustrated in Figure 3.21 when compressed by RUNLENC.C would display '261 Total Characters Eliminated from 326', instead of '309', since the BASICA program does not count line feeds, carriage returns and the end-of-file character.

Modifications to consider

The ASCII 125 character was used as a compression-indicating character due to its representation as a right brace on most printers. Normally, if your source data does not include characters beyond ASCII 127, then a character in the extended ASCII character set, such as ASCII 129 or another beyond ASCII 127, should be used to represent the occurrence of run-length encoding. For the preceding example, ASCII 129 was purposely excluded because its display on a monitor as the character ii will be printed on some printers as the £ (pound) character, while other printers simply ignore characters beyond ASCII 127. To correctly print characters beyond ASCII 127 using an IBM PC requires you to have a printer capable of printing the extended ASCII character set. In addition, a special disk operating system (DOS) program called GRAFTABL which is available under DOS 3.0 and higher versions of the operating system must be loaded into the computer prior to printing data. Due to this, the ASCII 125 character was used for illustrative purposes as the compression-indicating character.

If a character beyond ASCII 127 is used to indicate the occurrence of compression and that character naturally occurs in your data a false indication of compression will result. To prevent a receiving device from misinterpreting the character as an indication that run-length compression occurred, the program can be modified to send two such characters whenever a compression-indicating character occurs naturally in a data stream. Then, at the receiving device the decompression program would first examine each character for the occurrence of a compression-indicating character; however, when encountered it would not immediately signify run-length encoding had occurred. The program would then examine the next character to ascertain if that character is also a compression-indicating character. If it is, this would serve as an indicator that one compression-indicating character occurred naturally in the data, resulting in the removal of the second compression-indicating character by the receiver. The previously described technique is more formally referenced as the insert and delete character method. This method was previously described in Chapter 2 and readers are referred to that chapter for additional information concerning the use of that technique to alleviate a false indication of the occurrence of data compression.

Decompression

In Figure 3.23, readers will find the program listing of RUN-LEND.BAS, which is the program developed to decompress data previously compressed by the RUNLENC.BAS program. To as great an extent as possible, program variables and coding modules have been kept the same between compression and decompression programs presented in this book to facilitate their utilization and explanation.

Similar to the previously examined compression program, this program processes data on a line by line basis. The LINE INPUT statement in line 130 reads a line of data from the file used for input. Next, in line 140 the length of the line is determined.

The subroutine bounded by lines 180 and 320 is then invoked. In this subroutine the string representing one line from the input file is examined on a character by character basis, using the MID$ function in line 250 to extract one character at a time from the string. In line 260, each extracted character is compared to the character value of 125 which is the right brace character to determine if a compression-indicating character occurred. If so, a branch to line 360 occurs where the repeated count and the repeated character are extracted from the string in lines 370 and 380. Next, an index is obtained based upon the numerical value of K$, using the ASC function in line 390. This is followed by the FOR-NEXT loop bounded by lines 400 to 420, which place the repeated character in the output buffer the required number of times to match the count character. Then the J and I indexes are increased and the program branches back to line 250.

If a compression-indicating character does not occur in the data, line 290 is executed. This line causes the character extracted from the string to be placed directly into the output buffer. Next, the J index is incremented by 1 in line 300 and the boundary of the original FOR-NEXT loop checks to determine if the end of the loop was reached in line 310.

```c
/* RUNLENC.C - Program to demonstrate  run length compression     */

#include <stdio.h>
#define infile "RUNLEN.DAT"
#define outfile "RUNLENC.DAT"

char  o[256],               /* output buffer                        */
      filename[13],         /* store filename                       */
      buffer[256];          /* input buffer - max string length     */

int   t=0,                  /* initialize counters                  */
      t1=0,
      n=0,
      n1=0;


FILE  *input,
      *output;

main()
{
   fileop();
   while( fgets(buffer,256,input) != NULL )
   {
      n = strlen(buffer);                  /* get length of string    */
      tally( runlength() );
   }
   summary();
   return 0;
}

runlength()
{
   int    i=0                              /* set indices             */
          j=0,
          k=1;

   unsigned char   a,
                   b;

   for(i=0;i < n; i++)                     /*step through record      */
   {
      a = buffer[i];                       /* extract a character     */

      if (a == buffer[i+11] )              /* same as next?           */
      {
         b=a;
         k++;
         continue;
      }
      if  ( k > 3)                         /* have run length compression   */
      {
         o[j] = 125;                       /* set flag for run length     */
         o[j+1] = b;;                      /* insert repeated character   */
         0[j+2] = k;                       /* insert count                */
         j+=3;
         k=1;                              /* reset count                 */
      }
      else
      {
         if ( k== 3)                       /* when only three repeating   */
         {
            o[j]    = b;
            o[j+1] = b;
            j+=2;
            k=1;
         }
         else                              /* next character different    */
         {
            o[j] = a;
            j++;
         }
      }
   }
```

Figure 3.22 RUNLENC.C program listing

```
    }
    return(j);
}

tally(j)                                    /* tally the compression count   */
int    j;
{
    int    i=0;

    n1=n1+n;
    t=n-j;
    t1=t1+t;
    for(i=0; i<j; i++)
    {
        fputc(o[i],output);
    }
    return 0;
}

fileop()
{
    printf("Enter ASCII Filename.Eg, %s  :", infile);
    scanf("%13s",filename);
    if ( (input  = fopen(filename,"rb")) == NULL )
     {
     printf("ERROR: Cannot open the file: %s \n",filename);
     exit(-1);
     }
     if ( (output = fopen(outfile,"wb")) == NULL )
     }
     printf("ERROR: Cannot open the file %s for output\n",outfile);
     exit(-2);
     }
     return 0;
}

summary()
{
    fclose(input);
    fclose(output);

    if ( (input  = fopen(filename,"rb")) == NULL )
    {
        printf("ERROR: Cannot open the file %s for input\n",filename);
        exit(-1);
    }
    printf("File  %s Before Compression:\n",filename);
    while( fgets(buffer,256,input) != NULL )
    {
        printf("%s",buffer);
    }

    fclose(input);

    if ( (input = fopen("runlenc.dat","rb")) == NULL )
    {
        printf("ERROR: Cannot open the file %s for output\n",outfile);
        exit(-2);
    }

    printf("\nFile After Compression :\n");

    while( fgets(buffer,256,input) != NULL )
    {
        printf("%s",buffer);
    }

    fclose(input);

    printf("\n%d Total Characters Eliminated From %d Or %d%%\n\n",
    t1,n1,(int)((float)((float)t1/(float)n1)*100));
    return 0;
}
```

Figure 3.22   *Continued*

```
10 REM RUNLEND.BAS PROGRAM
20 DIM O$(132)
30 WIDTH 80:CLS
40 '**********MAIN ROUTINE***********************
50 '* THIS ROUTINE READS RECORDS FROM AN ASCII *
60 '* FILE INTO A STRING CALLED X$ WHICH IS     *
70 '* THEN PASSED TO DECOMPRESSION SUBROUTINE   *
80 '*********************************************
90 PRINT "ENTER ASCII FILENAME. EG, RUNLENC.DAT"
100 INPUT F$: OPEN F$ FOR INPUT AS #2
105 OPEN "RUNLEND.DAT" FOR OUTPUT AS #3
110 PRINT "PATIENCE - INPUT PROCESSING"
120 IF EOF(2) THEN GOTO 9000
130 LINE INPUT #2, X$
140 N= LEN(X$)
150 GOSUB 180
160 GOSUB 900
170 GOTO 120
180 '*****RUN LENGTH DECODING SUBROUTINE********
190 '* THIS ROUTINE PROCESSES RECORDS FROM X$  *
200 '* AND DECOMPRESSES RUN-ENCODED CHARACTERS *
210 '* USING O$ AS THE OUTPUT BUFFER.          *
220 '*******************************************
230 K=1:J=1                         'RESET INDICES
240 FOR I= 1 TO N                   'STEP THRU RECORD
250 A$= MID$(X$,I,1)                'EXTRACT A CHAR
260 IF A$= CHR$(125) THEN 360       'COMPRESSION FLAG?
290 O$(J)=A$                        'STUFF IN OUTPUT BUFFER
300 J=J+1                           'BUMP BUFFER INDEX
310 NEXT I                          'GO BACK FOR MORE
320 RETURN                          'END OF STRING
355 '*****************************************************
360 'DECODE COMPRESSION NOTATION TO OUTPUT BUFFER
365 '*****************************************************
370 K$= MID$(X$,I+2,1)             'GET REPEAT COUNT
380 A$= MID$(X$,I+1,1)             'GET REPEAT CHAR
390 K= ASC(K$)                     'SET UP INDEX
400 FOR L= J TO J+K                'SET OUTPUT LOOP
410 O$(L)= A$                      'STUFF REPEAT CHAR
420 NEXT L                         'KEEP GOING
430 J= L                           'BUMP OUTPUT INDEX
440 I= I+3                         'BUMP INPUT INDEX
450 GOTO 250                       'DONE
```

Figure 3.23 RUNLEND.BAS program listing

The statements from line 900 to the end of the program were included to tally the decompression count and display the before and after results of the program. Thus, this part of the program was included for illustrative purposes only.

Figure 3.24 contains a sample execution of the RUNLEND.BAS program. You will note that the data file RUNLENC.DAT was used as input to the program. This data file was created by the execution of the RUNLENC.BAS program and the top eight numbered lines in Figure 3.24 correspond to the lower eight numbered lines in Figure 3.21.

```
900 '*****TALLY THE DECOMPRESSION COUNT & WRITE BUFFER****
910 '* DISPLAY BEFORE & AFTER RESULTS OF DECOMPRESSION   *
920 '* AND SHOW THE NET RESULTS OBTAINED BY EACH METHOD  *
930 '*****************************************************
931 N1=N1+N                          'TALLY INPUT CHAR COUNT
932 T=N-J+1                          'NET DIFFERENCE IN BUFFERS
936 T1=T1-T                          'SAVE COUNT FOR SUMMARY
940 FOR I= 1 TO J-1
950 PRINT #3, O$(I);
960 NEXT I
965 PRINT #3, ""
970 RETURN
9000 CLOSE: OPEN F$ FOR INPUT AS #2
9010 PRINT "FILE ";F$;" BEFORE DECOMPRESSION:"
9020 LINE INPUT #2,X$
9030 IF EOF(2) THEN 9060
9040 PRINT X$
9050 GOTO 9020
9060 PRINT X$:OPEN "BYTED.DAT" FOR INPUT AS #3
9070 PRINT "FILE ";F$;" AFTER DECOMPRESSION:"
9080 LINE INPUT #3,O$
9090 IF EOF(3) THEN 9998
9100 PRINT O$
9110 GOTO 9080
9998 PRINT O$:PRINT T1;" TOTAL CHARACTERS INSERTED"
9999 CLOSE:END
```

Figure 3.23 *Continued*

```
ENTER ASCII FILENAME. EG. RUNLENC.DAT
? RUNLENC.DAT
PATIENCE - INPUT PROCESSING
FILE RUNLENC.DAT BEFORE DECOMPRESSION:
1 BEGIN}*#
2 }R!
3 }E"
4 }P#
5 }E$
6 }A%
7 }T&
8 }*%END
FILE RUNLENC.DAT AFTER DECOMPRESSION:
1 BEGIN*********************************
2 RRRRRRRRRRRRRRRRRRRRRRRRRRRRRRRRRR
3 EEEEEEEEEEEEEEEEEEEEEEEEEEEEEEEEEE
4 PPPPPPPPPPPPPPPPPPPPPPPPPPPPPPPPPP
5 EEEEEEEEEEEEEEEEEEEEEEEEEEEEEEEEEE
6 AAAAAAAAAAAAAAAAAAAAAAAAAAAAAAAAAA
7 TTTTTTTTTTTTTTTTTTTTTTTTTTTTTTTTTT
8 *********************************END
 276  TOTAL CHARACTERS INSERTED
Ok
```

Figure 3.24 Sample execution of RUNLEND.BAS program

Since the decompression program returns the compressed data to its original format, the eight numbered lines at the bottom of Figure 3.24 are exactly the same as the eight numbered lines at the top of Figure 3.21.

RUNLEND.C *program*

Based upon the previously discussed naming conventions, RUNLEND.C represents the C language program developed to demonstrate run-length decompression. Since RUNLEND.BAS was developed for the prior edition of this book, variable names, when possible, were carried over to the C language program. In addition, the structure of the C language program was based upon the structure of the previously developed BASICA program.

Since the program statements in RUNLEND.C closely resemble the statements contained in RUNLENC.C, we will limit our examination of the code of the program to its key functions—'main', 'runlength' and 'tally', whose statements are listed in Figure 3.25. Readers are referred to the file RUNLEND.C on the diskette accompanying this book to obtain a complete program listing.

Using our previous conventions, the two define statements in the program header, which is not shown in Figure 3.25, assign the string 'RUNLENC.DAT' to the variable 'infile' and the string 'RUNLEND.DAT' to the variable 'outfile'. The function 'runlength' performs the actual decompression. The 'FOR i' statement cycles through each character in the buffer, searching for the occurrence of the ASCII character whose value is 125. When a match occurs, the position of the next character is recognized as the compressed character and assigned to the variable 'a', while the character offset by two positions in the buffer represents the character count and is assigned to the variable 'k'. Next, the 'for 1' loop places 'k' occurrences of the compressed character into appropriate positions in the output buffer through the use of the statement 'o[j++]=a;'. Since the occurrence of ASCII 125 results in the use of the two succeeding characters in the data stream, the statement 'i += 2;' bumps the index to the position in the array buffer by 2. The remainder of coding in the function 'tally' represents coding very similar to the coding used in RUNLENC.C and should be self-explanatory to persons with a limited background in C. Thus, we will pass over a discussion of that function in the program.

The execution of RUNLEND.C using the file RUNLENC.DAT for input produces the same result as the use of RUNLEND.BAS. Thus, readers executing RUNLEND.C will obtain a display equivalent to that shown in Figure 3.23.

```
                    main()
                    {
                       fileop();

                       while( fgets(buffer,256,input) != NULL )
                       {
                          n = strlen(buffer);
                          tally( runlength() );
                       }
                       summary();
                       return 0;
                    }

                    runlength()
                    {
                       int   i=0,
                             j=0,
                             k=0,
                             l=0;

                       unsigned char  a,
                                      b;

                       for(i=0; i < n; i++)
                       {
                          if ( buffer[i] == 125 )
                          {
                             a = buffer[i+1];
                             k = buffer[i+2];

                             for (l=0; l<k; l++)
                             {
                                o[j++] = a;
                             }
                             i+=2;
                          }
                          else
                          {
                             o[j++] = buffer[i];
                          }
                       }
                       return(j);
                    }

                    tally(j)
                    int   j;
                    {
                       int   i=0;

                       n1=n1+n;
                       t=n-j;
                       t1=t1-t;
                       for(i=0; i<j; i++)
                       {
                          fputc(o[i],output);
                       }
                       return 0;
                    }
```

Figure 3.25 RUNLEND.C functions 'main', 'runlength' and 'tally'

3.4 HALF-BYTE PACKING

This data-compression technique can be viewed as a derivative of the bit mapping process. It can be successfully used under several data structure conditions; however, unlike one version of the bit mapping technique, it will never result in a compression ratio of less than unity.

As originally developed, half-byte packing takes advantage of the structure of certain characters in a character set. This technique is effective when a portion of the bit pattern used to represent those characters becomes repetitive. As an example of this type of situation, consider the EBCDIC character set where the first four bit positions used to represent numerics are all set to binary ones as illustrated in Table 3.6.

If a non-compressed data string contains 8-level EBCDIC coded characters, then run-length encoding does not permit compression of a sequence of digits that does not repeat by character. Since the first four bits, however, do repeat, compression can be accomplished if you can pack two numerics into one character. In a similar way to several versions of run-length encoding, a special character is required to indicate that half-byte packing has occurred. Again, like versions of runlength encoding that use a special compression-indi-cating character, this character should be selected from one of the unassigned characters in the character set.

As an alternative, you can consider the use of the insert and delete technique or the use of an extended character set obtained by the use of the shift out (SO) character. However, since most sequences of numerics are short in comparison to runs of the same character resulting from headings and spaces between columns, the use of the

Table 3.6 EBCDIC numeric representation. When an 8-bit byte is used to contain numeric values coded in the EBCDIC character set, the first four bit positions are always set to all ones (1s)

Bit structure	Numeric character
1111 0000	0
1111 0001	1
1111 0010	2
1111 0011	3
1111 0100	4
1111 0101	5
1111 0110	6
1111 0111	7
1111 1000	8
1111 1001	9

alternate methods to obtain a special compression-indicating character can significantly reduce the ability of half-byte packing to compress data.

Financial applications

When data characters do not have a repetitive bit structure, half-byte
packing can still be successfully employed under certain predefined
conditions. One example would be to predefine the occurrence of the
dollar sign, all 10 numerics, the comma, asterisk and decimal point
characters in succession as suitable for compression by half-byte
packing. In Table 3.7, the bit structure of ASCII data characters commonly used for financial representations is listed. If the occurrence
of a string consisting of any numeric digit as well as a comma, decimal
point, dollar sign and asterisk is predefined as suitable for half-byte
packing, then the occurrence of such strings as '$123,456.78',
'123,456' or '$****123,456.78' can be compressed.

In examining the bit structure for the financial characters listed in
Table 3.7, note that when the three high level bits are 'stripped', to
be able to encode the reamining four bits into a half-byte, that bit
structure will have duplicate entries. As an example, both the dollar
sign ($) and the digit four (4) would have the same bit representation

Table 3.7 ASCII financial character representation. In this
data representation, the parity bit was ignored. If a parity bit
exists, it can be stripped along with the first three bits shown
prior to the packing of the last four bits into half-bytes

Bit structure	Character
011 0000	0
011 0001	1
011 0010	2
011 0011	3
011 0100	4
011 0101	5
011 0110	6
011 0111	7
011 1000	8
011 1001	9
010 0100	$
010 1100	,
010 1110	.
010 1010	*

of 0100. To alleviate this problem, you are required to redefine the resulting half-byte representation of each character.

Table 3.8 lists one possible redefinition of a half-byte bit structure. Note that when you redefine the half-bytes you can include up to six non-decimal characters as half-byte representations to be packed when encountered in a compressible string. In Table 3.8, we have added the plus (+) and minus (–) characters to the half-byte bit definitions as those characters commonly occur in strings of financial characters, such as $1,506,203– and 487.32+. Although we originally commenced our half-byte data representation using the ASCII character set, you can follow the same procedure to encode EBCDIC financial characters into a sequence of up to 16 half-bytes. In doing so you can select to use the same six non-decimal characters listed in Table 3.8 or you could select any mixture of up to six non-decimal characters that you expect to encounter in strings representing financial information.

Encoding format and technique efficiency

To compress data into half-bytes, several encoding formats can be considered. Each format provides a certain level of efficiency based upon the sequence of characters encountered in the original data string. One typical format is illustrated in Figure 3.26. Using this format, up to 15 sequential numeric or predefined data characters in a string occurring sequentially can be compressed. The limit of 15

Table 3.8   Redefined half-byte bit composition

Bit structure	Character
0000	0
0001	1
0010	2
0011	3
0100	4
0101	5
0110	6
0111	7
1000	8
1001	9
1010	$
1011	,
1100	.
1101	*
1110	+
1111	–

Half-byte counter

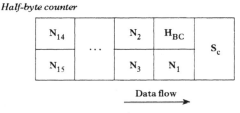

Data flow

Full-byte counter

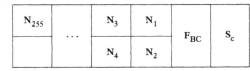

Figure 3.26 Half-byte encoding format
S   = Special character indicating half-byte encoding.
H_{BC} = Half-byte counter. Four bits are used to denote the number of numerics that have been packed.
       Number $\leq$ 15, or if a count of 0 indicates five packed characters then Number $\leq$ 20.
F_{BC} = Full-byte counter. Number $\leq$ 255.
N_1 to N_{255} = Up to 255 numerics packed 2 per 8-bit character

characters results from the use of a 4-bit, half-byte counter to denote the number of characters compressed.

Since, by definition, the special compression-indicating character indicates that half-byte encoding has occurred, you can assign a value of zero in the half-byte counter to indicate a count of one. Then, all four bits would indicate a run of 16 compressed characters. As an alternative, since half-byte encoding is only effective when five or more compressible characters occur in sequence, you can assign the half-byte count value of zero to indicate five half-bytes are encoded. Then, a counter value of 15 (all four bits set) would indicate that 20 half bytes are encoded.

If, instead of a half-byte counter, a full byte is used to indicate the half-byte packing count, up to 2^8 (or 255) numerics can be packed or 256 if the counter starts at zero to indicate one packed character. Similarly, if a count value of zero is used to indicate a string of five encoded half-bytes the full-byte counter becomes capable of indicating up to 260 packed characters. Since an extra half-byte is required to increase the counter capacity, only when the average number of characters in sequence is expected to exceed 15 (or 20 when a counter value of zero indicates five packed characters) should the full-byte counter be employed. Alternatively, you can use both a half-byte and a full-byte compression format and switch between the two depending upon the number of characters susceptible to half-byte packing that are encountered.

To examine the efficiency of half-byte packing, let us first explore the binary pattern of a sample data stream and the resulting compressed data stream. In Figure 3.27, the numeric sequence in the top part of the illustration consists of seven 8-bit characters or 56 bits. Through the use of the half-byte packing technique employing a half-byte (4-bit) counter, the resultant number of bits in the compressed data string is reduced to 40. In this example, the original data stream has been reduced by 28% (56 – 40)/56) for seven sequential numerics. It should be noted that 40 bits would also be required to represent six sequentially encountered characters susceptible to half-byte packing if transmission is on a character by character basis. Thus, any even number of sequentially encountered characters suitable for packing with a half-byte counter requires the transmission of four additional null bits when data is transferred on a character by character basis.

In Table 3.9, the original numeric data stream and its compressed format are compared when a 4-bit counter is used. Here, the number of continuous numerics was varied from 1 to 15. Since the number of bits in the original data stream is less than or equal to the number of bits in the compressed data stream, until the number of continuous numerics exceeds four, half-byte packing should not occur until five or more sequential numerics or predefined characters are encountered in a data stream.

The preceding can be represented mathematically as follows. For a sequence of S compressible characters, $S \geq 4$, the number of bits in the uncompressed string is $8S$. The number of bits in the compressed string when a half-byte counter is used is

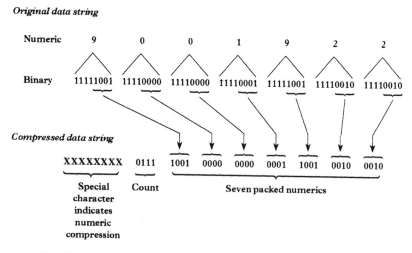

Figure 3.27 Half-byte encoding example. For 8-level character transmission, a multiple of eight bits of compressed data is transferred. Thus, a half-byte counter with an even number of packed characters will require four trailing null bits

Table 3.9 Half-byte compression efficiency using a 4-bit counter

Number of sequential compressible characters	Non-compressed bits	Compressed bits	Bit reduction (%)
1	8	16	N/A
2	16	24	N/A
3	24	24	N/A
4	32	32	0.00
5	40	32	20.00
6	48	40	16.66
7	56	40	28.00
8	64	48	25.00
9	72	48	33.33
10	80	56	30.00
11	88	56	36.36
12	96	64	33.33
13	104	64	38.46
14	112	72	35.71
15	120	72	40.00

$$12 + 4*\left\lceil\frac{S}{2}\right\rceil$$

giving a compression ratio of

$$\left(\frac{1}{8S}*\left\{12 + 4*\left\lceil\frac{S}{2}\right\rceil\right\}\right)^{-1}.$$

When a full-byte counter is used, the number of bits in the uncompressed string remains 8S. However, the number of bits in the compressed string now becomes:

$$16 + 4*\left\lceil\frac{S}{2}\right\rceil$$

resulting in a compression ratio of:

$$\left(\frac{1}{8S}*\left\{16 + 4*\left\lceil\frac{S}{2}\right\rceil\right\}\right)^{-1}.$$

Encoding process

A half-byte packing procedure for compressing numeric characters is illustrated in flow-chart format in Figure 3.28. After the numeric character counter is initialized, zero (1), a character is obtained from the original data string. If the character is numeric (3), the counter is incremented by one (4) and the next character in the original data string is examined (2). If the character comparison (3) shows that the character is not numeric, the counter is compared with four (5). If the counter is less than or equal to four, as previously discussed there is nothing to be gained by compression and the counter is reinitialized to zero (1). If the counter is greater than four (5), this means that our string of sequential numerics has ended with a sufficient number of such characters that half-byte compression is effective. At this point in time, we can set the compression format (6). Although the counter in Figure 3.28 does not have a limit, if a half-byte counter is employed, the maximum number of characters that can be packed is 15 unless you modify the assignment of values to the half-byte counter such that a value of zero indicates the compression of five half bytes, a value of one indicates the compression of six half bytes, and

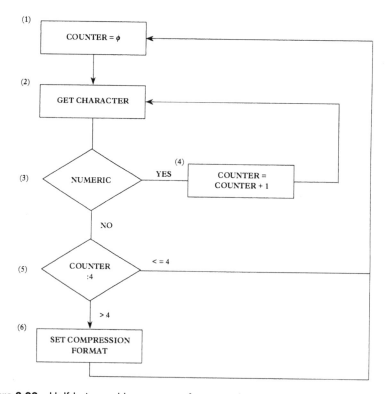

Figure 3.28 Half-byte packing process for numerics

so on. Thus, another counter comparison would be required between symbols (5) and (6).

If we desire to compress sequentially encountered strings of predefined characters to include the dollar sign, comma, period, etc., we would test for those characters in place of testing for numerics.

Buffer considerations

When a full character or multiple characters are used as a counter, buffer memory limitations must be considered in determining the maximum number of sequential characters that can be compressed, two to a byte. In Figure 3.29 half-byte packing buffer considerations are illustrated. As the original data stream is examined, sequential characters suitable for packing two per byte are placed into a buffer as illustrated in the top portion of that figure. When the counter exceeds four and the next character is not suitable for packing, the data in the first buffer can be operated upon. One half of each character is then transferred to its proper location in the compressed data string buffer as illustrated in the lower portion of Figure 3.29. Since the special character used to indicate half-byte compression and a count character can be preplaced in the contiguous compressed data string buffer, this technique of double buffering is suitable if you wish to employ a direct memory access (DMA) feature of the computer or microprocessor used for compression. Through the use of the DMA, data transfers can be effected independently of program control and data blocks are transferable on a word basis (bit parallel) to and from

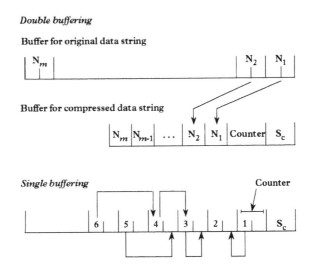

Figure 3.29 Half-byte encoding buffer considerations. Although single buffering requires additional processing, it eliminates the necessity of maintaining a separate buffer for compressed data. S_c = Special character indicating half-byte encoding

portions of main memory and peripheral devices. Thus, once the buffer in the lower portion of Figure 3.29 is completed, it can be set up for transmission through the use of a DMA transfer while the computer clears the original data string buffer and continues processing the incoming data stream. For an example of buffer size, consider the use of an 8-bit counter. In this situation, the buffer for the original data stream would have to be set up to hold up to 256 characters while the buffer for the compressed data stream would have to hold up to 130 characters, 256 compressed characters packed two per byte, a character count and the special character used to indicate half-byte packing.

Although double buffering is illustrated in the top part of Figure 3.29 for half-byte packing, single buffering can also be used. This is shown in the lower part of that illustration. In this situation, sequential characters suitable for packing are first placed into a buffer and once a non-compressible character is encountered in the original data stream and the counter exceeds four, the data elements in the buffer are manipulated as shown. In contrast to double buffering, this technique requires much more processing; however, it eliminates the necessity of having a separate buffer for compressed data. To determine total buffer requirements, the interrelationship of all data buffers must be examined as illustrated in Figure 3.30. In this example, the data to be operated upon is first read into a data-stream buffer where several different types of processing may be performed, depending upon the processing power and memory area availability of the computer being utilized. This data-stream buffer can be as

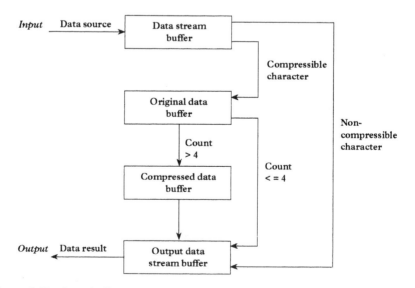

Figure 3.30 Data buffer relationships. To determine total buffer requirements, the interrelationship of all data buffers must be examined

small as one character or as large as a data block used for transmission. The buffer can be examined for compressible characters in several ways. First, a search can be made for any character suitable for half-byte packing; if none are encountered, the data-stream buffer can be directly transferred to the output data-stream buffer. Another method is to examine the data-stream buffer character by character. Non-compressible characters can then be sent to the output data-stream buffer, while compressible characters are transferred to the original data buffer. If less than five compressible characters are in the original data buffer when a non-compressible character is encountered in the data-stream buffer, the contents of the original data buffer are transferred to the output data-stream buffer. If there are five or more characters in the original data buffer when a non-compressible character is encountered in the data-stream buffer, the compression operation causes the contents of the original data buffer to be transferred in compressed format to the compressed data buffer. Finally, the contents of the compressed data buffer are transferred to an appropriate location in the output data-stream buffer.

Decoding

Decoding data compressed according to the half-byte packing technique is a relatively simple procedure. The decoding routine searches for the special character that is used to indicate that half-byte packing has occurred. Once that character is encountered, the next character or the following half byte will contain the count of the number of packed characters that follows. The special compression-indicator character itself can be used to inform the decoding software whether a full- or a half-byte counter is employed. Through the use of the buffering techniques previously discussed, the packed characters can be unpacked and the original data stream reconstructed.

Encoding application

Since strings of non-repeating numerics are not compressible by run-length encoding, the use of half-byte packing can be very advantageous when data files contain many numerical sequences. If predefined characters to include the dollar sign, comma, decimal point and asterisk are added to the numerics, half-byte packing becomes a very appropriate technique for compressing financial data.

Programming examples

Two different examples of half-byte encoding of data will be presented in this section. The first set of programming examples utilizes only the digits 0 to 9 for the encoding of data, following the classical approach of half-byte packing of numeric data. The second set of programming examples extends the number of characters that can be packed two per byte by including such characters as the comma, decimal point, asterisk and dollar sign as previously discussed in this section.

Encoding

The BASIC program BYTEC.BAS is listed in Figure 3.31. This program contains the coding required to perform simple half-byte encoding of strings containing five or more digits in sequence. The ASCII 126 character was used in this programming example to indicate the occurrence of half-byte encoding. This character is printed as the tilde (˜) on most printers and, therefore, provides a good visual indication of the use of a special compression-indicating character. In addition, the natural occurrence of this character in documents, correspondence, and programs written in different programming languages is minute. This also makes the tilde character suitable for the insertion and deletion method used to generate a special compression-indicating character.

Referencing the listing contained in Figure 3.31, the array 0$ is the output buffer into which each line input from an ASCII file is placed after it is first analyzed and compressed according to the half-byte encoding scheme, if so compressible. Each line from the file is read in line 130 and its length determined in line 140. Next, a branch to the subroutine starting at line 180 occurs. This subroutine steps through the record obtained from the file in increments of two character positions in line 240. The record is examined in increments of two character positions since the statements in lines 250 and 260 compare character I and character I + 1 to the range between and including the digits 0 and 9. If either the Ith or Ith + 1 character is in that range a branch to line 290 occurs.

To extend half-byte encoding to the characters $,. and * you could include them in the comparisons occurring in lines 250 and 260. This would be both tedious and slow, due to the time required to execute a group of MID functions joined together by many OR operators. A more elegant and speedier solution could be obtained by the creation of a one-dimensional array containing the characters to be encoded by half-byte compression. As an example, the following BASIC statements would initialize the array HBYTE, so each of its 14 elements would contain one of the characters that would be suitable for half-byte compression:

```
DIM HBYTE (14)
FOR I = I TO 14
```

```
READ HBYTE (I)
NEXT I
DATA "$","*",".",",","0","1","2","3",
DATA "4","5","6","7","8","9"
```

An interesting and practical assignment for the reader prior to examining the second version of this program presented in this section would be the modification of the half-byte encoding subroutine to include the compression of strings containing the characters $, . and * as well as the ten numerics.

Returning to the listing illustrated in Figure 3.31, if the Ith or Ith + 1 character is not a digit the counter is incremented by two in line 270 and the subroutine continues processing the line of input obtained from the file.

When the Ith or Ith + 1 character in the string is a numeric, a branch to line 290 in the program will occur. At this location, a comparison occurs to determine if there are enough numeric characters in sequence to encode. When K is greater than four a branch to line 350 occurs. At this program location, the subroutine actually performs the half-byte encoding of the data. In line 350 the ASCII character represented by the value 126 is placed into the Jth element of the array 0$. This character is used as the compression indicating character and, as previously discussed, will be displayed as a tilde (˜). In line 360, the length of the string is placed into the next element of the 0$ array and the output index is then incremented by 2 in line 370. Lines 380 to 420 perform the actual encoding of two bytes of non-compressed data into their half-byte representation and join two half bytes into a single byte.

Prior to examining the technique employed in line 400, let us first examine a conventional method to pack two numeric bytes into one byte in BASIC. In line 390, the VAL function is used to obtain the numeric part of the L and L + 1 characters contained in the X$ string. Thus, X represents one numeric character, while Y represents the second numeric character. Suppose X was 6 and Y was 9. Their byte composition would appear as follows:

0000 0110	X = 6
0000 1001	Y = 9

Packing two numeric into one byte can be accomplished by multiplying one character by 16 to shift it four bit positions to the left and either add it or OR it with the second character. Assuming Y is multiplied by 16, 9 * 16 is 144 and its bit composition becomes:

10010000	Y = 144

Then, adding X and Y results in a value of 150, whose byte composition is:

```
10 REM BYTEC.BAS PROGRAM
20 DIM O$(132)
30 WIDTH 80:CLS
40 '**********MAIN ROUTINE***********************
50 '* THIS ROUTINE READS RECORDS FROM AN ASCII *
60 '* FILE INTO A STRING CALLED X$ WHICH IS     *
70 '* THEN PASSED TO SUBROUTINES FOR COMPRESSION
80 '*********************************************
90 PRINT "ENTER ASCII FILENAME. EG, BYTE.DAT"
100 INPUT F$: OPEN F$ FOR INPUT AS #2
105 OPEN "BYTEC.DAT" FOR OUTPUT AS #3
110 PRINT "PATIENCE - INPUT PROCESSING"
120 IF EOF(2) THEN GOTO 9000
130 LINE INPUT #2, X$
140 N= LEN(X$)
150 GOSUB 180
160 GOSUB 900
170 GOTO 120
180 '*****HALF-BYTE ENCODING SUBROUTINE*********
190 '* THIS ROUTINE PROCESSES RECORDS FROM X$  *
200 '* AND ENCODES NUMERIC STRINGS OF DATA INTO*
210 '* HALF-BYTE OR 4 BIT REPRESENTATION USING *
215 '* DOUBLE BUFFERING WITH O$ AS OUTPUT BUFF.*
220 '*********************************************
230 K=1:J=1                      'RESET INDICES
240 FOR I=1 TO N STEP 2          'STEP THRU RECORD
250 IF (MID$(X$,I,1)<"0") OR (MID$(X$,I,1)>"9") THEN 290
260 IF (MID$(X$,I+1,1)<"0") OR (MID$(X$,I+1,1)>"9") THEN 290
270 K=K+2                        'BOTH NUMERIC-BUMP COUNT
280 NEXT I                       'GO BACK FOR MORE
285 RETURN                       'END OF STRING
290 IF K > 4 THEN GOSUB 350      'ENOUGH TO ENCODE
300 IF K > 1 THEN GOSUB 440      'DON'T ENCODE
310 O$(J) = MID$(X$,I,1)         'OUTPUT 1ST CHAR.
320 O$(J+1) = MID$(X$,I+1,1)     'OUTPUT 2ND CHAR.
330 J=J+2:K=1                    'BUMP OUTPUT-RESET COUNT
340 GOTO 280                     'AND GO FOR MORE
345 '***** SUBROUTINE TO PERFORM HALF-BYTE ENCODING *****
350 O$(J)=CHR$(126)              'FLAG FOR HALF-BYTE ENCODE
360 O$(J+1)=CHR$(K-1)            'INSERT LENGTH OF STRING
370 J=J+2                        'BUMP OUTPUT INDEX
380 FOR L=I-K+1  TO K STEP 2     'ENCODE 2 BYTES INTO 1
390 X= VAL(MID$(X$,L+1,1)):Y=VAL(MID$(X$,L,1))
400 O$(J)=CHR$(X+(Y*10))         'STUFF BYTE IN OUTPUT
410 J=J+1                        'BUMP OUTPUT INDEX
420 NEXT L                       'GO BACK FOR MORE
430 K=1:RETURN                   'RESET COUNT AND RETURN
435 '***** SUBROUTINE FOR STRING NOT WORTH ENCODING *****
440 FOR L=I-K+1 TO K             'PICKUP SHORT STRING
450 O$(J)=MID$(X$,L,1)           'STUFF IN OUTPUT BUFFER
460 J=J+1                        'BUMP OUTPUT INDEX
470 NEXT L                       'GO BACK FOR MORE
480 K=1:RETURN                   'RESET COUNT AND RETURN
```

Figure 3.31 BYTEC.BAS program listing

```
900 '*****TALLY THE COMPRESSION COUNT & WRITE BUFFER*****
910 '* DISPLAY BEFORE & AFTER RESULTS OF COMPRESSION     *
920 '* AND SHOW THE NET RESULTS OBTAINED BY EACH METHOD  *
930 '****************************************************
931 N1=N1+N                          'TALLY INPUT CHAR COUNT
932 T=N-J+1                          'NET DIFFERENCE IN BUFFERS
936 T1=T1+T                          'SAVE COUNT FOR SUMMARY
940 FOR I=1 TO J-1
950 PRINT #3, O$(I);
960 NEXT I
965 PRINT #3, ""
970 RETURN
1000 PRINT
1020 RETURN
9000 CLOSE: OPEN F$ FOR INPUT AS #2
9010 PRINT "FILE ";F$;" BEFORE COMPRESSION:"
9020 LINE INPUT #2,X$
9030 IF EOF(2) THEN 9060
9040 PRINT X$
9050 GOTO 9020
9060 PRINT X$:OPEN "BYTEC.DAT" FOR INPUT AS #3
9070 PRINT "FILE ";F$;" AFTER COMPRESSION:"
9080 LINE INPUT #3,O$
9090 IF EOF(3) THEN 9998
9100 PRINT O$
9110 GOTO 9080
9998 PRINT O$:PRINT T1;" TOTAL CHARACTERS ELIMINATED FROM ";
9999 PRINT N1;"OR ";INT((T1/N1)*100);"%":CLOSE:END
```

Figure 3.31 *Continued*

| 10010110 | X + Y packed = 150 |

A second method to accomplish the stuffing of the two numerics into one byte was used in line 400 of the program listing contained in Figure 3.31. In this method, the numeric value of Y was first multiplied by 10 and then added to the numeric value of X. Then, the character representing the numeric value of the addition of X to Y multiplied by 10 is placed into the O$ array as a single byte. Returning to the previous example where X was 6 and Y was 9, multiplying 9 by 10 and adding 6 results in the packing of the character that has an ASCII code of 96 into the appropriate element in the O$ array. Thus, if this half-byte encoding routine encounters the numerical sequence of 6 followed by a 9 and there is a sufficient run of numerics to pack those two characters together they would be displayed as an apostrophe ('), since that character is represented by an ASCII 96. In this technique the ASCII codes from 00 to 99 can be employed to directly represent the 100 possible combinations of two digits.

To determine the original data you can divide the received ASCII code by 10 to obtain one numeric and use the remainder of the division process for the second numeric. Unfortunately, this technique is not applicable if the additional characters previously dis-

cussed are included in the string of characters defined as susceptible
to half-byte encoding.

Again returning to the program listing contained in Figure 3.31,
note that whenever the count of characters suitable for half-byte
encoding is less than 5 or a non-numeric character is encountered a
branch to the subroutine located at line 440 occurs. This subroutine
simply takes the character from its appropriate position in the X$
string and places it in its appropriate position in the output buffer.

The last subroutine in this program was included to print a comparison of each line read from the file used for input and the half-
byte encoded version of the line. In addition, the subroutine creates
a file containing compressed data that will be used as an input file
to test the decompression routine that will be discussed next. Starting
at line 900, this subroutine also counts the characters' input and
output and computes and prints the percentage of characters eliminated as a result of half-byte encoding.

Figure 3.32 illustrates the execution of the BYTEC.BAS half-byte
encoding program, showing the original lines of data contained in the
input file followed by its resulting compressed data. Note that for clarity of illustration the input data was structured to ensure that certain
numeric pairs of characters were excluded. This was done to eliminate, as an example, two encoded half-bytes representing an ASCII 31
character or below, since such characters are non-printable and
would not be appropriate for illustrative purposes.

C language program

The program BYTEC.C was developed to provide readers with a C language version of half-byte encoding previously developed using
BASIC. The entire program is contained on the file BYTEC.C on the
disk accompanying this book. In this section we will focus our atten-

```
ENTER ASCII FILENAME. EG. BYTE.DAT
? BYTE.DAT
PATIENCE - INPUT PROCESSING
FILE BYTE.DAT BEFORE COMPRESSION:
 1  '+434567898765433345678987654333456789898769
 2  '-98357257894538629657398577526457497356872Ə
 3  '$434445464748495051525354555657585960612
FILE BYTE.DAT AFTER COMPRESSION:
 1  '+~&+-CYWA+!-CYWA+!-CY69
 2  '-~(b#H9Y-&>'9'UM42911I829
 3  '$~$+,-./0123456789:;19
   54  TOTAL CHARACTERS ELIMINATED FROM  132 OR  40 %
Ok
```

Figure 3.32 Sample execution of BYTEC.BAS program

tion upon the program header and several functions used in the program.

Figure 3.33 lists the statements in the header of the program BYTEC.C. Note that the DEFINE statements are now used to associate the string 'BYTE.DAT' to the variable 'infile' and the string 'BYTEC.DAT' to the string 'outfile'. Also note that the naming convention for the input and output buffers and input filename was retained.

Figure 3.34 lists the statements in the functions 'main', 'bytec', 'halfbyte', and 'tally', which represent the major functions of the program. The function 'main' first calls the function 'fileop' to prompt the user for the appropriate input file and then uses the 'fgets' functions in a 'while' statement to cycle through the input file, first obtaining the length of each line in the file and then invoking the functions 'bytec' and 'tally'. The function 'bytec' steps through the file record, calling the function 'halfbyte' which either compresses data into the output buffer or recognizes that data cannot be compressed and moves input data into its appropriate location in the output buffer.

In examining the statements contained in Figure 3.32, readers should note a minor but program-significant difference between the C language program and the previously discussed BASIC program. The indices in the C language program begin at zero, whereas they were set to unity in the BASIC program. Thus, this action results in the C language program performing count tests at one less count than the BASIC program.

The execution of BYTEC.C using the file BYTE.DAT as input will be similar to the sample execution shown in Figure 3.31. As previously mentioned, the C language program will read two extra characters

```
/* BYTEC.C program to demonstrate half-byte encoding           */

#include <stdio.h>
#define infile "BYTE.DAT"
#define outfile "BYTEC.DAT"

unsigned char   o[256],
                filename[13],
                buffer[256];

int    i=0,                    /*   pointer in record        */
       j=0,                    /*   output buffer pointer    */
       k=0,                    /*   numeric counter          */
       n=0,                    /*   line length              */
       n1=0,                   /*   input character counter  */
       t=0,                    /*   buffer difference        */
       t1=0;                   /* sum of buffer differences  */

FILE   *input,
       *output;
```

Figure 3.33 BYTEC.C header statements

```
main()
{
    fileop();

    while( fgets(buffer,256,input) != NULL )
    {
        n = strlen(buffer);
        bytec();
        tally();
    }
    summary();
    return 0;
}


bytec()
{
    k=0;
    j=0;

    for(i=0;i < n-1; i+=2)                          /* step thru record   */
    {
        if ( buffer[i]   < '0' || buffer[i]   > '9' )  /* no digit        */
        {
            halfbyte();
            continue;
        }
        if ( buffer[i+1] < '0' || buffer[i+1] > '9' ) /* next no digit */
        {
            halfbyte();
            continue;
        }
        k+=2;                                       /* both numeric - bump count */
    }
    return 0;
}


halfbyte()
{
    int    l=0;

    char   x=0,
           y=0;
```

Figure 3.34 BYTEC.C functions 'main', 'bytec', 'halfbyte' and 'tally'

per line and the End-of-File character. Thus, executing the program BYTEC.C using the file BYTE.DAT as input will result in the program reading 139 characters instead of 132.

```
    if ( k > 3 )                           /* have enough to encode  */
    {
       o[j]   = 126;                        /* set compression flag   */
       o[j+1] = k;                          /* insert string length - */
       j+=2;                                /* bump output index      */
       for (l=(i-k);l<=k;l+=2)              /* encode 2 bytes into 1  */
       {
          x = buffer[l+1];
          y = buffer[l];
          o[j] = x+(y*10) - 16;             /* stuff into output      */
          j++;                              /* bump output index      */
       }
    }
    else
    {
       if ( k > 0 )
       {
          for (l=(i-k);l<=k;l++)            /* pickup short string    */
          {
             o[j] = buffer[l];              /* stuff in output buffer */
             j++;                           /* bump output index      */
          }
       }
    }
    o[j]   = buffer[i];                     /* output first character */
    o[j+1]= buffer[i+1];                    /* output second character*/
    j+=2;                                   /* bump output index      */
    k=0;                                    /* reset count            */
    return 0;
}


tally()
{
    int    i=0;

    n1=n1+n;                                /* tally input character count */
    t=n-j;                                  /* net difference in buffers   */
    t1=t1+t;                                /* save count for summary      */
    for(i=0; i<j; i++)
    {
       fputc(o[i],output);                  /* output character            */
    }
       return 0;
}
```

Figure 3.34 *Continued*

Decompression

The program BYTED.BAS listed in Figure 3.35 was written to decode
or decompress data previously compressed by the BYTEC.BAS program.

Since the BYTEC.BAS program used the ASCII 126 character as a half-byte compression indicator, the BYTED.BAS program was written to search for the occurrence of this character. After a line of data is obtained from a file in line 130 of the program, the length of the line is determined in line 140. Then the subroutine at line 180 is invoked to scan the line character by character, looking for the occurrence of an ASCII 126. The FOR-NEXT loop bounded by lines 240 through 320 accomplishes this, extracting a character from the string through the use of the MID$ function in line 250 and then comparing the extracted character to ASCII 126 in line 260.

If the extracted character does not equal ASCII 126, the character is simply placed into the output buffer in line 290, the index is incremented by 1 in line 300 and the processing of the data in the loop continues. If the character is equal to ASCII 126, a branch line to 360 occurs and the decoding of the compressed data commences. First the repeat count which is the next character in the string is obtained in line 370. This character is then converted into a numeric value in line 390 since it will control the loop index for decompressing the following characters in the string that were previously encoded two per byte. This decoding is controlled by the FOR-NEXT loop bounded by lines 400 through 460. First the numeric value of the byte following the repeat count is obtained the first time line 410 is executed. In line 420, the value obtained in the preceding line is multiplied by 0.1, which, in effect, functions as a right shift. By taking the integer of the multiplication of the byte's numeric value by 0.1 we obtain a numeric between 0 and 9. This numeric represents the value of Y when X and Y were previously encoded in the BYTEC.BAS program by multiplying Y by 10 and adding the value of X to the result. Since we are working with characters based upon their ASCII values, 48 is added to the value of Y in line 430 to obtain the appropriate ASCII value of the digit. This value is then an ASCII character between 0 and 9 that represents the 10s position of the previously encoded data. In line 440, the value of Y multiplied by 10 is subtracted from the value of X to obtain the numeric value representing the unit's position in the packed data. Similar to line 430, line 450 adds 48 to the value of Z to obtain the appropriate ASCII character code that represents the decoded digit.

Figure 3.36 illustrates the execution of the BYTED.BAS program, using the file BYTEC.DAT as input to the program. Since the half-byte compression program, BYTEC.BAS, previously created this file it should be of no surprise that lines 1 to 3 at the top of Figure 3.36 are equal to lines 1 through 3 at the bottom of Figure 3.32, while lines 1 to 3 at the bottom of Figure 3.36 are equal to lines 1 to 3 at the top of Figure 3.32.

BYTED.C program

The program BYTED.C represents the C language version of the program BYTED.BAS. The full program listing of BYTED.C can be found in the file of that name on the convenience diskette accompanying this book. Since the C language programs developed for inclusion in this chapter were structured to be as modular as possible, only the DEFINE statements were changed from the statements in BYTEC.C. That is, the string 'BYTEC.DAT' is now associated with the variable 'infile', while the string 'BYTED.DAT' is now associated with the variable 'outfile'. Figure 3.37 lists the contents of the header of BYTED.C.

Figure 3.38 lists the statements in the functions 'main', 'byted', and 'tally' of the program BYTED.C. The 'main' function first calls the function 'fileop' (not listed) to prompt the user to enter the name of the compressed file and display appropriate error messages if input or output files cannot be opened. Next, the 'while' statement cycles through each line in the input file, obtaining the string length and then calling the functions 'byted' and 'tally'. The function 'byted' performs decompression and the comments in the function listing indicate the operations performed by statements in that function. Once the line is decompressed, the function 'tally' is called. That function keeps a running total of characters inserted and outputs the contents of the output buffer to the output file. Once the end of file is encountered, the function 'summary' is invoked which displays the contents of the compressed input file, the contents of the decompressed file, and the number of characters inserted into the file.

Extended half-byte encoding

A second example of half-byte encoding results from the inclusion of additional characters beyond the 10 numerics into half bytes when such characters occur sequentially. In Figure 3.39, you will find the program listing of the PACKC.BAS program that was developed to compress a string containing numerics and/or the dollar sign ($), comma (,), decimal point (.) and asterisk (*).

BASIC compression program

Similar to the previously described BYTEC.BAS program, a line of input is obtained from a file in line 130, the length of the line is determined in line 140 and a branch to the half-byte encoding subroutine occurs in line 150 of the program.

The subroutine bounded by lines 180 and 550 processes the line of input and encodes sequences of numerics and the special characters previously mentioned into half bytes. The FOR-NEXT loop bounded

```
10 REM BYTED.BAS PROGRAM
20 DIM O$(132)
30 WIDTH 80:CLS
40 '***********MAIN ROUTINE**********************
50 '* THIS ROUTINE READS RECORDS FROM AN ASCII *
60 '* FILE INTO A STRING CALLED X$ WHICH IS     *
70 '* THEN PASSED TO DECOMPRESSION SUBROUTINE   *
80 '*********************************************
90 PRINT "ENTER ASCII FILENAME. EG, BYTEC.DAT"
100 INPUT F$: OPEN F$ FOR INPUT AS #2
105 OPEN "BYTED.DAT" FOR OUTPUT AS #3
110 PRINT "PATIENCE - INPUT PROCESSING"
120 IF EOF(2) THEN GOTO 9000
130 LINE INPUT #2, X$
140 N= LEN(X$)
150 GOSUB 180
160 GOSUB 900
170 GOTO 120
180 '*****HALF BYTE  DECODING SUBROUTINE********
190 '* THIS ROUTINE PROCESSES RECORDS FROM X$  *
200 '* AND DECOMPRESSES BYTE-ENCODED CHARACTERS*
210 '* USING O$ AS THE OUTPUT BUFFER.          *
220 '*********************************************
230 K=1:J=1                           'RESET INDICES
240 FOR I= 1 TO N                     'STEP THRU RECORD
250 A$= MID$(X$,I,1)                  'EXTRACT A CHAR
260 IF A$= CHR$(126) THEN 360         'COMPRESSION FLAG?
290 O$(J)=A$                          'STUFF IN OUTPUT BUFFER
300 J=J+1                             'BUMP BUFFER INDEX
310 NEXT I                            'GO BACK FOR MORE
320 RETURN                            'END OF STRING
355 '**************************************************
360 'DECODE COMPRESSION NOTATION TO OUTPUT BUFFER
365 '**************************************************
370 K$= MID$(X$,I+1,1)               'GET REPEAT COUNT
380 M= I+2                           'SETUP INPUT INDEX
390 K= ASC(K$)                       'SET UP LOOP INDEX
400 FOR L= J TO J+K-1 STEP 2         'SET OUTPUT LOOP
410 X= ASC(MID$(X$,M,1))             'GET ONE BYTE
420 Y= INT(X* .1)                    'SHIFT RIGHT
430 O$(L)= CHR$(Y+48)                'DECODE TENS POS
440 Z= INT(X-(Y* 10))                'SUBTRACT TENS POS
450 O$(L+1)= CHR$(Z+48)              'DECODE UNITS POS
455 M= M+1                           'BUMP INPUT INDEX
460 NEXT L                           'KEEP GOING
470 J= L+1                           'RESET OUTPUT INDEX
480 I= M                             'RESET INPUT INDEX
490 GOTO 250                         'DONE
```

Figure 3.35 BYTED.BAS program listing

by lines 240 and 280 searches through the character positions in the string X$ that represents a line of input data. In line 242, the C(I) array flag is reset, while lines 243 and 244 extract two bytes from the string. The C(I) array flag is then set to a value between 1 and 4 if the first byte of the string (A$) is one of the special characters. If A$ is a

```
900 '*****TALLY THE DECOMPRESSION COUNT & WRITE BUFFER****
910 '* DISPLAY BEFORE & AFTER RESULTS OF DECOMPRESSION   *
920 '* AND SHOW THE NET RESULTS OBTAINED BY EACH METHOD   *
930 '****************************************************
931 N1=N1+N                          'TALLY INPUT CHAR COUNT
932 T=N-J+1                          'NET DIFFERENCE IN BUFFERS
936 T1=T1-T                          'SAVE COUNT FOR SUMMARY
940 FOR I= 1 TO J-1
950 PRINT #3, O$(I);
960 NEXT I
965 PRINT #3, ""
970 RETURN
9000 CLOSE: OPEN F$ FOR INPUT AS #2
9010 PRINT "FILE ";F$;" BEFORE DECOMPRESSION:"
9020 LINE INPUT #2,X$
9030 IF EOF(2) THEN 9060
9040 PRINT X$
9050 GOTO 9020
9060 PRINT X$:OPEN "BYTED.DAT" FOR INPUT AS #3
9070 PRINT "FILE ";F$;" AFTER DECOMPRESSION:"
9080 LINE INPUT #3,O$
9090 IF EOF(3) THEN 9998
9100 PRINT O$
9110 GOTO 9080
9998 PRINT O$:PRINT T1;" TOTAL CHARACTERS INSERTED"
9999 CLOSE:END
```

Figure 3.35 *Continued*

```
ENTER ASCII FILENAME. EG, BYTEC.DAT
? BYTEC.DAT
PATIENCE - INPUT PROCESSING
FILE BYTEC.DAT BEFORE DECOMPRESSION:
1 '+~&+-CYWA+!-CYWA+!-CY6ə
2 '-~(b#H9Y-&>'9'UM4ə911182ə
3 '$~$+,-./0123456789::1ə
FILE BYTEC.DAT AFTER DECOMPRESSION:
1 '+4345678987654333456789876544333456789543ə
2 '-983572578945386296573985775264574973565Q∂
3 '$434445464748495051525354555657585949554ə
  54  TOTAL CHARACTERS INSERTED
Ok
```

Figure 3.36 Sample execution of BYTED.BAS program

digit between 0 and 9 the C(I) array flag is then set to 5 in line 256. Otherwise, the C(I) array flag remains set to zero and a branch to line 258 occurs where the second byte represented by B$ is processed. Next, lines 258 to 266 process the second byte, assigning the C(I + 1) flag a value between 1 and 5 depending upon whether one of four special characters or a numeric is encountered. If either C(I) or C(I + 1) equal zero and four or more bytes containing numerics or special characters have been encountered in sequence there is enough to encode and a branch to the subroutine starting at line 350 occurs. If either C(I) or C(I + 1) equals zero and between one and three bytes

```
/* BYTED.C program to demonstrate half-byte decoding

#include <stdio.h>
#define infile "BYTEC.DAT"
#define outfile "BYTED.DAT"

unsigned char  o[256],
               filename[13],
               buffer[256];

int    i=0,
       j=0,
       k=0,
       n=0,
       t=0,
       t1=0;


FILE   *input,
       *output;
```

Figure 3.37 Statements in the header of BYTED.C

were encountered a branch to the subroutine beginning at line 500 occurs. This subroutine simply takes the encountered characters from the input string and places them into their appropriate positions in the output buffer.

When two bytes are extracted from the input string and no previous bytes were numeric or special characters C(I) and C(I + 1) are zero and line 268 causes a branch to line 290 to occur. Since K is zero, lines 310 to 330 are then executed, resulting in the two bytes just extracted from the input string being placed into their appropriate position in the output buffer.

Lines 350 to 480 contain the coding for generating the compression indicating character which, for this example, is ASCII 129 and then packing the characters eligible for half-byte compression into half bytes. Lines 352 and 354 enable two mask flags that will enable upper or lower half-bytes to be generated by ANDing the numerical value of a byte by the mask flag. Line 360 inserts the length of the string into the output buffer, while line 372 examines the C flag and encodes the byte (lines 376 to 382) based upon the type of special character in the byte. If the byte is numeric, line 384 is executed. Here, the numeric value of the byte is extracted. In line 386, it is multiplied by 16 which is equivalent to a shift 4-bit positions to the left while line 388 ANDs the value of the newly formed character flag or shifted byte by the first mask. Similarly, lines 390 to 420 perform the same operation on the second byte by first examining the second C flag. Finally, line 440 adds the two half bytes into one byte by the use of the OR

```
main()
{
    fileop();

    while( fgets(buffer,256,input) != NULL )
    {
        n = strlen(buffer);
        byted();
        tally();
    }
    summary();
    return 0;
}
tally()
{
    int    i=0;

    t=(n-1)-j;
    t1=t1-t;
    for(i=0;i<j;i++)
    {
        fputc(o[i],output);
    }
    return 0;
}
byted()
{
    int    m=0,                         /* input index               */
           l=0,                         /* output loop counter       */
           x=0,                         /* first byte                */
           y=0,                         /* right shift of x          */
           z=0;                         /* position shift            */

    k=0;
    j=0;

    for(i=0;i < n;)                     /* step thru record          */
    {
        if ( buffer[i] == 126 )         /* compression flag?         */
        {
            k = buffer[i+1];            /* get repeat count          */
            m = i+2;                    /* setup input loop          */
            for (l=j;l<j+k;l+=2)        /* setup output loop         */
            {
                x = buffer[m];          /* get a byte                */
                y = x * .1;             /* shift right               */
                o[l] = y+48;            /* decode tens position      */
                z = x-(y*10);           /* subtract tens position    */
                o[l+1] = z+48;          /* decode units position     */
                m++;                    /* bump input index          */
            }
            j=l;                        /* reset output index        */
            i=m;                        /* reset input index         */
        }
        else
        {
            o[j] = buffer[i];           /* no decode - just stuff    */
            j++;                        /* in output buffer & bump   */
            i++;                        /* increment input buffer ptr*/
        }
    }
    return 0;
}
```

Figure 3.38 'Main', 'byted' and 'tally' functions of the program BYTED.C

```
10 REM PACKC.BAS PROGRAM
20 DIM O$(132),C(132)
30 WIDTH 80:CLS
40 '*********MAIN ROUTINE*********************
50 '* THIS ROUTINE READS RECORDS FROM AN ASCII *
60 '* FILE INTO A STRING CALLED X$ WHICH IS    *
70 '* THEN PASSED TO SUBROUTINES FOR COMPRESSION
80 '*******************************************
90 PRINT "ENTER ASCII FILENAME. EG, PACK.DAT"
100 INPUT F$: OPEN F$ FOR INPUT AS #2
105 OPEN "PACKC.DAT" FOR OUTPUT AS #3
110 PRINT "PATIENCE - INPUT PROCESSING"
120 IF EOF(2) THEN GOTO 9000
130 LINE INPUT #2, X$
140 N= LEN(X$)
150 GOSUB 180
160 GOSUB 900
170 GOTO 120
180 '****HALF-BYTE ENCODING SUBROUTINE*********
190 '* THIS ROUTINE PROCESSES RECORDS FROM X$  *
200 '* AND ENCODES  MIXED  STRINGS OF DATA INTO*
210 '* HALF-BYTE OR 4 BIT REPRESENTATION USING *
215 '* DOUBLE BUFFERING WITH O$ AS OUTPUT BUFF.*
220 '*******************************************
230 K=1:J=1                          'RESET INDICES
240 FOR I=1 TO N STEP 2              'STEP THRU RECORD
242 C(I)=0:C(I+1)=0                  'RESET ENCODE FLAGS
243 A$= MID$(X$,I,1)                 'GET 1ST BYTE
244 B$= MID$(X$,I+1,1)               'GET 2ND BYTE
246 IF A$= "$" THEN C(I)= 1          'SET 1ST ENCODE FLAG
248 IF A$= "," THEN C(I)= 2
250 IF A$= "." THEN C(I)= 3
252 IF A$= "*" THEN C(I)= 4
254 IF A$< "0" OR A$> "9" THEN 258   'SKIP OTHERS
256 C(I)= 5
258 IF B$= "$" THEN C(I+1)= 1        'SET 2ND ENCODE FLAG
260 IF B$= "," THEN C(I+1)= 2
262 IF B$= "." THEN C(I+1)= 3
263 IF B$= "*" THEN C(I+1)= 4
264 IF B$< "0" OR B$> "9" THEN 268   'SKIP OTHERS
266 C(I+1)= 5
268 IF C(I)= 0 OR C(I+1)= 0 THEN 290 'NOT CANDIDATE
270 K=K+2                            'BOTH NUMERIC-BUMP COUNT
280 NEXT I                           'GO BACK FOR MORE
285 RETURN                           'END OF STRING
290 IF K > 4 THEN GOSUB 350          'ENOUGH TO ENCODE
300 IF K > 1 THEN GOSUB 500          'DON'T ENCODE
310 O$(J) = MID$(X$,I,1)             'OUTPUT 1ST CHAR.
320 O$(J+1) = MID$(X$,I+1,1)         'OUTPUT 2ND CHAR.
330 J=J+2:K=1                        'BUMP OUTPUT-RESET COUNT
340 GOTO 280                         'AND GO FOR MORE
350 O$(J)=CHR$(129)                  'FLAG FOR BYTE PACKING
352 MASK1= &HF0                      '11110000
354 MASK2= &HF                       '00001111
360 O$(J+1)=CHR$(K-1)                'INSERT LENGTH OF STRING
```

Figure 3.39 PACKC.BAS program listing

```
370 J=J+2                          'BUMP OUTPUT INDEX
371 FOR L=I-K+1 TO K STEP 2        'SETUP ENCODE LOOP
372 ON C(L) GOTO 376,378,380,382,384 'USE FLAG TO ENCODE
376 X=&HA0:GOTO 388                '10100000
378 X=&HB0:GOTO 388                '10110000
380 X=&HC0:GOTO 388                '11000000
382 X=&HD0:GOTO 388                '11010000
384 X=VAL(MID$(X$,L,1))            'GET NUM VALUE OF BYTE 1
386 X=X*16                         'SHIFT 4 BITS LEFT
388 X=X AND MASK1                  'MASK LOWER HALF-BYTE
390 ON C(L+1) GOTO 394,396,398,400,410 'USE ENCODE FLAG
394 Y=&HA:GOTO 420                 '00001010
396 Y=&HB:GOTO 420                 '00001011
398 Y=&HC:GOTO 420                 '00001100
400 Y=&HD:GOTO 420                 '00001101
410 Y=VAL(MID$(X$,L+1,1))          'GET NUM VALUE OF BYTE 2
420 Y=Y AND MASK2                  'MASK UPPER HALF-BYTE
440 Z= X OR Y                      'OR THE TWO TOGETHER
450 O$(J)= CHR$(Z)                 'OUTPUT BYTE TO BUFFER
460 J=J+1                          'BUMP OUTPUT INDEX
470 NEXT L                         'GO BACK FOR MORE
480 K=1:RETURN                     'RESET COUNT AND RETURN
500 '***** SUBROUTINE FOR STRING NOT WORTH ENCODING *****
510 FOR L=I-K+1 TO K               'PICKUP SHORT STRING
520 O$(J)=MID$(X$,L,1)             'STUFF IN OUTPUT BUFFER
530 J=J+1                          'BUMP OUTPUT INDEX
540 NEXT L                         'GO BACK FOR MORE
550 K=1:RETURN                     'RESET COUNT AND RETURN
900 '*****TALLY THE COMPRESSION COUNT & WRITE BUFFER*****
910 '* DISPLAY BEFORE & AFTER RESULTS OF COMPRESSION    *
920 '* AND SHOW THE NET RESULTS OBTAINED BY EACH METHOD *
930 '****************************************************
931 N1=N1+N                        'TALLY INPUT CHAR COUNT
932 T=N-J+1                        'NET DIFFERENCE IN BUFFERS
936 T1=T1+T                        'SAVE COUNT FOR SUMMARY
940 FOR I=1 TO J-1                 'OUTPUT FILE LOOP
950 PRINT #3, O$(I);               'BUFFER CHAR STRING
960 NEXT I
965 PRINT #3, ""                   'NOW WRITE TO FILE
970 RETURN                         'DONE
9000 CLOSE: OPEN F$ FOR INPUT AS #2
9010 PRINT "FILE ";F$;" BEFORE COMPRESSION:"
9020 LINE INPUT #2,X$
9030 IF EOF(2) THEN GOTO 9060
9040 PRINT X$
9050 GOTO 9020
9060 PRINT X$: OPEN "PACKC.DAT" FOR INPUT AS #3
9070 PRINT "FILE ";F$;" AFTER COMPRESSION:"
9080 LINE INPUT #3,O$
9090 IF EOF(3) THEN 9998
9100 PRINT O$
9110 GOTO 9080
9998 PRINT O$:PRINT T1;" TOTAL CHARACTERS ELIMINATED FROM ";
9999 PRINT N1;"OR ";INT((T1/N1)*100);"%":CLOSE:END
```

Figure 3.39 *Continued*

operator and the newly formed character that now represents two characters is placed into the output buffer.

Like the other programs previously discussed, lines 900 to 9999 keep track of the compression count and generate a file named PACKC.DAT which represents the compressed data contained in the file PACK.DAT. Later the extended half-byte decompression program called PACKD.BAS will use the PACKC.DAT file as input to perform extended half-byte decompression.

Figure 3.40 illustrates the execution of the PACKC.BAS program using a three-line data file whose contents are listed at the top of the figure. Since the packing of some half bytes resulted in the generation of a full byte whose ASCII code was below 31 and therefore unprintable, the first two lines of compressed data may appear odd due to the effect these characters have on the printer used by the author.

C language program

The program PACKC.C was created to illustrate the C language code required to perform expanded half-byte compression. In developing this program, an attempt was made as far as possible to follow the structure and coding of the previously developed BASIC language program as well as to use a liberal amount of comments to document the program. The complete PACKC.C program is included in the file of that name on the diskette accompanying this book. Due to the length of the program, we will examine the statements of key portions of the program as individual entities. Similar to previously developed C language programs, the statements in the functions 'fileop' and 'summary' will not be examined due to their commonality with previously developed C language programs.

```
ENTER ASCII FILENAME. EG, PACK.DAT
? PACK.DAT
PATIENCE - INPUT PROCESSING
FILE PACK.DAT BEFORE COMPRESSION:
1 '+$43,456,789.87‡‡65433345678987654333456789872ə
2 '-$9835$72.57$89.45$386,296,573.8577526457497356782ə
3 '$434445464748495051525354555657585987əə
FILE PACK.DAT AFTER COMPRESSION:
1 '+,ñ;Ekx▌eC3Eg          eC3Eg      əə

2 '-0⌐Zr┽z          —Z8k)kW‹wRdWIsVəə

3 '$$CDEFGHIPQRSTUVWXYəə
  61   TOTAL CHARACTERS ELIMINATED FROM  146 OR  41 %
Ok
```

Figure 3.40 Sample execution of the PACKC.BAS program

Figure 3.41 lists the statements in the header and function 'main' of the program PACKC.C. Note that the two DEFINE statements were used to associate the string 'PACK.DAT' to the variable 'infile' and the string 'PACKC.DAT' to the variable 'outfile'. This results in the file PACK.DAT representing the input file, while the file PACKC.DAT will contain the compressed data generated by the program. Also note that the header includes the initialization of the two masks while the function 'main' operates in a manner very similar to 'main' in previously developed C language programs. That is, after invoking the function 'fileop', the function 'main' cycles through each line of the

```
/* PACKC.C program to perform expanded half byte compression        */

#include <stdio.h>
#define infile "PACK.DAT"
#define outfile "PACKC.DAT"

char  o[256],                           /* output buffer           */
      c[256],
      filename[13],
      buffer[256];                      /* input buffer            */

int   i=0,                              /* record position pointer */
      j=0,                              /* output pointer          */
      k=0,                              /* count of half-bytes      */
      n=0,                              /* length of record/string  */
      n1=0,                             /* input character count    */
      t=0,                              /* difference in buffers    */
      t1=0;                             /* total characters elimin  */

char  mask1 = 0xf0,                     /* set 11110000 mask        */
      mask2 = 0x0f;                     /* set 00001111 mask        */

FILE  *input,
      *output;

main()
{
    fileop();                           /* prompt user for file     */

    while( fgets(buffer,256,input) != NULL )
    {
        n = strlen(buffer);             /* get length of string     */
        halfbyte();
        tally();
    }
    summary();
    return 0;

}
```

Figure 3.41 Statements in header and function 'main' of the program PACKC.C

input file, obtaining the length of each line and then invoking the functions 'halfbyte' and 'tally'.

The statements in the function 'halfbyte' are listed in Figure 3.42. Those statements essentially represent the C language code for lines 180 through 360 of PACKC.BAS. Instead of using GOSUB statements, the function 'halfbyte' calls the functions 'encode' and 'noencode'. The function 'encode' is invoked when four or more appropriate half-bytes are encountered. Although the test to call the function 'encode' uses the statement 'if(k>73),' note that in the C language program k is initialized at zero, whereas in the PACKC.BAS program k was initialized at k=1.

The statements in the function 'encode' are listed in Figure 3.43 and essentially represent the C language equivalent of lines 350 through 480 of the program PACKC.BAS. Note that the computed GOTO blocks of code in the BASIC language program were replaced by the use of a switch block of code. This action enabled the values of x and y to be obtained based upon the flag values of c[1] and c[1+1], respectively.

The last two functions we will examine from the program PACKC.C are 'noencode' and 'tally', whose statements are listed in Figure 3.44. The function 'noencode' represents the C language coding of the statements contained in lines 500 through 550 of the BASIC program, while the function 'tally' represents the C language coding of the statements contained in lines 900 through 970 of the BASIC program. Readers should note that the statement N1=N1+N in the BASIC program was changed to N1=N1+(N−1) in the subroutine 'tally' to eliminate the counting of the null character added to each string read by the C language program.

BASIC decompression program

Figure 3.45 contains the program listing of the PACKD.BAS program that was developed to decompress data previously compressed by the PACKC.BAS program.

Similar in construction to the PACKC.BAS program, PACKD.BAS obtains a line of data from a file in line 130, determines the length of the line in line 140 and then branches to the subroutine starting at line 180 to perform the required decoding. The FOR-NEXT loop bounded by lines 240 and 320 extracts one character at a time from the input string, searching for ASCII 129 which is the compression indicating character used to denote the occurrence of extended half-byte compression.

When the compression flag is encountered in line 260, a branch to line 330 occurs which is the beginning of the routine that decompresses the compressed data. After initializing the masks in lines 330 and 335 the length of the string is obtained in line 340 while the FOR-

```
halfbyte()
{
    unsigned     char  a,            /*  first byte holder         */
                       b;            /*  second byte holder        */

    k=0;                             /*  count of bytes packable   */
    j=0;                             /*  position in output buffer  */

    for(i=0;i < n; i+=2)             /*  step thru record          */
    {
        c[i]   = 0;                  /*  reset encode flags        */
        c[i+1] = 0;

        a = buffer[i];               /*  get first byte            */
        b = buffer[i+1];             /*  get next byte             */

        if(a=='$') c[i]=1;           /*  set 1st encode flag       */
          else if(a==',') c[i]=2;
          else if(a=='.') c[i]=3;
          else if(a=='*') c[i]=4;

        if ( a >= '0' && a <= '9' )  /*  when a digit              */
        {
            c[i] = 5;                /*  set flag to 5             */
        }
        if(b=='$') c[i+1]=1;         /*  set 2nd encode flag       */
          else if(b==',') c[i+1]=2;
          else if(b=='.') c[i+1]=3;
          else if(b=='*') c[i+1]=4;

        if ( b >= '0' && b <= '9' )  /*  when a digit              */
        {
            c[i+1] = 5;              /*  set flag to 5             */
        }
        if ( c[i] == 0 || c[i+1] == 0 )  /*  nothing worth encoding */
        {
            if ( k > 3 )             /*  enough to encode          */
            {
                encode();            /*  yes, call encode          */
            }
            if ( k > 0 )             /*  not enough to encode      */
            {
                noencode();          /*  call noencode             */
            }
            o[j]   = buffer[i];      /*  output first character    */
            o[j+1] = buffer[i+1];    /*  output second character   */
            j+=2;                    /*  bump output buffer index  */
            k=0;                     /*  reset counter             */
            continue;
        }
        k+=2;                        /*  both numeric - bump count */
    }
    return 0;
}
```

Figure 3.42 Function 'halfbyte' in the program PACKC.C

```
encode()                          /* function to perform half  */
{                                 /* byte packing              */

    int   l=0,
          x=0,
          y=0,
          z=0;

    o[j]   = 129;                 /* flag for byte packing      */
    o[j+1] = k;                   /* insert length of string    */
    j+=2;                         /* bump output by 2           */

    for (l=(i-k); l<=k; l+=2)     /* set up encoding loop       */
    {
        switch( c[l] )            /* set x based on flag in     */
        {                         /* c[l]                       */
            case 1 : x = 0xa0;    /* 10100000                   */
                    break;
            case 2 : x = 0xb0;    /* 10110000                   */
                    break;
            case 3 : x = 0xc0;    /* 11000000                   */
                    break;
            case 4 : x = 0xd0;    /* 10100000                   */
                    break;
            case 5 : x = (buffer[l]<<4);  /* num val, shift left 4bits */
                    break;
        }
        x=x & mask1;              /* mask lower half byte       */

        switch( c[l+1] )          /* do the same for the        */
        {                         /* next byte                  */
            case 1 : y = 0xa;     /* 00001010                   */
                    break;
            case 2 : y = 0xb;     /* 00001011                   */
                    break;
            case 3 : y = 0xc;     /* 00001100                   */
                    break;
            case 4 : y = 0xd;     /* 00001101                   */
                    break;
            case 5 : y = buffer[l+1];  /* get num value of byte 2 */
                    break;
        }
        y = y & mask2;            /* mask upper half-byte       */
        z = x | y;                /* OR the two together        */
        o[j] = z;                 /* output byte to buffer      */
        j++;                      /* bump output index          */
    }
    k=0;                          /* reset counter              */
    return 0;                     /* and return to calling fn   */
}
```

Figure 3.43   Function 'encode' in the program PACKC.C

```
noencode()
{
   int    l;

   for(l=(i-k); l<=k; l++)            /* pick up short string    */
   {
      o[j]=buffer[l];                 /* stuff in output buffer   */
      j++;                            /* bump output index        */
   }
   k=0;                               /* reset count              */
   return 0;
}


tally()                              /* tally character counts   */
{
   int    i=0;

   n1=n1+(n-1);
   t=n-j+1;
   t1=t1+t;
   for(i=0; i<j-1; i++)
   {
      fputc(o[i],output);
   }
   return 0;
}
```

Figure 3.44   Statements in the functions 'noencode' and 'tally' from the program PACKC.C

NEXT loop bounded by lines 350 to 498 breaks up each byte into the original two characters that were previously compressed. First line 370 takes a byte and ANDs it with the first mask and divides by 16 which is equivalent to a right shift of 4 bit positions. In line 375, the character is tested to determine if it is numeric. If so, a branch to line 430 occurs where 48 is added to the character to obtain its appropriate ASCII value. If the character is not numeric, lines 380 to 410 test to determine what special character the character represents by examining its code value and then based upon its code value the character is reset to its original value. Next, lines 440 to 490 perform the same operation on the second half byte in the received character.

Program execution

Figure 3.46 illustrates the execution of the PACKD.BAS program using PACKC.DAT as the input data file to decompress. The reader will note that the first three lines in Figure 3.46 are identical to the last three lines of Figure 3.40 while the last three lines of Figure 3.46 that represents the decompressed data are identical to the top three

```
10 REM PACKD.BAS PROGRAM
20 DIM O$(132)
30 WIDTH 80:CLS
40 '*********MAIN ROUTINE*********************
50 '* THIS ROUTINE READS RECORDS FROM AN ASCII *
60 '* FILE INTO A STRING CALLED X$ WHICH IS    *
70 '* THEN PASSED TO DECOMPRESSION SUBROUTINE  *
80 '*******************************************
90 PRINT "ENTER ASCII FILENAME. EG, PACKC.DAT"
100 INPUT F$: OPEN F$ FOR INPUT AS #2
105 OPEN "PACKD.DAT" FOR OUTPUT AS #3
110 PRINT "PATIENCE - INPUT PROCESSING"
120 IF EOF(2) THEN GOTO 9000
130 LINE INPUT #2, X$
140 N= LEN(X$)
150 GOSUB 180
160 GOSUB 900
170 GOTO 120
180 '*****HALF BYTE  DECODING SUBROUTINE********
190 '* THIS ROUTINE PROCESSES RECORDS FROM X$  *
200 '* AND DECOMPRESSES BYTE-ENCODED CHARACTERS*
210 '* USING O$ AS THE OUTPUT BUFFER.          *
220 '*******************************************
230 J=1                           'RESET INDEX
240 FOR I= 1 TO N                 'STEP THRU RECORD
250 A$= MID$(X$,I,1)              'EXTRACT A CHAR
260 IF A$= CHR$(129) THEN 330     'COMPRESSION FLAG?
290 O$(J)=A$                       'STUFF IN OUTPUT BUFFER
300 J=J+1                          'BUMP BUFFER INDEX
310 NEXT I                         'GO BACK FOR MORE
320 RETURN                         'END OF STRING
322 '*************************************************
324 'DECODE COMPRESSION NOTATION TO OUTPUT BUFFER    *
326 '*************************************************
330 MASK1= &HF0                    '11110000
335 MASK2= &HF                     '00001111
340 K= ASC(MID$(X$,I+1,1))         'GET STRING LENGTH
345 M= I+(K/2)                     'SET END OF STRING
350 FOR L=I+2 TO M                 'SETUP LOOP TO DECODE
362 Z= ASC(MID$(X$,L,1))           'GET BYTE
370 X= (Z AND MASK1)/16            'MASK LOWER HALF-BYTE
375 IF X< 10 THEN 430              'ITS NUMERIC
380 IF X= 10 THEN O$(J)= "$"       'SPECIAL
390 IF X= 11 THEN O$(J)= ","       'SPECIAL
400 IF X= 12 THEN O$(J)= "."       'SPECIAL
410 IF X= 13 THEN O$(J)= "*"       'SPECIAL
415 GOTO 440                       'SKIP IF SPECIAL
430 O$(J)= CHR$(X+48)              'OUTPUT 1ST NUMERIC
440 Y= Z AND MASK2                 'MASK UPPER HALF-BYTE
445 IF Y< 10 THEN 490              'ITS NUMERIC
450 IF Y= 10 THEN O$(J+1)= "$"     'SPECIAL
460 IF Y= 11 THEN O$(J+1)= ","     'SPECIAL
470 IF Y= 12 THEN O$(J+1)= "."     'SPECIAL
480 IF Y= 13 THEN O$(J+1)= "*"     'SPECIAL
485 GOTO 495                       'SKIP IF SPECIAL
490 O$(J+1)= CHR$(Y+48)            'OUTPUT 2ND NUMERIC
495 J= J+2                         'BUMP OUTPUT BY TWO
498 NEXT L:I= M                    'CONTINUE, BUMP INPUT INDEX
499 GOTO 310                       'GO BACK FOR MORE
```

Figure 3.45 PACKD.BAS program listing

```
900 '*****TALLY THE DECOMPRESSION COUNT & WRITE BUFFER****
910 '* DISPLAY BEFORE & AFTER RESULTS OF DECOMPRESSION   *
920 '* AND SHOW THE NET RESULTS OBTAINED BY EACH METHOD  *
930 '**********************************************************
931 N1=N1+N                          'TALLY INPUT CHAR COUNT
932 T=N-J+1                          'NET DIFFERENCE IN BUFFERS
936 T1=T1-T                          'SAVE COUNT FOR SUMMARY
940 FOR I= 1 TO J-1
950 PRINT #3, O$(I);
960 NEXT I
965 PRINT #3, ""
970 RETURN
9000 CLOSE: OPEN F$ FOR INPUT AS #2
9010 PRINT "FILE ";F$;" BEFORE DECOMPRESSION:"
9020 LINE INPUT #2,X$
9030 IF EOF(2) THEN 9060
9040 PRINT X$
9050 GOTO 9020
9060 PRINT X$:OPEN "PACKD.DAT" FOR INPUT AS #3
9070 PRINT "FILE ";F$;" AFTER DECOMPRESSION:"
9080 LINE INPUT #3,O$
9090 IF EOF(3) THEN 9998
9100 PRINT O$
9110 GOTO 9080
9998 PRINT O$:PRINT T1;" TOTAL CHARACTERS INSERTED"
9999 CLOSE:END
```

Figure 3.45 *Continued*

```
ENTER ASCII FILENAME. EG, PACKC.DAT
? PACKC.DAT
PATIENCE - INPUT PROCESSING
FILE PACKC.DAT BEFORE DECOMPRESSION:
1 '+,ñ;Ekx█eC3Eg          eC3Eg       әә

2 '-O┌Zr┼z        ─Z8k)kW<wRdWIsVәә

3 '$$CDEFGHIPQRSTUVWXYәә
FILE PACKC.DAT AFTER DECOMPRESSION:
1 '+$43,456,789.87**65433345678987654333456789әә
2 '-$9835$72.57$89.45$386,296,573.857752645749735620ә
3 '$434445464748495051525354555657585920ә
 55  TOTAL CHARACTERS INSERTED
Ok
```

Figure 3.46 Sample execution of the PACKD.BAS program

lines of Figure 3.40. Again, this is no surprise since the decompression program simply reconstructs the compressed data into its original form. The reader should also note that the 61 characters denoted as eliminated by half-byte compression in Figure 3.40 do not take into account the additional compression characters required to indicate each occurrence of half-byte encoding. If this was done, then a total of 55 characters would have been eliminated which matches the 55 character insertion count in Figure 3.46.

C language program

The program PACKD.C contained on the diskette accompanying this book represents the C language version of the previously developed PACKD.BAS expanded half-byte decompression program. Since the functions 'fileop' and 'summary' are essentially duplicates of the same functions used in PACKC.C, I will leave it to you to list those functions from the file PACKD.C. Instead, let's focus our attention upon the program header and the functions 'main', 'halfbyte' and 'tally'.

Figure 3.47 lists the statements in the header and the function

```
/* PACKD.C program to perform expanded half byte decompression      */

#include <stdio.h>
#define infile "PACKC.DAT"                /* input file              */
#define outfile "PACKD.DAT"               /* output file             */

char   o[256],                            /* output buffer           */
       c[256],
       filename[13],
       buffer[256];                       /* input buffer            */

int    i=0,                               /* record position pointer */
       j=0,                               /* output pointer          */
       k=0,                               /* count of half-bytes     */
       m=0,                               /* end of string           */
       n=0,                               /* length of record/string */
       n1=0,                              /* input character count   */
       t=0,                               /* difference in buffers   */
       t1=0;                              /* total characters elimin  */

char   mask1 = 0xf0,                      /* set 11110000 mask        */
       mask2 = 0x0f;                      /* set 00001111 mask        */

FILE   *input,
       *output;

main()
{
   fileop();

   while( fgets(buffer,256,input) != NULL )
   {
      n = strlen(buffer);
      halfbyte();
      tally();
   }
   summary();
   return 0;
}
```

Figure 3.47 Statements in the header and function 'main' of the program PACKD.C

'main' from the program PACKD.C. Note that the DEFINE statements associate the string 'PACKC.DAT' with the variable 'infile' and the string 'PACKD.DAT' with the variable 'outfile'. The variables 'infile' and 'outfile' are then used in the functions 'fileop' and 'summary' to perform required user file prompts and file I/O operations. The statements contained in Figure 3.47 are very similar to the statements in Figure 3.41 due to our desire to maintain a degree of consistency between programs.

The actual decompression performed by the program PACKD.C occurs in the function 'halfbyte' whose statements are listed in Figure 3.48. The statements in the function 'halfbyte' essentially represent the statements in lines 290 through 498 of the program PACKD.BAS. Since the statements in the C language function are fully documented through the use of comments and closely follow the coding of the previously developed BASIC program, it is left to the reader to review the statements in that function.

The statements in the last function of the program PACKD.C we will examine are those that tally the decompression count. Figure 3.49 lists the statements in the function 'tally'. Those statements essentially are equivalent to the statements contained in lines 900 through 970 of the BASIC program PACKD.C.

Encoding variations and efficiency considerations

The coding examples previously covered in this section have illustrated the use of special compression-indicating characters by assuming that each character was available for selection. Suppose your half-byte encoding and decoding algorithms must work on 7-bit ASCII data or non-standard 8-bit ASCII as defined by the character set used by the IBM PC and compatible personal computers. Would we want to modify our selection of a compression-indicating character and, if so, how might we do so?

Let us first assume we will transmit or store 7-level ASCII data. Since computer systems operate upon 8-bit bytes, it becomes possible to use the eighth bit position as a compression indicator, similar to the alternative method described for the bit mapping encoding technique. The top of Figure 3.50 illustrates how we could define one character by the use of the parity bit (bit 8) in a 7-bit ASCII character set to represent both a compression-indicating character as well as a count of the number of compressed half bytes. In this example, the setting of bit position 8 serves as an indicator that half-byte compression has occurred, while bit positions 1 through 7 contain the binary count of the number of half bytes that were compressed. Here the 7-bit position counter can represent a sequence of up to 127 compressed half bytes.

Since we previously used this technique to indicate the occurrence

```
halfbyte()                              /* function to decompress    */
{
    unsigned char   a,
                    z;

    int    i=0,
           l=0,
           x=0,
           y=0;

    j=0;                                /* reset index               */

    for(i=0; i<= n; i++)                /* step thru record          */
    {
        a = buffer[i];                  /* extract a character       */
        if ( a == 129 )                 /* compression flag?         */
        {
            k = buffer[i+1];            /* get string length         */
            m = i+(k/2);                /* set end of string         */
            l = i+2;

            for(; l<=m; l++)            /* setup loop to decode      */
            {
                z = buffer[l];          /* get byte                  */
                x = (z & mask1) >> 4;   /* mask lower half-byte      */

                if (x < 10)             /* its numeric               */
                {
                    o[j] = x+48;        /* output first numeric      */
                }
                else if (x==10) o[j] = '$';  /* or appropriate special   */
                else if (x==11) o[j] = ',';  /* character                */
                else if (x==12) o[j] = '.';
                else if (x==13) o[j] = '*';

                y = z & mask2;          /* mask upper half-byte      */

                if ( y < 10 )           /* its numeric               */
                {
                    o[j+1] = y+48;      /* output second numeric     */
                }
                else if (y==10) o[j+1] = '$';  /* or appropriate special   */
                else if (y==11) o[j+1] = ',';  /* character                */
                else if (y==12) o[j+1] = '.';
                else if (y==13) o[j+1] = '*';

                j+=2;                   /* bump output by two        */
            }
            i=m;                        /* bump input index          */
        }
        else
        {
            o[j] = a;                   /* stuff in output buffer    */
            j++;                        /* bump buffer index         */
        }
    }
    return 0;
}
```

Figure 3.48 Statements in the function 'halfbyte' of the program PACKD.C

```
tally()                                  /* function to tally the  */
{                                        /* decompression count    */
    int   i=0;

    n1=n1+n;
    t=n-j+1;
    t1=t1-t;
    for(i=0;i<j-1;i++)
    {
        fputc(o[i],output);
    }
    return 0;
}
```

Figure 3.49   Statements in the function 'tally' of the program PACKD.C

Combining a compression indicator and count

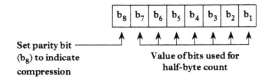

Set parity bit
(b_8) to indicate
compression

Value of bits used for
half-byte count

Defining multiple compression techniques with one character

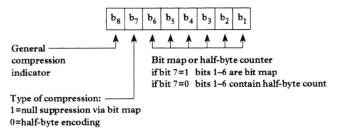

General
compression
indicator

Bit map or half-byte counter
if bit 7 =1 bits 1–6 are bit map
if bit 7 =0 bits 1–6 contain half-byte count

Type of compression:
1=null suppression via bit map
0=half-byte encoding

Figure 3.50   Using 7-level ASCII for half-byte encoding and decoding

of bit mapping, suppose we desire to implement both bit mapping of repeated nulls and half-byte encoding of financial data using a 7-level ASCII character set. To accomplish this task we would use the setting of bit 8 as a 'general' compression indicator. Then, we could use the value of bit 7 to denote two types of compression. Finally, bit positions 1 through 6 would either represent a 6-position bit map or the count of compressed half bytes based upon the value of bit 7. The lower portion of Figure 3.50 illustrates how an 8-bit character format can be used to define two types of data compression when a 7-bit data code is encapsulated in eight bits. Of course, in using the eighth bit you eliminate the ability to use that bit for parity checking. However, since most communications software programs now include a block checking mechanism for error control purposes, the loss of a parity checking capability for most persons will be considered as a non-event.

The format illustrated in the lower portion of Figure 3.50 is only applicable for defining two types of compression. You can extend this technique further by using bit positions 6 and 7 to indicate up to four types of compression. Doing so then reduces the bit map to five bits and the maximum half-byte count to 31.

If we are using a true 8-bit character set we would not want to use the previously described technique. This is because the previous scheme could eventually generate all characters whose ASCII values exceed 127. Instead, we would probably use the insert and delete technique to obtain two characters that infrequently appear in data to use as compression-indicating characters—perhaps the characters whose ASCII values are 126 and 129!

3.5 DIATOMIC ENCODING

As the name implies, diatomic encoding is a data-compression process whereby a pair of characters is replaced by a special character. The bit structure of the special character represents the encoded pair of characters and, thus, permits a 50% data reduction or a 2:1 compression ratio.

Since the number of special characters that can be employed to represent different types of compression is limited, the theoretical potential of obtaining a 50% data reduction by substituting one character for every pair of characters cannot be obtained. To maximize potential compression requires a prior understanding of the composition of your data. Once you know the expected frequency of occurrence of pairs of characters, then the most commonly encountered pairs can be selected as candidates for diatomic encoding. The actual number of pairs selected will depend upon the number of special characters available to represent those pairs of frequently occurring characters.

Operation

A block diagram representation of the diatomic encoding process will be found in the top portion of Figure 3.51. In the lower portion of that illustration is a flowchart denoting the major processes required to

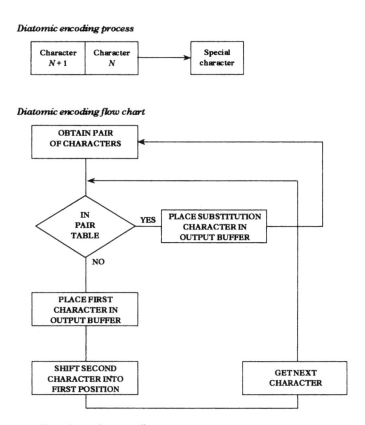

Figure 3.51 The diatomic encoding process

encode data diatomically. Note that the flowchart assumes that a continuous input data stream occurs. In actuality, the input and output buffers would be of finite length. Since the output buffer will always be less than or equal to the character size of the input buffer, you may be able to assign a pointer which will be incremented through the input buffer. Upon reaching the end of that buffer, the contents of the output buffer will be transmitted while the input buffer will be refilled with additional non-compressed data.

Pair frequency of occurrence

The major problem in the implementation of diatomic encoding is in determining what pairs should be represented by special characters. To perform diatomic encoding and obtain a meaningful compression ratio requires the assignment of special characters to represent the most frequently occurring pairs of characters that are expected to be encountered in the original data stream. This means you must have some prior knowledge concerning the type of data to be operated upon so that you can base the assignment of special characters in a meaningful manner (Snyderman and Hunt, 1970).

To assist readers in selecting the appropriate character pairs to replace with special characters, several tables of pair combinations are presented in this section. In Table 3.10, you will find a table containing the first 25 most frequently encountered pairs of characters in a 12 198-character English language text (Aronson, 1977). This table, prepared by Jewell, denotes the rank, pair combination, number of occurrences of the pair and the occurrences per thousand data characters (Jewell, 1976).

Since many users of data transmission will transfer program files in addition to textual data, an analysis of the paired-character composition of BASIC, COBOL and FORTRAN programs is presented. The analysis of these programs was obtained by the execution of the DAT-ANALYSIS program written by 4-Degree Consulting located in Macon, Georgia. This program performs a compression susceptibility analysis upon data files and the paired-character analysis listed in Tables 3.11

Table 3.10 Jewell character-combination pairing

Rank	Combination	Occurrences	Occurrences per thousand
1	E_	328	26.89
2	_T	292	23.94
3	TH	249	20.41
4	_A	244	20.00
5	S_	217	17.79
6	RE	200	16.40
7	IN	197	16.15
8	HE	183	15.00
9	ER	171	14.02
10	_I	156	12.79
11	_O	153	12.54
12	N_	152	12.46
13	ES	148	12.13
14	_B	141	11.56
15	ON	140	11.48
16	T_	137	11.23
17	TI	137	11.23
18	AN	133	10.90
19	D_	133	10.90
20	AT	119	9.76
21	TE	114	9.35
22	_C	113	9.26
23	_S	113	9.26
24	OR	112	9.18
25	R_	109	8.94

Note: __ represents a space character

to 3.14 is but one of several compression algorithms analysed by that software package. The listing of the software statements in the DAT-ANALYSIS program will be found in Appendix A. Its use will facilitate the selection of one or more compression algorithms based upon an analysis of the susceptibility of your anticipated or actual data traffic to several compression algorithms.

Table 3.11 shows the paired-character compression analysis results based upon an examination of a 9322-character BASIC program. In general, most BASIC language programs contain a high proportion of input messages and prompts as well as output headings. This structure makes the paired-character consistency form a modified English text paired-character consistency. Normally, the degree of deviation from normal English textual data pairs results from the ratio of computation statements to input/output statements in the program. In Table 3.11, note that '_P' 'NT' and 'RI' are the most commonly encountered pairs. All three pairs come from PRINT statements in the program with the pair '_P' resulting from a programmer using a space to precede each PRINT statement. Similarly, the BASIC language statement of the form 'IF X:Y THEN' can be denoted by the frequently encountered pairs '_I', 'F_' and 'HE'.

Table 3.11 Paired-character compression analysis, BASIC data file of 9322 characters

Pair/count		Pair/count		Pair/count		Pair/count		Pair/count		Pair/count	
_P	13	NT	13	RI	13	_T	12	_	11	_I	8
O_	7	HE	7	_B	6	_S	6	T_	5	N_	5
_A	5	F_	5	_E	4	AB	4	BU	4	EX	4
_L	4	IN	4	IO	4	NI	4	_N	4	_U	4
TO	4	TS	4	_F	3	R_	3	S_	3	ND	3
E_	3	_R	3	OF	3	UR	3	OU	3	_C	3
SE	3	ET	3	_O	3	_D	2	NP	2	NS	2
EL	2	AC	2	ON	2	IR	2	OT	2	IT	2
LI	2	RO	2	LL	2	AT	2	NG	2	UT	2
IM	1	AL	1	AN	1	_K	1	BO	1	IU	1
LA	1	LD	1	BR	1	M_	1	LO	1	LU	1
MB	1	CK	1	NE	1	CO	1	CT	1	NO	1
DO	1	ED	1	_X	1	NU	1	EN	1	OL	1
EQ	1	ER	1	OS	1	ES	1	Y_	1	PP	1
PS	1	PT	1	_Y	1	GH	1	_G	1	SO	1
SS	1	ST	1	TA	1	TE	1	TH	1	TI	1
HI	1	HT	1	TU	1	UG	1	UI	1	UL	1
UN	1	IA	1	VA	1	XX	1	YE	1	YI	1
ZE	1	**	0	**	0	**	0	**	0	**	0

Total combinations found: 288

Note: __ represents a space character

Table 3.12 Paired-character compression analysis, FORTRAN data file of 20 465 characters

Pair/count		Pair/count		Pair/count		Pair/count		Pair/count		Pair/count	
_I	167	_F	116	TE	106	UT	105	OR	99	OU	99
RI	96	_W	87	MA	86	_C	86	IN	86	TP	81
IR	69	HA	66	_S	61	O_	60	ER	60	C_	57
IT	51	HE	48	EN	46	RA	44	_D	42	E_	42
AL	40	RE	39	SI	38	IX	38	ON	37	_T	36
HS	35	HD	34	TO	34	SU	33	_R	32	T_	31
IM	30	_B	30	TA	30	HB	30	HF	30	HN	30
HC	30	HO	29	HR	29	HG	29	HU	29	HV	29
TI	29	HH	29	HL	29	AR	29	HJ	28	HT	28
_N	28	HM	28	HW	28	HX	28	HY	28	HZ	28
IA	28	CT	28	HI	28	IP	28	HP	28	HQ	28
HJ	28	L_	27	IO	26	_G	26	EQ	25	NY	25
_O	25	UB	25	IY	25	LE	24	_E	24	_P	23
AN	23	__	23	N_	23	ND	22	CO	22	SE	22
_A	22	Y_	21	AT	21	R_	20	S_	20	IS	20
RO	20	IC	20	NC	20	PR	20	ED	19	TR	18
G_	18	UR	18	ES	18	OT	17	ET	16	NG	16
NA	16	LY	16	TH	15	AC	15	PE	14	D_	14
PU	14	UM	14	NU	14	CH	14	BL	13	IV	13
LI	13	_J	13	FI	12	PF	12	GO	12	_K	12
OW	12	ST	12	_M	11	IL	11	SS	11	LA	11
AI	11	EP	11	NS	11	DA	11	EA	10	EC	10
TS	10	TY	10	FO	10	UE	10	F_	10	UN	10

Total combinations found: 4 370

Note: _ represents a space character

In Table 3.12, the results of a similar analysis of a 20 465-character FORTRAN program is presented while Table 3.13 denotes the pairs encountered when a 54 417-character COBOL program was analyzed. In the FORTRAN program analysis, common pairs result from such frequently used statements as 'FORMAT', 'WRITE' and 'READ'. Similarly, common pairs encountered in the COBOL program are normally a result of the 'PICTURE IS' statement. Finally, Table 3.14 shows the results of an analysis of the merger of the individual BASIC, FORTRAN and COBOL programs into one entity. Here, the 230 paired characters represent 16 266 total combinations. Since the file contained a total of 84 204 data characters, diatomic compression of the 230 most frequently encountered pairs would result in a 19.3% (16 266 / 84 204) data reduction. Note that the 12 most frequently encountered pairs represent a potential data reduction of 4388 characters or approximately 25% of the theoretical reduction obtainable by diatomically encoding the 230 most frequently encountered

Table 3.13 Paired-character compression analysis, COBOL program containing 54 417 characters

Pair/count		Pair/count		Pair/count		Pair/count		Pair/count		Pair/count	
_P	542	_F	391	IC	342	AL	316	IN	309	RE	297
E_	286	_V	251	_X	243	_W	239	LE	235	R_	235
_T	231	IL	229	UE	229	TE	221	PG	212	_O	211
NT	204	M_	190	AR	186	RO	186	O_	182	UT	178
CN	164	_C	164	CT	156	LN	153	RI	152	_M	151
CH	149	_L	148	OV	144	CI	141	FO	139	TY	137
_B	136	OR	129	_S	129	_I	122	_A	118	TO	110
ER	109	G_	108	NG	106	RM	102	EF	101	CE	97
AD	91	_	91	VA	90	PA	88	ES	87	AC	84
NC	84	T_	82	F_	79	AN	79	TA	67	DV	66
FI	63	S_	61	WR	60	TI	59	ST	57	_E	55
Y_	54	PU	52	_R	50	BU	49	QU	48	OM	46
ON	44	OT	43	AT	43	OF	42	CO	40	RK	40
LS	36	CD	35	D_	35	NE	35	AS	34	OU	33
_Z	33	_G	33	ME	32	IV	30	RP	30	EN	30
ND	29	_U	28	BE	27	OP	27	L_	25	H_	24
EL	24	TP	23	SE	23	CA	22	_H	22	_D	22
TH	22	DD	21	TR	20	ET	20	VE	20	VI	20
EC	20	YI	20	_N	19	GS	19	IT	19	C_	19
GE	18	WA	18	UA	18	SH	18	_K	17	IM	17
RA	17	RS	17	MO	16	DE	16	GI	15	EX	15
ED	15	MP	15	CC	15	WO	15	EQ	15	UN	15
LI	14	IO	14	_Q	14	UR	14	DI	13	FL	13

Total combinations found: 12 509

Note: _ represents a space character

pairs. From this, it is apparent that diatomic encoding can be effectively used in conjunction with other compression techniques by selecting only a portion of the most frequently expected pairs of characters for representation by special compression-indicator characters.

Communications hardware implementation

The use of a diatomic data-compression technique was implemented by Infotron Systems (now Gandalf Corporation) in combination with several other compression algorithms on their TL780 statistical multiplexer.

In a conventional time-division multiplexer, data from each input channel is assigned to a slot on the high-speed multiplexed output line, regardless of whether or not the bandwidth is used. Since each input line is assigned a corresponding time slot, implementing com-

Table 3.14 Paired-character compression analysis, combined 84 204 character file

Pair/count		Pair/count		Pair/count		Pair/count		Pair/count		Pair/count	
_P	578	_F	510	IN	399	IC	362	AL	357	RE	336
E_	331	TE	328	_W	326	_I	297	UT	285	_T	279
RI	261	LE	259	R_	258	_V	255	_C	253	_X	252
O_	249	NY	242	IL	240	UE	239	_O	239	OR	231
AR	215	PG	212	RO	208	_S	196	M_	194	CT	185
_B	172	ER	170	CN	164	CH	163	_M	162	_L	159
LN	153	FO	149	TO	148	TY	147	_A	145	OV	144
CI	141	OU	135	G_	126	__	125	NG	124	T_	118
CE	107	ES	106	RM	105	TP	104	NC	104	AN	103
EF	102	AC	101	PA	98	TA	98	MA	94	F_	94
AD	91	VA	91	TI	89	_R	85	S_	84	_E	83
ON	83	EN	77	HA	77	C_	76	Y_	76	FI	75
IT	72	IR	71	ST	70	DV	66	PU	66	AT	66
_D	66	CO	63	OT	62	HE	62	RA	61	_G	60
WR	60	QU	56	ND	54	OF	4	BU	54	OM	53
L_	52	_N	51	D_	49	SE	48	IM	48	IO	44
IV	43	SI	41	NE	41	EQ	41	RK	40	N_	40
ET	39	IX	38	TH	38	TR	38	ME	38	HD	37
AS	37	LS	36	HR	36	UB	35	HS	35	CD	35
_U	35	ED	35	_Z	34	SU	33	HO	33	HP	32
HY	32	UR	32	IA	31	IS	31	HI	31	OP	31
HF	30	HB	30	HN	30	PR	30	HC	30	RP	30
_K	30	EC	30	LI	29	HV	29	HH	29	BE	29

Total combinations found: 16 266

Note: __ represents a space character

pression on the high-speed link will not increase any individual line efficiency. If compression is implemented on the low-speed line side, normally referenced as the channel side or level, the efficiency of only each low-speed compressed link will be increased, since each link is reserved a fixed slot on the high-speed side. This is illustrated in the upper portion of Figure 3.52.

In a statistical multiplexer, the bandwidth for a particular channel on the high-speed link is used only when the channel is transmitting data or control signals. Therefore, compression of one or more low-speed links permits the statistical multiplexer to utilize less of the bandwidth of the high-speed line for the low-speed link being compressed. The compression of the low-speed link at the channel side will then result in a lower, high-speed line rate or permit more low-speed channels to be added since compression reduces the total number of data characters transmitted over the high-speed line. Conversely, if compression is performed at the high-speed line level, the number of characters transmitted on that link will be reduced. This

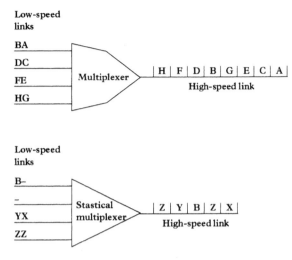

Figure 3.52 Multiplexing and compression. If compression occurs on one or more low-speed links, the effective information transfer ratio of those individual links will increase when a conventional time-division multiplexer is employed. When statistical multiplexers are employed, data may be compressed at the individual channel level or overall at the high-speed line level

will permit a lower composite high-speed operating data rate or permit additional low-speed channels to be serviced. While some vendors have elected to compress the high-speed link, Infotron uses a diatomic encoding process combined with additional data compression techniques on its low-speed channel adapters to perform compression at the channel level. This technique permits the user to select which channels, if any, should be compressed.

In the Infotron technique, statistical multiplexer compression occurs through the use of multiple-space codes, repeated character codes, common character pair codes (diatomic encoding) and packed decimal codes (half-byte encoding). In addition, since the data must be queued at the channel level to compress it, it becomes necessary to transmit control signals through the data path of the high-speed link to preserve the time relationship between data and control signals. The addition of these signals reduces the overall compression efficiency. Since each channel adapter on the multiplexer requires a buffer area and a microprocessor to effect compression, compression of a large number of low-speed channels becomes more expensive from a hardware standpoint than compressing data at the high-speed line level where only one buffer area and a single microprocessor can be used.

The Infotron channel adapter that performs compression only operates on asynchronous ASCII coded data. To obtain a sufficient number of special compression-indication codes, the parity bit in the normal 8-bit ASCII code is stripped for transmission. This results in 128 character codes that can be used to represent and indicate compressed information. The stripping of parity by the microprocessor within the multiplexer has no effect on errors since the multiplexer employs an HDLC-like frame transmission on the high-speed link level to include generating a cyclic redundancy check of transmitted frames.

In the Infotron system, codes are assigned to represent groups of two to seven consecutive spaces for various multiple-space compression code schemes. These codes are most effective when transmitted data has been formatted in columns separated by groups of spaces or for textual information that contains paragraph indentations and margin justification through the use of spaces.

To represent repeated characters, 16 codes were assigned to represent groups of three to 18 consecutive identical characters. This code is followed by the character to be repeated, in a similar way to run-length encoding, and results in a 2-byte code. To represent common character pairs, 48 codes have been assigned. The characters pairs used by Infotron are listed in Table 3.15. With the exception of the decimal point, space and carriage return–line feed pairs, all other pairs include both upper and lower case characters.

Lastly, 16 codes are assigned to specify when four to 19 characters are in packed decimal (half-byte) format. Here, characters are represented by 4-bit codes packed two per 8-bit byte. In addition to numerics, the dollar sign, period, comma, per cent, diagonal sign and space are stripped of the leading four bits if they occur in the string and are included in the packed format.

Although the effectiveness of the compression technique employed obviously depends upon the data to which the technique is applied, using multiple techniques increases the possibility of being able to use one technique effectively upon a portion of the data stream. During channel adapter compression tests, a compression ratio of up to 1.8 was noted by Infotron, indicating that only 55% of the input data stream was actually transmitted.

The Infotron compression-performing channel adapter was developed in an era where low speed asynchronous transmission was

Table 3.15 Common character pair codes compressed by Infotron: both upper and lower case

S␣	␣T	IN	TE	AN
T␣	␣A	ED	ER	TI
E␣	␣N	AT	RE	·ON
R␣	␣O	ES	TH	CRLF
D␣	␣I	SE	HE	
␣				

Note: ␣ represents a space character. CRLF denotes carriage return followed by line feed

primarily accomplished by the use of 7-level ASCII terminal devices. The growth in the use of 8-level ASCII personal computer transmission resulted in the Infotron technique being unsuitable for supporting the general personal computer communications environment; however, it is still suitable for special applications, such as compressing the data flow from certain types of badge readers, banking devices and 7-level ASCII terminals. In addition, the Infotron compression method indicates an important compression concept applicable to most applications of this technology. That is, more effective compression techniques are based upon the use of two or more compression algorithms, a technique referred to as multi-level compression.

Programming examples

The BASIC program PAIRC.BAS listed in Figure 3.53 was developed to perform diatomic compression based upon the Jewell character-combination pairing previously listed in Table 3.10. Although this example of diatomic compression was programmed to use the Jewell character-combination pairing, it is easily modified to compress data based upon the use of other character pairs that may more appropriately reflect the reader's data.

Similar to other BASIC language compression routines previously presented in this chapter, the diatomic compression program was developed using subroutines linked together to provide distinct code modules that can be easily analyzed by the reader. After the data file is opened in line 100, the subroutine commencing at line 400 is invoked. This subroutine initializes the P$ array elements to the Jewell character combination pairing, resulting in 25 character pairs assigned to the array P$. You can change the data pairs contained in lines 410 and 420 of the subroutine; however, if the number of data pairs is changed from 25, the appropriate indices in the program must be changed to reflect the actual number of pairs. In addition, the dimension size of the P$ array must be changed to reflect the new number of pairs to be used in the diatomic compression routine. Thus, lines 400 and 425 would require modification in the subroutine previously discussed when a new set of character pairs are entered in lines 410 and 420 whose sum differs from 25.

After a line of data is read in line 130, its length is determined in line 140. The subroutine invoked in line 150 processes the line of data read from the file commencing in line 230. After the indices are reset in line 230, the FOR-NEXT loop bounded by lines 240 to 330 steps through the record, extracting pairs of data in line 250. The inner FOR-NEXT loop bounded by lines 260 and 280 compares the pair extracted from the record in line 250 to the pairs contained in the pair table previously set up by the subroutine in line 400. Note

```
10 REM PAIRC.BAS PROGRAM
20 DIM O$(132)
30 WIDTH 80:CLS
40 '**********MAIN ROUTINE**********************
50 '* THIS ROUTINE READS RECORDS FROM AN ASCII *
60 '* FILE INTO A STRING CALLED X$ WHICH IS    *
70 '* THEN PASSED TO SUBROUTINES FOR COMPRESSION
80 '*********************************************
90 PRINT "ENTER ASCII FILENAME. EG, PAIR.DAT"
100 INPUT F$: OPEN F$ FOR INPUT AS #2
105 OPEN "PAIRC.DAT" FOR OUTPUT AS #3
110 PRINT "PATIENCE - INPUT PROCESSING"
115 GOSUB 400                    'PAUSE TO SET UP TABLE
120 IF EOF(2) THEN GOTO 9000
130 LINE INPUT #2, X$
140 N= LEN(X$)
150 GOSUB 180
160 GOSUB 900
170 GOTO 120
180 '*****DIATOMIC COMPRESSION SUBROUTINE*******
190 '* THIS ROUTINE PROCESSES RECORDS FROM X$  *
200 '* AND COMPRESSES OUT COMMON PAIRS         *
210 '* USING O$ AS THE OUTPUT BUFFER.          *
220 '*********************************************
230 I=1                          'RESET INDICES
240 FOR J= 1 TO N-1              'STEP THRU RECORD
250 A$= MID$(X$,J,2)             'EXTRACT A PAIR
260 FOR K = 1 TO 25              'SETUP PAIR TABLE LOOP
270 IF A$=P$(K) THEN GOSUB 350   'IS INPUT PAIR IN TABLE?
280 NEXT K                       'NO - TRY NEXT
290 IF M = 1 THEN 310            'IF MATCH FLAG SET?
300 O$(I) = MID$(A$,1,1)         'NO-STUFF 1ST CHAR IN BUFFER
310 I=I+1                        'BUMP INPUT STRING INDEX
320 M=0                          'RESET MATCH FLAG
330 NEXT J                       'GO BACK FOR MORE
340 RETURN                       'DONE
350 M=1                          'SET PAIR MATCH FLAG
355 '*****************************************************
360 'INSERT COMPRESSION NOTATION IN OUTPUT BUFFER
365 V = K + 224                  'INDEX OUT TO SUBSTITUTE CHAR
370 O$(I)=CHR$(V)                'INSERT PAIR SUBSTITUTION
380 J=J+1                        'FORCE INPUT SHIFT 2 OVER PAIR
390 K = 25                       'FORCE END OF PAIR SEARCH
395 RETURN                       'GO BACK FOR MORE
400 DIM P$(25)                   'JEWELL CHAR. COMBINATION PAIRS
410 DATA "E "," T",TH," A","S ",RE,IN,HE,ER," I"," O","N ",ES,
420 DATA " B",ON,"T ",TI,AN,"D ",AT,TE," C"," S",OR,"R "
425 FOR I = 1 TO 25              'SETUP PAIR TABLE
430 READ Z$                      'GET COMMON PAIR
440 P$(I) = Z$: NEXT I           'AND STUFF INTO PAIR TABLE
450 RETURN                       'DONE - TABLE COMPLETE
```

Figure 3.53 PAIRC.BAS program listing

```
900 '*****TALLY THE COMPRESSION COUNT & WRITE BUFFER******
910 '* DISPLAY BEFORE & AFTER RESULTS OF COMPRESSION     *
920 '* AND SHOW THE NET RESULTS OBTAINED BY EACH METHOD  *
930 '*****************************************************
931 N1=N1+N                          'TALLY INPUT CHAR COUNT
932 T=N-I+1                          'NET DIFFERENCE IN BUFFERS
936 T1=T1+T                          'SAVE COUNT FOR SUMMARY
940 FOR I=1 TO J-1
950 PRINT #3, O$(I);
960 NEXT I
965 PRINT #3, ""
970 RETURN
1000 PRINT
1010 PRINT "**RUN-LENGTH ENCODING SAVED ";T;" CHARACTERS"
1020 RETURN
9000 CLOSE: OPEN F$ FOR INPUT AS #2
9010 PRINT "FILE ";F$;" BEFORE COMPRESSION:"
9020 LINE INPUT #2,X$
9030 IF EOF(2) THEN 9060
9040 PRINT X$
9050 GOTO 9020
9060 PRINT X$:OPEN "PAIRC.DAT" FOR INPUT AS #3
9070 PRINT "FILE ";F$;" AFTER COMPRESSION:"
9080 LINE INPUT #3,O$
9090 IF EOF(3) THEN 9998
9100 PRINT O$
9110 GOTO 9080
9998 PRINT O$:PRINT T1;" TOTAL CHARACTERS ELIMINATED FROM ";
9999 PRINT N1;"OR ";INT((T1/N1)*100);"%":CLOSE:END
```

Figure 3.53 *Continued*

that the outer limit of 25 in line 260 should also be changed if the number of pairs used in the program changes from that value.

If a pair of characters extracted from the record matches a pair in the pair table, the subroutine in line 350 is invoked. Line 350 uses the variable M to denote that a match occurred. In line 365, the variable V is set to K + 224. Here the value of K is the position in the pair table where the pair extracted from the record matched a predetermined pair. The reason 224 was added to this value was for clarity of display of the results of this compression routine. That is, italics are printed from ASCII 225 upward on many printers including one printer used by the author. Thus, the pair 'E space' is represented by an italic 'a' when printed, and so on.

The pair substitution character is inserted into the appropriate element of the O$ array as indicated in line 370. Note that J is incremented by 1 in line 380 to force a shift over the current position in the input record. Next, line 390 sets K to 25 to terminate the pair comparison in the FOR-NEXT loop bounded by lines 260 and 280, from which the compression routine was called and to which it returns upon execution of line 395.

Since the variable M was set to 1 to indicate a pair match occurred, the termination of the FOR K loop causes the execution of line 290 to result in a branch to line 310. Here the index used for the 0$ array is increased by one and the match flag is rest to zero prior to the loop terminating.

Figure 3.54 illustrates how the execution of the diatomic compression routine will appear on your monitor, while Figure 3.55 illustrates the screen image after it has been 'dumped' to a printer that outputs ASCII values from 225 upward as italics. Thus, some readers may prefer to use the execution illustrated in Figure 3.55 to compare the compression characters in italics with respect to the original data and the Jewell character combination pairs used in the program. Since an italic lower case 'a' represents the first combination pair while an italic 'b' represents the second pair and so on, it should be easier to use the second example of the PAIRC.BAS program execution for readers who wish to follow the logical flow of the program in detail.

```
ENTER ASCII FILENAME. EG. PAIR.DAT
? PAIR.DAT
PATIENCE - INPUT PROCESSING
FILE PAIR.DAT BEFORE COMPRESSION:
1 TO BE OR NOT TO BE THAT IS THE QUESTION
2 THE RAIN IN SPAIN FALLS MAINLY IN THE PLAIN
FILE PAIR.DAT AFTER COMPRESSION:
1ΓΟΠß· NO±TOΠßπJΩσπßQUØ≥≡
2Γ§ RAτΩ∞SPAτ FALLσMAτLYΩ∞πßPLAτ
  30  TOTAL CHARACTERS ELIMINATED FROM  91 OR  32 %
Ok
```

Figure 3.54 Sample execution of PAIRC.BAS program as displayed on a monitor

```
ENTER ASCII FILENAME. EG, PAIR.DAT
? PAIR.DAT
PATIENCE - INPUT PROCESSING
FILE PAIR.DAT BEFORE COMPRESSION:
1 TO BE OR NOT TO BE THAT IS THE QUESTION
2 THE RAIN IN SPAIN FALLS MAINLY IN THE PLAIN
FILE PAIR.DAT AFTER COMPRESSION:
1bOoay NOqTOoacujecaQUarp
2bh RAgjlSPAg FALLeMAgLYjlcaPLAg
  30  TOTAL CHARACTERS ELIMINATED FROM  91 OR  32 %
```

Figure 3.55 Sample execution of PAIRC.BAS program when printed using a printer that displays characters greater than ASCII 224 as italics

C language program

Although many readers executing BASIC and C language versions of previously developed programs may not notice any appreciable difference in the speed of execution, a comparison of PAIRC.BAS and PAIRC.C will certainly be noticeable. The program PAIRC.C, which was developed to provide readers with a C language program to demonstrate diatomic compression, executes considerably faster than PAIRC.BAS. Thus, for many readers the speed of program execution obtained by coding in C will justify using this programming language.

Similar to previously developed C language programs, PAIRC.C was developed using modular functions that, when possible, replicate the operation of subroutines in PAIRC.BAS. In our examination of the code used in PAIRC.C we will focus our attention upon three groups of program code. Readers are referred to the diskette accompanying this book for the entire program.

Figure 3.56 lists the program statements in the header and function 'main' of the program PAIRC.C. In examining the header portion of the program, note the use of the array p which through C's automatic dimensioning capability forms 25 three-character arrays, with the third character being the null character. The remaining program statements are very similar to those of previously developed program headers and functions named 'main', using DEFINE statements to associate file names to the variables 'infile' and 'outfile' and initialize characters and integers. The function 'main' first invokes the function 'fileop', then it uses a 'while' loop to read its line from the input file into the array 'buffer', determines the length of 'buffer', and invokes the subroutines 'diatomic' and 'tally'. After all lines are processed, the function 'main' invokes the function 'summary' and terminates.

The function 'diatomic' whose statements are listed in Figure 3.57 is equivalent to the code in lines 180 through 395 of the program PAIRC.BAS. Since C does not have a statement equivalent to BASIC's MID statement, the array 'a' was used to perform an equivalent storage location to the variable A as pairs of characters are extracted from the array buffer. However, a three-element array was required to add the null character used to terminate each string. Once the array 'a' is formed, the function 'strcmp' (string compare) is used within the 'for k' loop to compare the contents of the array 'a' against each pair of characters in the pair table. Since the variable 'i' denotes the element position of the array '0' holding the encoded or compressed line, the 'return' statement is used to pass the value of 'i' to the next function. This will enable the function 'tally' to use that value in a loop to control a 'fputc' function.

```
/* PAIRC.C - program to demonstrate diatomic (pair) encoding          */

#include <stdio.h>
#define infile "PAIR.DAT"
#define outfile "PAIRC.DAT"

char   o[256],                                  /* output buffer          */
       filename[13],
       buffer[256],                             /* input buffer           */
       p[][3]  =  { "E "," T","TH","A ","S ",   /* form 25 three          */
                    "RE","IN","HE","ER"," I",   /* character arrays to    */
                    " O","N ","ES"," B","ON",   /* include null at end    */
                    "T ","TI","AN","D ","AT",   /* of each character      */
                    "TE"," C"," S","OR","R "    /* pair                   */
                  };

int    t=0,                                     /* buffer difference      */
       t1=0,                                    /* sum of differences     */
       n=0,                                     /* length of string       */
       n1=0;                                    /* sum of lengths         */


FILE   *input,
       *output;

main()
{
    fileop();                                   /* do file processing     */

    while( fgets(buffer,256,input) != NULL )    /* get line of input      */
    {
        n = strlen(buffer);                     /* get length of line     */
        tally( diatomic());                     /* compress & count       */

    }
    summary();                                  /* display results        */
}
```

Figure 3.56 Header and function 'main' of program PAIRC.C

Figure 3.58 lists the statements in the function 'tally'. Note that the function 'diatomic' passes the value of 'i' to the function 'tally', where the variable 'j' then obtains the value of 'i'. Thus, the 'for' loop in the function 'tally' outputs the elements of the array '0' from 0 to the value of i passed by the subroutine 'diatomic'.

To provide readers with another example of the variance of output among printers when printing characters whose ASCII values exceed 127, the output of the program PAIRC.C was printed on an Epson 850 printer. Figure 3.59 illustrates the execution of PAIRC.C obtained on an Epson 850 printer.

```
diatomic()                                  /* search for pairs and  */
{                                           /* replace when in table */
   int   i=0,
         j=0,
         k=0,                               /* index to output buffer*/
         pairs=0,
         m=0;                               /* flag  character        */

   unsigned char  a[3];
   pairs= sizeof(p)/sizeof(p[0]);           /* nbr pairs in line      */
   for(j=0; j < n; j++)                     /* step through record    */

   {
      a[0]=buffer[j];                       /* extract a pair of      */
      a[1]=buffer[j+1];                     /* characters             */
      a[2]='\0';                            /* and add null           */
      m=0;
      for (k=0; k<pairs; k++)               /* setup pair loop        */
      {
         if (strcmp(a,p[k])==0)             /* is pair in table?      */
         {
            m=1;                            /* set pair match flag    */
            o[i++] = k+225;                 /* insert pair sub char   */
            j++;                            /* force input shift      */
            break;                          /* force end of search    */
         }
      }
      if(!m )                               /* match flag set ?       */
      {
         o[i++]=a[0];                       /* no-stuff 1st char      */
      }
   }

   return(i);
}
```

Figure 3.57 Statements in function 'diatomic' of the program PAIRC.C

Decompression

The program listing of PAIRD.BAS is listed in Figure 3.60. As indicated by the naming conventions used in this book, this program performs decompression upon previously compressed pairs of characters.

From an examination of the program coding listed in Figure 3.60, you will note that the construction of the code modules for decompression closely resemble the previously examined BASIC language compression program. Although our programming goal was to do this to facilitate a comparison between programs, due to the relationship

```
tally(j)                                    /* tally count & write   */
  int j;
{
    int i=0;                                /* dont count null       */
    n1=n1+(n-1);
    t=n-j;
    t1=t1+t;

    for(i=0;i<j;i++)
    {
       fputc(o[i],output);

    }
    return 0;
}
```

Figure 3.58 Statements in the function 'tally' of the program PAIRC.C

```
Enter ASCII Filename.Eg, ΓAIR.DAT : pair.dat
File pair.dat Before compression:
1 TO BE OR NOT TO BE THAT IS THE QUESTION
2 THE RAIN IN SΓAIN FALLS MAINLY IN THE ΓLAIN

File ΓAIRC.DAT After compression:
1ΓOЄß˙ NO≡TOЄßπ⌈ΩσπßQUø±∩
2ΓΦ RAτΩ∞SΓAτ FALLσMAτLYΩ∞πßΓLAτ
29 Total characters eliminated from 87 or 33%
```

Figure 3.59 Execution of PAIRC.C printed on an Epson 850 printer

between compression and decompression such modular coding relationships will normally be the rule and not the exception.

After opening files for input and output, the subroutine beginning at line 500 is invoked by line 115 of the program. This subroutine simply builds the P$ table that will contain the Jewell character combination pairs that the program will search for. In line 130, the familiar LINE INPUT statement is used to obtain a record from the input file. Next, line 140 is employed to determine the length of the record while line 150 invokes the subroutine beginning at line 180 which performs the actual decompression of data.

The FOR-NEXT loop bounded by lines 240 and 310 searches through the record previously extracted from the input file on a character by character basis. This is accomplished by the use of the MID$ function in line 250. If the character extracted from the record exceeds a value of 224, it is assumed that diatomic or paired compression has occurred. This assumption is based upon the selection of each character beyond ASCII 224 to represent a pair of characters in this coding example. If the character extracted from the record equals or is less than ASCII 224, that character does not represent a previously compressed pair of characters. Thus, line 290 simply places the extracted character into its appropriate position in the output buffer.

When an ASCII character greater than 224 is encountered, the branch to line 360 in the program results in the actual decompression of a previously compressed pair of characters. In line 370, the numerical value of the character that actually represents a pair of characters is obtained. Next, line 380 subtracts 224 from the numerical value of the character to obtain the appropriate index in the paired table (P\$(25)). Line 390 places the pair of characters that was previously represented by one character into the output buffer while lines 400 and 405 increment the index position in the output buffer and the variable V which is only employed to compute the difference in size between the input and output buffers and is not required for decompression.

Figure 3.61 illustrates the execution of the PAIRD.BAS program as it would appear on our monitor using the data file PAIRC.DAT as input. PAIRC.DAT was created by the PAIRC.BAS program. Thus, it is of no surprise that the two compressed lines of data at the top of Figure 3.61 match lines 1 and 2 in the lower part of Figure 3.55, while lines 1 and 2 at the bottom of Figure 3.61 match those lines at the top of Figure 3.55.

C language program

Continuing our naming convention, the program PAIRD.C was developed to demonstrate diatomic decoding. Figure 3.62 lists the statements in the program header and function 'main' of PAIRD.C. The key differences between the program header of PAIRD.C and PAIRC.C are in the use of the DEFINE statements and use of the unsigned character definition. As might be expected, PAIRD.C reverses the association of PAIRC.DAT from that used in the program PAIRC.C. That is, the file PAIRC.DAT is now used as input to the program. Similarly, the program PAIRD.C uses the file PAIRD.DAT for output instead of the file PAIRC.DAT. The other difference between the two programs is the assignment of the elements in the array 0 as unsigned characters. This results in each element having a value of 0 to 255.

Figure 3.63 lists the statements in the functions 'diatomic' and 'tally' from the program PAIRD.C. Note that in the function 'diatomic' the 'for' loop cycles through each element of the array 'buffer'. When the value of an element in the array exceeds 224, this indicates that diatomic compression occurred. Thus, the ith element of the output buffer (array 0) is set to the appropriate characters in the two-dimen-

```
10 REM PAIRD.BAS PROGRAM
20 DIM O$(132)
30 WIDTH 80:CLS
40 '**********MAIN ROUTINE**********************
50 '* THIS ROUTINE READS RECORDS FROM AN ASCII *
60 '* FILE INTO A STRING CALLED X$ WHICH IS    *
70 '* THEN PASSED TO DECOMPRESSION SUBROUTINE  *
80 '*********************************************
90 PRINT "ENTER ASCII FILENAME. EG, PAIRC.DAT"
100 INPUT F$: OPEN F$ FOR INPUT AS #2
105 OPEN "PAIRD.DAT" FOR OUTPUT AS #3
110 PRINT "PATIENCE - INPUT PROCESSING"
115 GOSUB 500
120 IF EOF(2) THEN GOTO 9000
130 LINE INPUT #2, X$
140 N= LEN(X$)
150 GOSUB 180
160 GOSUB 900
170 GOTO 120
180 '*****DIATOMIC    DECODING SUBROUTINE********
190 '* THIS ROUTINE PROCESSES RECORDS FROM X$   *
200 '* AND DECOMPRESSES PAIR-ENCODED CHARACTERS *
210 '* USING O$ AS THE OUTPUT BUFFER.           *
220 '*********************************************
230 K=1:J=1:V=0                    'RESET INDICES
240 FOR I= 1 TO N                  'STEP THRU RECORD
250 A$= MID$(X$,I,1)               'EXTRACT A CHAR
260 IF A$> CHR$(224) THEN 360      'COMPRESSED PAIR?
290 O$(J)=A$                       'STUFF IN OUTPUT BUFFER
300 J=J+1                          'BUMP BUFFER INDEX
310 NEXT I                         'GO BACK FOR MORE
320 RETURN                         'END OF STRING
355 '*********************************************************
360 'DECODE COMPRESSION NOTATION TO OUTPUT BUFFER
365 '*********************************************************
370 K= ASC(A$)                     'GET ORDINAL EQUIV.
380 K= K-224                       'SUBTRACT FOR INDEX
390 O$(J)= P$(K)                   'STUFF PAIR IN BUFFER
400 J= J+1                         'BUMP OUTPUT INDEX
405 V= V+1                         'SUM VARIABLE COUNT
410 GOTO 310                       'DONE
500 DIM P$(25)                     'JEWELL CHAR. COMBINATION PAIRS
510 DATA "E "," T",TH," A","S ",RE,IN,HE,ER," I"," O","N ",ES,
520 DATA " B",ON,"T ",TI,AN,"D ",AT,TE," C"," S",OR,"R "
530 FOR I = 1 TO 25                'SET UP PAIR TABLE
540 READ Z$                        'GET COMMON PAIR
550 P$(I) = Z$: NEXT I             'AND STUFF INTO PAIR TABLE
560 RETURN                         'DONE - TABLE COMPLETE
900 '*****TALLY THE DECOMPRESSION COUNT & WRITE BUFFER****
910 '* DISPLAY BEFORE & AFTER RESULTS OF DECOMPRESSION   *
920 '* AND SHOW THE NET RESULTS OBTAINED BY EACH METHOD  *
930 '*********************************************************
931 N1=N1+N                        'TALLY INPUT CHAR COUNT
932 T=N-J+1+V                      'NET DIFFERENCE IN BUFFERS
936 T1=T1-T                        'SAVE COUNT FOR SUMMARY
```

Figure 3.60 PAIRD.BAS program listing

```
940 FOR I= 1 TO J-1
950 PRINT #3, O$(I);
960 NEXT I
965 PRINT #3, ""
970 RETURN
9000 CLOSE: OPEN F$ FOR INPUT AS #2
9010 PRINT "FILE ";F$;" BEFORE DECOMPRESSION:"
9020 LINE INPUT #2,X$
9030 IF EOF(2) THEN 9060
9040 PRINT X$
9050 GOTO 9020
9060 PRINT X$:OPEN "PAIRD.DAT" FOR INPUT AS #3
9070 PRINT "FILE ";F$;" AFTER DECOMPRESSION:"
9080 LINE INPUT #3,O$
9090 IF EOF(3) THEN 9998
9100 PRINT O$
9110 GOTO 9080
9998 PRINT O$:PRINT ABS(T1);" TOTAL CHARACTERS INSERTED"
9999 CLOSE:END
```

Figure 3.60 *Continued*

```
ENTER ASCII FILENAME. EG, PAIRC.DAT
? PAIRC.DAT
PATIENCE - INPUT PROCESSING
FILE PAIRC.DAT BEFORE DECOMPRESSION:
1ΓΟΠß• NO±TOΠßπJΩσπßQUØ≥≡
2Γ⌠ RAτΩ∞SPAτ FALLσMAτLYΩσπßPLAτ
FILE PAIRC.DAT AFTER DECOMPRESSION:
1 TO BE OR NOT TO BE THAT IS THE QUESTION
2 THE RAIN IN SPAIN FALLS MAINLY IN THE PLAIN
 29  TOTAL CHARACTERS INSERTED
Ok
```

Figure 3.61 Sample execution of PAIRD.BAS program

sional array 'p'. The value of p[buffer[j]–225] represents the position in the array 'p' or its first dimension. The second element of the array 'p' [0] or [1], represents the first and second characters stored in the two-dimensional array. For example, if buffer [j] had the value 227, then 227–225 is 2. Thus p[2][0] is the character T, while p[2][1] is the character H.

Similar to the program PAIRC.C, PAIRD.C includes the functions 'fileop' and 'summary'. Both functions are very similar to the previously described functions having those names. Thus, you are referred to the diskette accompanying this book to obtain a listing of those functions.

The execution of PAIRD.C using the file PAIRC.DAT as input results in a display similar to that previously illustrated in Figure 3.61. The only difference is that the display of the message concerning 'patience' during input processing was removed due to the enhanced processing speed of the C language program.

```
C:\TC\EXAMPLES>
/* PAIRD.C - program to demonstrate diatomic (pair) decoding          */

#include <stdio.h>
#define infile  "PAIRC.DAT"
#define outfile "PAIRD.DAT"

unsigned char  o[256],                    /* output buffer value 0-255 */
        filename[13],
        buffer[256],                      /* input buffer             */
        p[][3]  =  { "E "," T","TH","A ","S ",   /* form 25 three       */
                     "RE","IN","HE","ER"," I",   /* character arrays to */
                     " O","N ","ES"," B","ON",   /* include null at end */
                     "T ","TI","AN","D ","AT",   /* of each character   */
                     "TE"," C"," S","OR","R "    /* pair                */
                 };

int    t=0,                               /* buffer difference        */
       t1=0,                              /* sum of differences       */
       n=0,                               /* length of string         */
       n1=0;                              /* sum of lengths           */


FILE   *input,
       *output;

main()
{
   fileop();                              /* do file processing       */

   while( fgets(buffer,256,input) != NULL )   /* get line of input    */
   {
      n = strlen(buffer);                 /* get length of line       */
      tally( diatomic());                 /* compress & count         */

   }
   summary();                             /* display results          */
   return ;
}
```

Figure 3.62 Statements in the header and function 'main' of the program PAIRD.C

Encoding considerations

The ability to perform diatomic compression is based upon the replacement of two characters with a special compression-indicating character. Thus, you must either use a character set from which undefined characters are available as substitutions for pairs of characters or you must consider the frequency of occurrence of pairs of characters versus the frequency of occurrence of the selected character using the insert and delete technique. As a third option, if

```
diatomic()                              /* decompression        */
{
   int   i=0,
         j=0;

   for(j=0;j < n;j++)                   /* set up loop = # char */
   {
      if ( buffer[j] > 224 )            /* is character val>224 */
      {                                 /* have compressed pair */
         o[i++] = p[buffer[j]-225][0];  /* get first character  */
         o[i++] = p[buffer[j]-225][1];  /* get second character */
      }
      else
      {                                 /* not compressed       */
         o[i++] = buffer[j];            /* move to buffer        */
      }
   }
   return(i);
}


tally(j)
int   j;
{
   int   i=0;

   n1=n1+(n-1);
   t=n-j;
   t1=t1-t;
   for(i=0;i<j;i++)
   {
      fputc(o[i],output);
   }
   return 0;
}
```

Figure 3.63 Statements in the functions 'diatomic' and 'tally' of the program PAIRD.C

you preprocess a data file you may note that certain characters from the file's character set are not used. Those characters can be used to represent compression-indicating characters, in effect providing you with the ability to perform diatomic compression based upon the distribution of characters used within a file. This technique, more formerly referred to as byte pair encoding (BPE), is described in the next section in this chapter.

Concerning the frequency of occurrence of characters, assume you wish to apply diatomic compression and will transmit or store 8-bit ASCII characters. Since all characters in that character set are defined, the only way that diatomic compression can be effective if you do not preprocess a file is if the frequency of occurrence of a pair of characters exceeds twice the frequency of occurrence of the character selected to represent the character pair. The reason for this is that the insert and delete technique doubles the frequency of occurrence of the selected character, because each natural occurrence of the character selected to represent the pair is doubled, since the insert portion of the insert and delete technique adds an extra character whenever the selected character is encountered.

3.6 BYTE PAIR ENCODING

A most interesting variation to the previously described diatomic coding technique was developed by Philip Gage (Gage, 1994). Referred to as Byte Pair Encoding (BPE), this technique replaces pairs of adjacent bytes in a file with a byte that was not encountered in the file. Thus, this technique avoids the problems associated with only a limited number of unused or non-defined characters available as replacement pair identifiers in most codes. While this technique can overcome the limitations associated with conventional diatomic compression, it is normally not suitable for use on large binary files where few, if any, of the 256 characters in a character set are unused and hence available for pair substitution. Two additional limitations of BPE involve its design requirements for data to be stored in memory. This results in the inability of the algorithm to support data streams and the potential of files to be too big to fit into memory. In spite of those problems, BPE provides a near-equivalent to slightly better compression ratio on ASCII text files as the popular Lempel–Ziv string compression technique described in Chapter 5 while requiring less code, making it very suitable for applications with limited memory.

Gage's work in developing the BPE compression method was originally described in his article 'A New Algorithm for Data Compression' published in *The C Users Journal*, February 1994 (volume 12, issue 2). *The C Users Journal* is an R&D Publication of Lawrence, Kansas, and the author of this book appreciates permission granted by *The C Users Journal* and Mr Gage to extract information from the previously referenced article.

Figure 3.64 lists the pseudocode for the BPE compression algorithm while Figure 3.65 lists the pseudocode for the BPE expansion algorithm. In examining the expansion pseudocode you will note another advantage associated with BPE in that it provides an easy and relatively fast expansion process.

The C language BPE compression program called COMPRESS.C by Mr Gage is listed in Figure 3.66. Although Mr Gage graciously provided permission for me to include his BPE compression and expansion program listings in this book, readers should note that he holds the copyright to those programs. Thus, the program listings are presented as compression educational tools and are not included on the diskette accompanying this book. Readers can contact Mr Gage via

```
While not end of file
        Read block of data into buffer and
        enter all pairs into hash table with
        counts of their occurrence
    While compression possible
        Locate most frequent byte pair
        Replace pair with an unused byte
        IF substitution deletes a pair from buffer
            decrease its count in the hash table
        IF substitution adds a new pair to the buffer
            increase its count in the hash table
    End while
    Write pair table and packed data
End while
```

Figure 3.64   BPE Compression algorithm pseudocode


```
While not end of file
    Read pair table from input
    While more data in block
        IF stack empty, read byte from input
        Else pop byte from stack
        IF byte in table, push pair on stack
        Else write byte to output
    End while
End while
```

Figure 3.65   Expansion algorithm pseudocode

CompuServe at 70541,3645 for information concerning the use of these programs.

In examining the compression program listing in Figure 3.66 note that the hash table consists of the arrays left[ ], right[ ], and count[ ], while the variable HASHSIZE, which represents the size of the hash table, must be a power of 2. In addition, the hash table size should not be significantly smaller than the maximum block size (BLOCKSIZE) as this situation could result in an overflow condition. The maximum block size supported is 32 767 bytes. Another parameter that can be adjusted is THRESHOLD. That parameter specifies the minimum occurrence count of pairs to be compressed.

The algorithm developed by Mr Gage commences with the most frequently occurring pair, replacing that pair with an unused character. As each pair is replaced, the algorithm updates the pair count in the hash table. Once the program completes the compression of a buffer, the pair table will contain a record of those pairs of bytes that were replaced by single bytes within the buffer.

To illustrate how the pair table is updated, Figure 3.67 shows a sample pair table resulting from the compression of a string of nine characters. In this example it was assumed that the character set was limited to [A. .H], representing a set of eight characters. Note that

```
/* compress.c */
/* Copyright 1994 by Philip Gage */

#include <stdio.h>

à  fine BLOCKSIZE 5000    /* Maximum block size */
#define HASHSIZE  4096     /* Size of hash table */
#define MAXCHARS   200     /* Char set per block */
#define THRESHOLD    3     /* Minimum pair count */

unsigned char buffer[BLOCKSIZE]; /* Data block */
unsigned char leftcode[256];     /* Pair table */
unsigned char rightcode[256];    /* Pair table */
unsigned char left[HASHSIZE];    /* Hash table */
unsigned char right[HASHSIZE];   /* Hash table */
unsigned char count[HASHSIZE];   /* Pair count */
int size;          /* Size of current data block */

/* Function prototypes */
int lookup (unsigned char, unsigned char);
int fileread (FILE *);
void filewrite (FILE *);
void compress (FILE *, FILE *);

/* Return index of character pair in hash table */
/* Deleted nodes have count of 1 for hashing */
int lookup (unsigned char a, unsigned char b)
{
  int index;

  /* Compute hash key from both characters */
  index = (a ^ (b << 5)) & (HASHSIZE-1);

   * Search for pair or first empty slot */
  while ((left[index] != a || right[index] != b) &&
         count[index] != 0)
    index = (index + 1) & (HASHSIZE-1);

  /* Store pair in table */
  left[index] = a;
  right[index] = b;
  return index;
}

/* Read next block from input file into buffer */
int fileread (FILE *input)
{
  int c, index, used=0;

  /* Reset hash table and pair table */
  for (c = 0; c < HASHSIZE; c++)
    count[c] = 0;
  for (c = 0; c < 256; c++) {
    leftcode[c] = c;
    rightcode[c] = 0;
  }
  size = 0;

  /* Read data until full or few unused chars */
  while (size < BLOCKSIZE && used < MAXCHARS &&
         (c = getc(input)) != EOF) {
    if (size > 0) {
      index = lookup(buffer[size-1],c);
      if (count[index] < 255) ++count[index];
    }
    buffer[size++] = c;
```

Figure 3.66 The Byte Pair Encoding COMPRESS.C compression program

```
            /* Use rightcode to flag data chars found */
            if (!rightcode[c]) {
              rightcode[c] = 1;
              used++;
            }
          }
        return c == EOF;
      }

      /* Write each pair table and data block to output */
      void filewrite (FILE *output)
      {
        int i, len, c = 0;

        /* For each character 0..255 */
        while (c < 256) {

          /* If not a pair code, count run of literals */
          if (c == leftcode[c]) {
            len = 1; c++;
            while (len<127 && c<256 && c==leftcode[c]) {
              len++; c++;
            }
            putc(len + 127,output); len = 0;
            if (c == 256) break;
          }

          /* Else count run of pair codes */
          else {
            len = 0; c++;
            while (len<127 && c<256 && c!=leftcode[c] ||
                   len<125 && c<254 && c+1!=leftcode[c+1]) {
              len++; c++;
            }
            putc(len,output);
            c -= len + 1;
          }

          /* Write range of pairs to output */
          for (i = 0; i <= len; i++) {
            putc(leftcode[c],output);
            if (c != leftcode[c])
              putc(rightcode[c],output);
            c++;
          }
        }

        /* Write size bytes and compressed data block */
        putc(size/256,output);
        putc(size%256,output);
        fwrite(buffer,size,1,output);
      }

      /* Compress from input file to output file */
      void compress (FILE *infile, FILE *outfile)
      {
        int leftch, rightch, code, oldsize;
        int index, r, w, best, done = 0;

        '* Compress each data block until end of file */
        hile (!done) {
          done = fileread(infile);
          code = 256;

          /* Compress this block */
```

Figure 3.66 *Continued*

```
        for (;;) {

            /* Get next unused char for pair code */
            for (code--; code >= 0; code--)
              if (code==leftcode[code] && !rightcode[code])
                break;

            /* Must quit if no unused chars left */
            if (code < 0) break;

            /* Find most frequent pair of chars */
            for (best=2, index=0; index<HASHSIZE; index++)
              if (count[index] > best) {
                best = count[index];
                leftch = left[index];
                rightch = right[index];
              }

            /* Done if no more compression possible */
            if (best < THRESHOLD) break;

            /* Replace pairs in data, adjust pair counts */
            oldsize = size - 1;
            for (w = 0, r = 0; r < oldsize; r++) {
              if (buffer[r] == leftch &&
                  buffer[r+1] == rightch) {
                if (r > 0) {
                  index = lookup(buffer[w-1],leftch);
                  if (count[index] > 1) --count[index];
                  index = lookup(buffer[w-1],code);
                  if (count[index] < 255) ++count[index];
                }
                if (r < oldsize - 1) {
                  index = lookup(rightch,buffer[r+2]);
                  if (count[index] > 1) --count[index];
                  index = lookup(code,buffer[r+2]);
                  if (count[index] < 255) ++count[index];
                }
                buffer[w++] = code;
                r++; size--;
              }
              else buffer[w++] = buffer[r];
            }
            buffer[w] = buffer[r];

            /* Add to pair substitution table */
            leftcode[code] = leftch;
            rightcode[code] = rightch;

            /* Delete pair from hash table */
            index = lookup(leftch,rightch);
            count[index] = 1;
          }
          filewrite(outfile);
       }
    }

    void main (int argc, char *argv[])
    {
      FILE *infile, *outfile;

       f (argc != 3)
         printf("Usage: compress infile outfile\n");
       else if ((infile=fopen(argv[1],"rb"))==NULL)
         printf("Error opening input %s\n",argv[1]);
       else if ((outfile=fopen(argv[2],"wb"))==NULL)
         printf("Error opening output %s\n",argv[2]);
       else {
         compress(infile,outfile);
         fclose(outfile);
         fclose(infile);
       }
    }
```

Figure 3.66 *Continued*

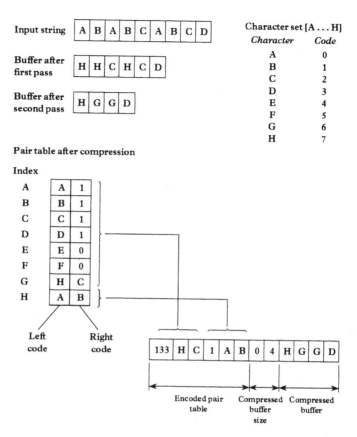

Figure 3.67 Developing the encoded pair table with a hypothetical character set

the last unused character in the character set with respect to the input string was 'H'. Therefore, after the first pass 'H' is used to replace the pair 'AB', which was the most frequently encountered pair in the input string. After the first pass, the pair 'HC' is the most frequently encountered pair in the buffer. Since the character 'G' is the next unused character with respect to the end of the character set, 'G' replaces each occurrence of the pair 'HC' during the second pass. In the lower portion of Figure 3.67 the creation of the pair table is shown. Note that a pair's position in the table is used to indicate the value of the replacement byte instead of storing the replacement byte. For example, consider the pair 'AB' which is located in the eighth entry in the pair table. This indicates that the character 'H' replaced that pair.

In examining the pair table note that an unused entry which does not contain a replaced pair has a left code whose value equals its index (index==leftcode[index]). In examining the pair table after compression, note that when the index is 'G' the left code is 'H'. Similarly, when the index is 'H' the left code is 'A'. Thus, the two pair replacements are indicated in the left code entries.

In comparison, the right code has a dual use. First, the right code[ ] is used by the compression algorithm in the function 'fileread' to flag used versus unused characters, assigning a '1' if the character was used and a '0' if the character was not used. Once the flagging is completed, right code[ ] then serves as the second half of the pair table. Thus, index 'G' results in a left and right code pair of 'HC', while index 'H' has a left and right code pair of 'AB'.

The operation of the BPE compression algorithm uses another trick when outputting pair substitution tables along with packed data. Instead of outputting the character code and pair it represents for each substitution, which would require three characters, the program first classifies each entry in the pair table. In doing so it steps through the table from the first entry to the last, classifying the entry as representing a replaced pair (index != leftcode[index]) or as not representing a replaced pair (index = leftcode[index]). A group of contiguous replaced pairs are encoded by outputting a positive byte count followed by the pairs. If a group of contiguous table entries do not represent replaced pairs, a negative byte count followed by one pair is output. The expansion program, labeled EXPAND.C, whose listing is contained in Figure 3.68 uses a positive byte count as an indicator of the number of following pairs to read. In comparison, a negative byte count tells the expansion program to skip a range of the character set and then read a single pair. As a result of this technique, many pairs can be stored using two bytes per code. To further increase pair table compression, Mr Gage modified the previously described algorithm so that an isolated single byte not representing a pair is not encoded. Instead, the revised algorithm writes the byte to output along with the pair data without the use of an accompanying right code. This trick is recognized by the expansion algorithm which recognizes that the byte does not represent pair data since it occurs at the position byte value == leftcode[byte value].

Expansion

A comparison of the code of the decompression-performing expansion program listed in Figure 3.68 with the code of the compression program indicates one of the prime advantages of this technique, its relatively fast decompression capability. Expansion in comparison to compression requires a single pass, with input to the algorithm occurring from either the input file or a stack. If the stack is empty, its original condition, the algorithm obtains its next input byte from the input file. If the stack contains data, the algorithm obtains its next input byte from the stack. Regardless of the source of the input byte, it is processed according to the following rule: if the byte is a literal it is passed 'as is' to the output; if the byte represents a pair the algorithm replaces it with a pair that is pushed onto the stack.

Once data is placed onto the stack, the expansion algorithm reads the top byte, removing it from the stack. If the byte is expandable, it expands the byte into a pair and pushes the pair onto the stack. The expansion algorithm continues this process of removing the top character in the stack, expanding it into a pair and pushing the new pair onto the stack until it reads a byte that does not expand. Then, that byte is output and the algorithm continues its repetitive operation on the contents of the stack until it is empty. At that time, the algorithm inputs a new character from the file and repeats the previously described process. Figure 3.69 illustrates an example of the expansion process.

Efficiency

The BPE algorithm is very efficient at compression patterns since both bytes in a pair can represent previously defined pair codes. In fact, the previously defined pair codes can represent nested codes which upon expansion represent long strings. For example, a run of 1024 identical bytes can be reduced to a single byte after 10 pair substitutions. Based upon Mr Gage's testing, he was able to compress the Windows 3.1 program WIN386.EXE from 544 789 bytes to 276 955 bytes employing the BPE program using the default parameters listed in Figure 3.66. In comparison, the use of 12-bit and 14-bit Lempel–Ziv string compression programs that he used to compare the efficiency of BPE reduced the same file to 299 118 and 292 588 bytes, respectively.

3.7 PATTERN SUBSTITUTION

This compression technique is basically a more sophisticated form of diatomic encoding. Here, a special character code is substituted for a predefined character pattern. The employment of the pattern substitution compression technique can be highly advantageous when you are transmitting program listings and other types of data files containing known repeating patterns. In addition, pattern substitution forms the basis for the development of dictionary-based compression techniques in which recognized patterns are replaced by numeric codes. Thus pattern substitution can be considered as the predecessor to modern dictionary-based compression techniques.

The advantage offered by pattern substitution is best understood by examining a higher-level language such as FORTRAN. In any FORTRAN program, a very high probability exists that one or more types of statements will be encountered containing common key words

```c
/* expand.c */
/* Copyright 1994 by Philip Gage */

#include <stdio.h>

/* Decompress data from input to output */
void expand (FILE *input, FILE *output)
{
  unsigned char left[256], right[256], stack[30];
  short int c, count, i, size;

  /* Unpack each block until end of file */
  while ((count = getc(input)) != EOF) {

    /* Set left to itself as literal flag */
    for (i = 0; i < 256; i++)
      left[i] = i;

    /* Read pair table */
    for (c = 0;;) {

      /* Skip range of literal bytes */
      if (count > 127) {
        c += count - 127;
        count = 0;
      }
      if (c == 256) break;

      /* Read pairs, skip right if literal */
      for (i = 0; i <= count; i++, c++) {
        left[c] = getc(input);
        if (c != left[c])
          right[c] = getc(input);
      }
      if (c == 256) break;
      count = getc(input);
    }

    /* Calculate packed data block size */
    size = 256 * getc(input);
    size += getc(input);

    /* Unpack data block */
    for (i = 0;;) {

      /* Pop byte from stack or read byte */
      if (i)
        c = stack[--i];
      else {
        if (!size--) break;
        c = getc(input);
      }

      /* Output byte or push pair on stack */
      if (c == left[c])
        putc(c,output);
      else {
        stack[i++] = right[c];
        stack[i++] = left[c];
      }
    }

  }
}

void main (int argc, char *argv[])
{
```

Figure 3.68 The Byte Pair Encoding EXPAND.C decompression program

```
        FILE *infile, *outfile;

        if (argc != 3)
          printf("Usage: expand infile outfile\n");
        else if ((infile=fopen(argv[1],"rb"))==NULL)
          printf("Error opening input %s\n",argv[1]);
        else if ((outfile=fopen(argv[2],"wb"))==NULL)
          printf("Error opening output %s\n",argv[2]);
        else {
          expand(infile,outfile);
          fclose(outfile);
          fclose(infile);
        }
    }
```

Figure 3.68 *Continued*

such as 'READ', 'WRITE' and 'FORMAT', among others. Instead of transmitting the characters of these key words on a character by character basis each time they appear, one of the unassigned characters from the employed character set can be substituted. When pattern substitution is applied to language text, common key words or phrases can similarly be replaced. For English text transmission, such commonly encountered words as 'and', 'the', 'that' and 'this' would be among the first candidates for substitution.

The pattern table

To employ pattern substitution, a pattern table is required. This table contains a set of list arguments and a set of function values. Each function value is a special compression-indicator character which represents the compressed value of a particular argument (Aronson, 1977). Figure 3.70 shows an example of the use of a pattern table. Although each list argument was of similar length, this table can be expanded to include many additional entries of various character length. Strings of four, five, six and more blanks, for example, could be assigned values represented by different special characters as well as patterns of alphanumeric data.

Encoding process

To obtain the compressed data stream, the source data must be broken down into distinct search arguments, initially equal to the smallest-sized argument in the pattern table. The search argument is matched with those list arguments of equal character width. If a match is obtained, the function value associated with the list argument then replaces that portion of the original data stream and results in data compression. If no match is obtained, the width of the search argument is increased to the width of the next larger list

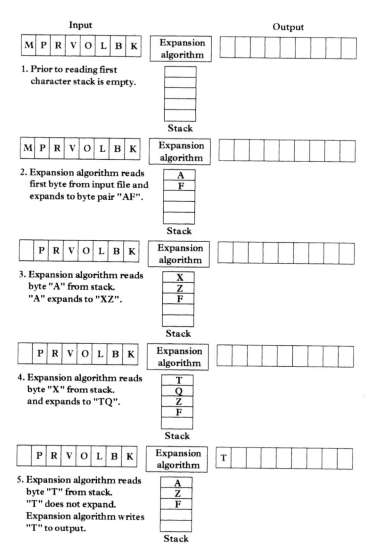

Figure 3.69 Expansion process example

argument or series of list arguments and the process is repeated. If after increasing the width of the search argument to the largest width of the list argument no match results, the first character of the original data string is passed to the compressed data string and the process is repeated, starting with the second character from the original data stream.

A second method of performing pattern substitution results from the use of blanks as delimiters. The binary or octal value of the characters between blanks can be generated and compared with the binary or octal values in the list argument portion of the pattern table.

NOW IS THE TIME FOR ALL GOOD MEN

Pattern table

List arguments	Function values
THE	Sc_1
FOR	Sc_2
ALL	Sc_3

Compressed data stream

NOW IS Sc_1 TIME Sc_2 Sc_3 GOOD MEN

Figure 3.70 Pattern table utilization. Upon a portion of the original data stream matching the list argument, the appropriate function value is substituted. In the above example, special compression indicator characters Sc_1, Sc_2 and Sc_3 are substituted for the words 'the', 'for' and 'all' as they are encountered

This process simplifies the searching of a long argument list and minimizes the processing time required to encode patterns.

Patterns in programming languages

Due to the utilization of keywords or reserved words in most programming languages, pattern substitution is often a very effective compression technique for storing or transmitting program files. Since the number of keywords or reserved words in a programming language can be as high as several hundred, a 2-byte sequence can be employed to represent each keyword pattern substitution. Here, the first byte or character would be used to indicate pattern substitution has occurred, while the following character would denote the actual pattern that was substituted for the keyword or reserved word. To illustrate this concept in additional detail, let us assume that the version of BASIC we are working with is limited to eight keywords. Table 3.16 lists these keywords and their equivalent function values contained in the pattern table that could be constructed.

For clarity of explanation the dollar sign ($) was employed as the compression-indicating character in Table 3.16, although obviously any character in the character set could be used. Preferably, you should select a character which is seldom or, better yet, never used. As an alternative, you can use the Byte Pair Encoding (BPE) technique described in Section 3.6 and first determine if there are unused characters in a file. If so, you can use those characters as pattern compression-indicating characters. However, this technique requires two passes and cannot be used for read-time applications to include compressing data streams that are being transmitted.

If you select characters seldomly used as pattern-indicating charac-

Table 3.16 BASIC language pattern table

Keywords	Function values
END	$1
GOTO	$2
IF	$3
INPUT	$4
LETPRINT	$5
REM	$6
THEN	$7
	$8

BASIC program	Compressed program
100 REM COMMISSION CALCULATION	100$7COMMISSION CALCULATION
110 PRINT "ENTER SALE PRICE"	110$6"ENTER SALE PRICE"
120 INPUT W	120$4W
130 PRINT "ENTER NUMBER SOLD"	130$6"ENTER NUMBER SOLD"
140 INPUT N	140$4N
150 LET C=W∗N∗.0875	150$5C=W∗N∗.0875
160 PRINT "COMMISSION=";C	160$6"COMMISSION=";C
170 PRINT "ANOTHER CALCULATION–Y/N"	170$6"ANOTHER CALCULATION–Y/N"
180 INPUT A$	180$4A$$
190 IF A$ ⟨⟩ "Y" THEN 210	190$3A$$⟨⟩"Y"$8210
200 GOTO 110	200$2110
210 END	210$1

Figure 3.71 Compressing a BASIC program

ters there is always the possibility that the character could occur in a BASIC program. You can replace each occurrence of the pattern compression-indicating character by duplicating that character when it is encountered. Then, the decompression routine would disregard each such duplication of a pattern compression-indicating character.

The compression of a short BASIC program is illustrated in Figure 3.71 based upon the employment of pattern substitution, which in actuality is the replacement of BASIC keywords. Note that the pattern table contained in Table 3.16 was used for the compression process. Since most BASIC languages require keywords to be delimited by spaces, we assumed that the keywords entered in Table 3.16 contained leading and trailing blanks, enabling the functional value substituted for the keyword to be a more effective substitution. Using this method of substitution, 25 spaces as well as 26 other characters are eliminated from the program while two characters are added. The additional characters are due to the replacement of the natural occurrence of the $ character in the program by the special sequence $$ in lines 180 and 190.

Although the overall data reduction, which in this example was approximately 20%, may not appear significant, it should be noted that the actual effort involved to compress data using pattern substitution may not be significantly demanding. To increase the data reduction resulting from compression usually requires the application of several compression techniques to data. In this particular example, you might first preprocess programming files through the utilization of pattern substitution compression. Then you could statistically encode the resulting compressed data. Since the statistical encoding process results in the replacement of frequently occurring characters by short bit sequences, statistically encoding data where keywords were previously replaced by short patterns is more effective than the statistical encoding of the original data. As an example, a 5-bit sequence might be required to represent the keyword PRINT; however, a short bit sequence would be required to represent the character sequence $6 that was substituted for the keyword. The reader is referred to Chapter Four for additional information concerning statistical encoding.

3.8 FORMS-MODE OPERATION

Forms-mode operation is a method of compression that can be employed when data is to be communicated to and from a CRT display in a predefined series of formats. When operated in the forms mode, the display can be used for a fill in the blank type of operation. In this mode of operation, two basic types of data are displayed—protected information and variable information. Fixed or protected information corresponds to the preprinted information of a data field in a standard printed form such as name, address, social security number and similar types of information. Such information when the operation is in the forms mode is not cleared when the screen is erased, is not transmitted to the central processor and is not alterable by accidental keyboard entries. Each fixed field is one-half of a field pair, the other half being the corresponding variable field. Thus, in the forms mode the fixed field can be viewed as the question while the information entered into the corresponding variable field can be considered as the answer.

An example of forms-mode data entry is illustrated in Figure 3.72. Here, the blank spaces indicate the additional positions available for data entry into the variable fields.

In using the forms mode of operation, the operator denotes the form he or she wishes to complete and that form is transmitted from the computer to the terminal display or is locally generated from terminal memory or from a peripheral device attached to the terminal. Fixed fields are preceded by an 'FS' (start fixed field) character while variable fields are preceded by a 'GS' (begin variable field) character and a parameter character. The parameter character is used to define the

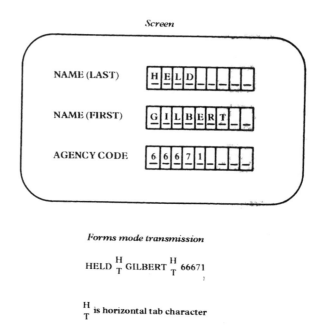

Figure 3.72   Forms-mode data entry

allowable operations within the variable field such as numeric only, alphabetic only, alphanumeric, inhibit transmission and so on. The exact sequence of the GS and FS characters as well as the bit configuration of the parameter character to define allowable operations depends upon the terminal's program. When in forms mode, certain keyboard operations are usually changed from those of the normal mode of operation. As an example, the TAB key on most displays permits the operator to move the cursor (positioning data entry marker) to the first character position of the next sequential variable field, permitting rapid skipping-over of variable fields for which no data is to be entered (Peterson, Bitner and Howard, 1978).

Transmission

The transmission of data in the forms mode is normally performed on a screen basis. When the operator depresses the TRANSMIT key, only the data previously entered into the variable fields is transmitted, with all trailing blanks eliminated.

Here, transmission can occur on-line to the computer or it can be to one of the peripheral units of the terminal. In the case of the latter, a large number of terminal screens may be batched onto a peripheral device such as a cassette or floppy disk for transmission to the computer at one time. Using this combination of forms-mode encoding and off-line storage for transmission by batching screens of information, computer system resources in the form of computer ports and line requirements can be reduced or used more effectively.

By reducing the amount of transmission time required to send batched screen information, a reduction in the number of computer ports required to support remote terminals may be possible. Concerning more effective line utilization, consider the situation where 10 terminals operate in a poll and select environment connected via a common modem sharing unit and modem to a central computer as illustrated in Figure 3.73. In the configuration illustrated, all terminals except the terminal transmitting or receiving data are locked out for the duration of the transmission. Normally, blocks of data up to the screen size, 1920 (80×24) characters or less, are transmitted. At 4800 bps, the transmission of a 1920 8-bit character block to completely fill a screen would require 3.2 s. If 10 terminals were connected to the modem sharing unit with a round robin polling sequence and each operator transmitted or received a full screen of data, it would take 32 seconds, ignoring transmission overhead, until the first terminal operator could again transmit or receive information. Thus, reducing the number of characters transmitted and received through the employment of forms-mode encoding can be used to decrease the response time of existing terminals or to permit additional terminals to be clustered without increasing overall response times.

Returning to the example in Figure 3.72, the 'HT' (horizontal tab) character is normally used as a variable field separator, resulting in the transmitted message indicated in the lower portion of that illustration. If a maximum of eight characters can be entered into each of the three variable fields shown in Figure 3.72, a maximum of 26 characters (24 data and two horizontal tab characters) will be transmitted to the computer for each form completed. This method of forms-mode data entry should be contrasted with conventional interactive line-by-line transmission as shown in Figure 3.74. Here, the message ENTER NAME (LAST), NAME (FIRST), AGENCY CODE serves

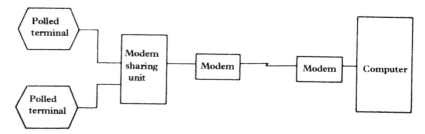

Figure 3.73 Forms-mode encoding increases line service. Forms-mode encoding reduces the poll and select time required per terminal, permitting more terminals to be connected on a shared line or an increase in throughput to existing terminals sharing the line

ENTER NAME (LAST), NAME (FIRST), AGENCY CODE

HELD, GILBERT, 6671 (C/R)

MORE?
YES (C/R)

ENTER NAME (LAST), NAME (FIRST), AGENCY CODE

Figure 3.74 Conventional interactive line-by-line data entry. In conventional time-sharing, the prompt messages requesting data as well as the user responses are transmitted

as a variable field indicator, denoting to the terminal operator the data to be entered. The carriage return (C/R) character acts as a line termination character; however, if data was entered incorrectly, such as alphabetic characters in an all-numeric field, data must first be sent to the computer for processing to determine that such an error has occurred. In such cases, the computer would transmit an error message to the terminal operator who would then hopefully retype the entire line correctly and retransmit the data. In contrast, using an intelligent terminal and forms-mode operation the data entry operation can be preprocessed and such errors corrected prior to transmission.

In comparison with the operator depressing the TRANSMIT key on the display and having the forms-mode method of operation transmit and clear the variable fields so new data can be entered, conventional interactive line-by-line transmission requires the program to prompt the operater to determine if more data is to be entered. The MORE? and YES (C/R) sequence in Figure 3.74 adds additional characters beyond the repeated message used as a variable field indicator. In comparing the sample forms-mode data entry with the conventional line-by-line data entry example, 18 characters are required for the former while 65 characters, excluding line feed and carriage return characters, are required for the latter.

Modified forms-mode method

We can implement a modified method of forms-mode compression by treating a screen display as a collection of forms. To do so we can subdivide a screen display into n segments or forms as illustrated in Figure 3.75. Then, we can operate upon each segment as if it were an individual form to reduce the transmission between a terminal device and a computer.

The key to the utilization of the previously described modified method of forms-mode operation is the assignment of a 'data tag' to

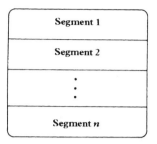

Figure 3.75 Treating the screen as a collection of forms. The screen display can be subdivided into *n* segments, with each segment operated upon as if it represents a form. Then, only if changes occurred to the segment will the contents of the segment be transmitted

1	2
3	4
5	6
7	8
9	10

Figure 3.76 Screen Segmentation Example. By subdividing the terminal screen into 10 segments of 40 characters by 5 lines screen updates can be implemented in increments of 200 characters

each screen segment. Modifying the data tag each time a screen segment changes allows the data tag to serve as an indicator as to whether or not the screen segment must be transmitted. To illustrate the potential for this method to reduce the data required to update screen displays, assume the 80×25 character display of the typical IBM PC screen is subdivided into segments of 40 characters by 5 lines. Doing so would result in the screen being subdivided into a total of 10 segments that could be numbered as indicated in Figure 3.76.

When a terminal operator updates the screen, the terminal or personal computer keeps track of changes made to each segment but does not transmit any data until an ENTER, RETURN, program function (PF), or program assistance (PA) key is pressed. At that time the terminal or personal computer transmits a data tag for each segment which indicates whether or not the segment has been modified. If the segment has been modified the data tag is then followed by the 200 characters contained in the segment. Through the implementation of this compression technique it becomes possible to transmit full-screen updates in as little as one segment of a few hundred characters, including a modified data tag which indicates both the segment which was modified as well as the fact that only that segment was modified.

To illustrate one method by which a display screen can be segmented, let us assume we are using a conventional 7-level ASCII character set. We could set bit 8 to indicate it is a data tag character, which, in effect, allows us to use up to 128 unique characters to represent different information concerning screen segments.

Figure 3.77 illustrates one possible format for the construction of a data tag character. Here the setting of bit position 7 would indicate whether or not the segment that the data tag represents had been changed since the last time the ENTER, RETURN, or a PF or PA key was pressed. Bits 3 through 6 would identify the segment that the data tag is associated with, while bit 1 would be used to indicate whether or not the segment identified by the data tag was the last segment changed.

If the end-of-change bit is set to 0 the characters in the segment would follow the data tag character. If the end-of-change bit is set to 1 this would indicate that no additional characters would follow the characters that define the changed segment identified by the data tag character.

Although the previous discussion concerned the use of a character set from which up to 128 characters could be defined to represent data tag information, how could you accomplish a similar compression scheme using EBCDIC, since most full-screen terminal operations use that character set? As previously discussed in Chapter 2, there are many EBCDIC 8-bit character representations that are undefined. Thus, you could implement the previously described modified forms-mode method of compression by selecting one undefined character to represent a data tag character. Then, you could define a format in which the data tag character is followed by another character which indicates whether or not the segment was modified. If the segment was

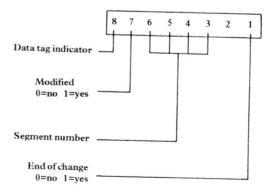

Figure 3.77 Possible data tag format

modified, you could then follow the character used to indicate a modified segment by the characters that represent the segment. Note that in using this method you might construct your format so that you would not transmit a segment number. Thus, if only segment 5 changed, you would have to transmit four data tags to represent segments 1 through 4 without each data tag character being followed by a character indicating that the segment was changed. In comparison, using the previous example you would transmit the data tag with bits 3 through 6 set to a value of 5 to indicate that segment 5 was changed. Regardless of the method used, the important aspect of this technique is that for most interactive terminal applications only a small portion of the screen is updated. Thus, subdividing the screen into segments, and only transmitting those segments that changed since a 'key' was pressed, can significantly increase the efficiency of the transmission system.

STATISTICAL ENCODING

One common element of the data compression techniques discussed in Chapter 3 is that they all operate upon character codes of a fixed bit size. In comparison with those compression methods statistical encoding takes advantage of the probabilities of occurrence of single characters and groups of characters, so that short codes can be used to represent frequently occurring characters or groups of characters while longer codes are used to represent less frequently encountered characters and groups of characters.

The statistical encoding process can be used to obtain a minimization of the average code length of the encoded data, in a manner similar to that in which Morse selected the dot and dash representations of characters so that a single dot was used to represent the letter E, which is the most frequently encountered character in the English language, while longer strings of dots and dashes were used to represent characters that appear less frequently. Included in the class of statistical compression methods is the Huffman coding technique, Shannon–Fano encoding and the arithmetic encoding method, among others.

In this chapter we will focus our attention upon a variety of data-compression methods that are based upon the statistical encoding of characters. Our examination will begin with the use of static compression tables and then investigate the construction and operation of dynamic tables. Prior to discussing these techniques in detail, a review of some basic information theory concepts is warranted. These concepts will provide an understanding of how redundancy can be statistically reduced.

4.1 INFORMATION THEORY

For a system capable of transmitting at n discrete levels at intervals of λ s, the number of different signal combinations in T s is $n^{T/\lambda}$. Since information is proportional to the length of time of transmission, we can take the logarithm of $n^{T/\lambda}$, to obtain the information transmitted in Ts being proportional to $(T/\lambda) \log n$.

The proportionality factor will depend upon the base of the logarithm used, the most common choice being the base 2. This results in the information unit H becoming

$$H = \frac{T}{\lambda} \log_2 n.$$

The unit of information defined in the preceding manner is known as the bit or binary digit. For the transmission of data over a 20 s period using two discrete levels (0 and 1) at 1 s intervals, the information content becomes:

$$H = \frac{20}{1} \log_2 2 = 20 \text{ bits.}$$

The capacity of a given system is defined as the maximum amount of information per second that a system can transmit and can be expressed in bits per second (bps). Thus, the capacity of the preceding example becomes:

$$C = \frac{H}{T} = \frac{1}{\lambda} \log_2 n = \frac{1}{1} \log_2 2 = 1 \text{ bps.}$$

The relative frequency of occurrence of any one combination or event is defined as the probability of occurrence, denoted symbolically as P, where

$$P = \frac{\text{number of times an event occurs}}{\text{total number of possibilities}}$$

If n possible events are specified to be the n possible signal levels, then $P = 1/n$ for events that are equally likely to occur. The information contained by the appearance of any one event in one time interval (H_1) becomes:

$$H_1 = \log_2 n = -\log_2 P \text{ bits/interval}$$

where P represents $1/n$. During t periods of time, consisting of periods λ s long, we should have t times as much information, or

$$H = tH_1 = -t \log_2 P \text{ bits in } t \text{ periods.}$$

Since the number of periods, t, equals the total time, T, divided by the number of intervals, λ, the information available in Ts becomes:

$$H = -\frac{T}{\lambda} \log_2 P = \frac{T}{\lambda} \log_2 n \text{ bits in } T \text{ s.}$$

With the preceding serving as a foundation, we can consider the case where different events or signal levels do not have equal probabilities of occurrence. Let us assume just two levels are to be transmitted, 0 or 1, the first with probability P and the second with probability Q, where $P + Q = 1$. Then:

$$P = \frac{\text{number of times 0 occurs}}{\text{total number of possibilities}}$$

$$Q = \frac{\text{number of times 1 occurs}}{\text{total number of possibilities}}.$$

The information content of a long message consisting of many 0s and 1s is thus dependent upon $P*\log_2 P + Q*\log_2 Q$ which is the information in bits per occurrence of a 0 or 1 times the relative frequency of occurrence of the bit value. We can let the frequency of occurrence of each possible signal level or signal be denoted by P_i, where $P_1 + P_2 + \ldots + P_n = 1$. Then each interval carries $-\log_2 P_i$ bits of information. In t periods of time, i will appear on the average $t*P_i$ times. By summing the information in bits contributed on the average by each symbol appearing $t*P_i$ times over the t intervals, we obtain:

$$H = -t* \sum_{i=1}^{n} P_i \log_2 P_i \text{ bits in } t \text{ periods.}$$

For the interval T, we then obtain:

$$H = -\frac{T}{\lambda}* \sum_{i=1}^{n} P \log_2 P_i \text{ bits in } T \text{ s.}$$

For a message with n possible symbols or levels with probability of occurrence P_i to P_n, the average information per single symbol interval of λ is:

$$H_{\text{avg}} = - \sum_{i=1}^{n} P_i \log_2 P_i \text{ bits/symbol interval.}$$

The above equation represents the mathematical definition of entropy, a term used in information theory to denote the average number of bits required to represent each symbol of a source alphabet. An equivalent definition of entropy is that it references the total information content of a source of information. The more likely an event the less information is conveyed by the event. In fact, as we will shortly note, a given event that always occurs conveys no information and has an entropy of zero.

Based upon the preceding, it becomes possible to compute the redundancy contained in information. Since the unit of information is $\log_2 n$ for a system capable of transmitting at n discrete levels, its redundancy becomes

$$R = \log_2 n - H_{avg}.$$

Then, when there is zero redundancy:

$$H_{avg} = \log_2 n.$$

Entropy examples

Since entropy represents the average number of bits required to represent each symbol of a source alphabet, let us probe deeper to verify its applicability to data compression. Let us first start with a simple coin toss experiment in which the probability of a head is equal to the probability of a tail. Then, $P_H = 0.5 = P_T$. How many bits are required to encode the result of this coin toss?

From the prior definition of entropy, $H_{avg} = -\Sigma^n_{i=1} P_i \log_2 P_i$. Here $\log_2 x$ for those that may have forgotten means $2^? = x$ or 2 to what power equals x. Substituting the probability of a head and the probability of a tail occurring into the equation for computing entropy we obtain:

$$H_{avg} = -[\tfrac{1}{2}\log_2\tfrac{1}{2} + \tfrac{1}{2}\log_2\tfrac{1}{2}].$$

One of the properties of logarithms is that $\log(1/X) = -\log X$. Thus $\log_2\tfrac{1}{2}$ is equivalent to $-\log_2 2$. Since $\log_2 2$ becomes -1, returning to our calculation of entropy for the one-coin toss, we obtain:

$$H_{avg} = -[\tfrac{1}{2}*(-1) + \tfrac{1}{2}*(-1)] = -[-1] = 1.$$

From the preceding, the computation of entropy tells us that one bit would be required to encode the results of a single-coin toss. Now let us expand the preceding to two coins.

The two sides of a coin, heads (H) and tails (T), correspond to members X_1 and X_2 from an alphabet X containing two symbols. If we toss two coins and encode the results so that $T = 0$ and $H = 1$, the coin toss result probabilities correspond to a four-symbol alphabet as tabulated in Table 4.1. The entropy or average number of bits required to represent each possible outcome or symbol from our four-symbol alphabet becomes:

$$H_{avg} = -\sum_{i=1}^{4} P_i \log_2 P_i = -4*0.25 \log_2 0.25.$$

Since $\log_2 0.25$ is -2, we can rewrite the preceding equation as follows:

$$H_{avg} = -4[0.25*(-2)] = 2.$$

For the coin toss experiment results listed in Table 4.1, two binary symbols were required to encode each alphabetic symbol. If for some reason the coin toss was fixed such that only tails (T) occurs, the only symbol required in our alphabet would be X_1. Under this condition, we would never have to do any coin tossing to determine the outcome since the result is always known in advance. The entropy of this one-symbol alphabet can be computed as follows:

$$H_{avg} = -\sum_{i=1}^{4} P_i \log_2 P_i = -\sum_{i=1}^{1} \log_2 1 = 0.$$

In this case, since the outcome is known in advance the symbol provides no information; hence, its entropy is zero.

We can again fix the coin toss experiment; however, this time we will fix it so the probability of tails (T) occurring is increased to 0.75, leaving a 0.25 probability of heads occurring. Under these circumstances, the tabular results of the coin toss outcomes representing a four-symbol alphabet would be as listed in Table 4.2. Although the representative code, number of coin toss outcomes and alphabet symbols have remained the same, the outcome probabilities have changed. Thus, the probability of two tails is now 0.75 times 0.75, or 0.5625, and so on. The entropy of this four-symbol alphabet is now:

Table 4.1 Coin toss representing four-symbol alphabet

Coin toss outcome	Alphabet symbol	Outcome probability	Representative code
TT	X_1	0.25	00
TH	X_2	0.25	01
HT	X_3	0.25	10
HH	X_4	0.25	11

Table 4.2 Fixed coin-toss representing four-symbol alphabet. Probability of head = 0.25; probability of tail = 0.75

Coin toss outcome	Alphabet symbol	Outcome probability	Representative code
TT	X_1	0.5625	00
TH	X_2	0.1875	01
HT	X_3	0.1875	10
HH	X_4	0.0625	11

$$H_{avg} = -\sum_{i=1}^{4} P_i \log_2 P_i = 0.5625 \log_2 0.5625 + 0.1875 \log_2 0.1875$$

$$+\ 0.1875 \log_2 0.1875 + 0.0625 \log_2 0.0625 = 1.62 \text{ bits per symbol.}$$

Based upon the preceding, let us compute the redundancy in the fixed coin toss experiment. Since a two-symbol event results in four discrete levels

$$R = \log_2 n - H_{avg} = \log_2 4 - 1.68 = 2 - 1.68 - 0.38.$$

In comparison with the first coin toss experiment, the average number of bits required to represent a symbol from the four-symbol alphabet has been reduced by 0.38. This indicates that using another type of coding scheme to represent the four-symbol alphabet could result in an approximate 20% reduction from the two bits per symbol previously used to represent the four-symbol alphabet. To obtain this reduction, we must assign short codes to the most frequently occuring symbols of the alphabet and longer codes to the less frequently encountered symbols. This method will result in a long string of data symbols having, on the average, fewer bits per symbol and is the foundation for what is known as Huffman coding (Dishon, 1977; Moilanen, 1978).

Huffman coding is one of several techniques where data is encoded based upon their probability of occurrence. Although the original method of Huffman coding was applied to characters, it can also be applied to strings of characters as well as previously compressed data, such as diatomic encoded pairs of characters. Due to Huffman coding forming the basis for a large number of data-compression techniques, we will examine this method as a separate entity in the next section of this chapter.

4.2 HUFFMAN CODING

Huffman coding is a statistical data-compression technique whose employment will reduce the average code length used to represent the symbols of an alphabet. The alphabet can be the English language alphabet or a type of data-coded alphabet such as the ASCII or EBCDIC character sets. It can also be any code set to include the 256 distinct codes in any file of eight-bit stored characters.

Prefix property of code

The Huffman code is an optimum code since it results in the shortest average code length of all statistical encoding techniques. In addition, Huffman codes have a prefix property which means that no short code group is duplicated as the beginning of a longer group. This means that if one character is represented by the bit combination 100, then 10001 cannot be the code for another letter since in scanning the bit stream from left to right the decoding algorithm would interpret the five bits as the 100-bit configuration character followed by a 01 bit configuration character.

The prefix property of the Huffman code ensures that the code is uniquely decipherable. To understand the importance of this property and how it relates to the construction of a Huffman code, consider a four-symbol alphabet X_1, X_2, X_3, X_4, which is to be encoded as follows:

$$X_1 = 0, \quad X_2 = 01, \quad X_3 = 11, \quad X_4 = 00.$$

If the received message is 0001, this bit sequence could represent $X_1X_1X_2$ or X_4X_2. Thus, the code is not uniquely decodable.

Using a decision tree diagram we can determine visually why the preceding encoded four-symbol alphabet is not instantly decodable. To construct a decision tree, let us start at an initial state and draw a branch to a node represented by the symbol X_1 and label the binary value assigned to the node on the branch. Next, since X_2 was assigned the value 01, let us draw a branch from the X_1 node to a node labeled X_2 and assign the value 1 to the branch between nodes X_1 and X_2. Similarly, we can route a branch from node X_1 to a node representing X_4 and assign that branch the binary value 0.

Since X_3 was assigned the value 11, we can draw a branch from the initial state to an intermediate node and another branch from the intermediate node to a node labeled X_3, assigning the binary value of 1 to each branch. Figure 4.1 illustrates the decision tree which corresponds to the assignment of values to the four-symbol alphabet.

In examining the decision tree illustrated in Figure 4.1, note that the route to node X_3 is through an intermediate node that is not

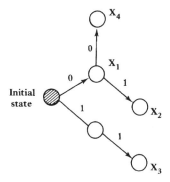

Figure 4.1 Decision tree to form the four-symbol alphabet X_1, X_2, X_3, X_4; where $X_1=0$, $X_2=01$, $X_3=11$, and $X_4=00$

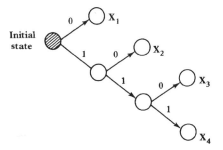

Figure 4.2 Revised decision tree. In constructing this tree each binary digit causes a branch, either to a terminal state or to a decision point

necessary as there is no route from the intermediate node to another node other than node X_3. Also note that node X_1 is actually an intermediate node and the path to X_1 does not represent a unique bit combination. Now suppose in forming a decision tree we add the rules that each branch either ends with a node as a terminal state or functions as a decision point which allows a route to a terminal state. Following those rules, let us redraw the decision tree and assign binary values to each branch to develop codes for a four-symbol alphabet. Figure 4.2 illustrates the revised decision tree. Note that X_1 still has the binary value of 0, but X_2 through X_4 are now assigned the values 10, 110, and 111, respectively. Also note that each bit is only examined once and each path represents a unique bit combination.

The rules we have just developed for constructing a decision tree that results in a code which is instantly decodable are the foundation from which the Huffman code is constructed. The Huffman code can be developed through the utilization of a tree structure as illustrated in Figure 4.3. Here, the symbols are first listed in descending order of frequency of occurrence. The groups with the smallest frequencies (X_3 and X_4) are combined into a node with a joint probability of occurrence of 0.25. Next, that node is merged with the next lowest probability of occurrence symbol or pair of symbols. In this illustration, the pair X_3X_4 is merged with X_2 to produce a node whose joint probability is 0.4375. Finally, the node representing the probabilities of occurrence of X_2,X_3 and X_4 is merged with X_1, resulting in a node whose probability of occurrence is unity. This master node represents the probability of occurrence of all four characters in the character set. By asssigning binary 0s and 1s to every segment emanating from each node, you can derive the Huffman code for each character. The code is obtained by tracing from the 1.0 probability node to each character symbol, noting the 1s and 0s encountered. If you construct the decision tree which represents the codes assigned to the characters X_1 through X_4 the tree is exactly the same as previously illustrated in Figure 4.2. Since each binary digit causes a branch to a terminal state or to a decision point, it represents a code that is uniquely decodable.

The average number of bits per symbol can be calculated by multiplying the Huffman code lengths by their probability of occurrence. Thus, the code uses:

$$1*0.5625 + 2*0.1875 + 3*0.1875 + 3*0.0625$$

or 1.63 bits per symbol. Note that the Huffman code result of 1.63 bits per symbol closely approaches the entropy of 1.62 bits per symbol (Dishon, 1977; Moilanen, 1978).

As previously explained, a key property of the Huffman code is that it can be instantaneously decoded as the coded bits in the com-

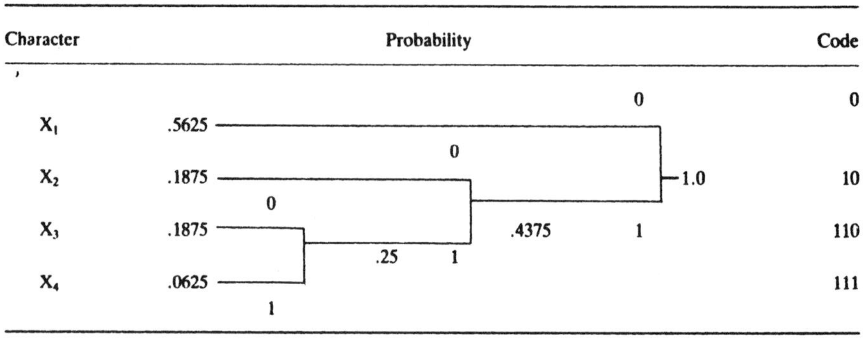

Figure 4.3 Huffman code development employing a tree structure. Huffman codes can be developed by employing a tree structure. The Huffman code resulting from this construction method is derived by tracing from the 1.0 probability node to each source character (symbol), noting 1s and 0s encountered

pressed data stream are encountered. An example of the instantaneous decoding property is illustrated in Figure 4.4. Here, the compressed data stream can be decoded immediately by reading left to right without waiting for the end of the block of data to occur.

The substitution of a number of bits representing a particular data character or group of characters is a fairly simple process when the number of substitutions is limited. As the number of substitutions increases, the complexity of the substitution process increases. In Figures 4.5 and 4.6, the development of a Huffman code for the English alphabet is illustrated. The second column in each figure indicates the approximate frequency of occurrence of each letter in the English language. If you compute the entropy for the 26 letters in the English language, you will obtain a value of approximately 4.1 bits per letter. This indicates that the application of a data compression technique that operates on single characters and results in an average bit length per character approaching entropy should provide a compression ratio of approximately 2:1. The tree structure used to develop the code shown in Figure 4.5 is produced as follows:

A. The character set is arranged in a column on the left in order of decreasing frequency of occurrence with the frequency placed in a column next to each character.
B. Commencing at the bottom of the table, lines are drawn horizontally from each character frequency. The lines with the two lowest frequencies of occurrence are merged and their associated frequencies are added to obtain a composite frequency. This composite frequency is entered on a single new line and reflects the combined frequency of the previously paired characters.
C. The process of combining the two lowest frequency lines into a single line containing combined frequencies is continued until all the lines have been merged.

After the tree has been developed, the Huffman code for each charac-

Encoded message	0	10	0	111	10	110	0	
Decoded message	X_1	X_2	X_1	X_4	X_2	X_3	X_1	

Figure 4.4 Instantaneous decoding property. One of the key properties of the Huffman technique is the fact that encoded data can be instantly decoded

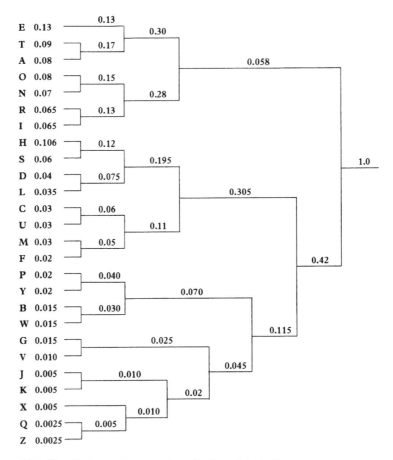

Figure 4.5 Developing a tree structure for the alphabet

ter can be assigned by placing a 0 bit to one side of each nodal point and a 1 bit to the other path emanating from that point towards the left-hand symbol. The assignment of 0 and 1 bits is arbitrary. However, the assignment must be consistent. Thus, if a binary 0 is selected to represent one upper route then it must be used for all upper routes. The appropriate bit sequence assigned to each data character is then determined by tracing the route from the master nodal point where the probability of all character frequencies of occurrence is unity back to the starting node for the appropriate character, noting the bits assigned to the path. The assignment of bits to the paths and the resulting Huffman coded values for the English alphabet are illustrated in Figure 4.6.

The number of bits required to encode a letter using the Huffman technique can be determined from the following formula:

$$b = f(-\log_2 P)$$

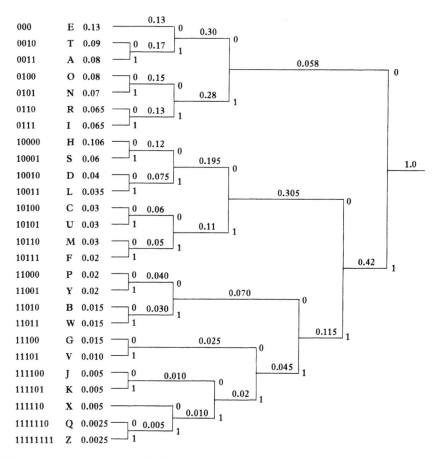

Figure 4.6 Assigning the Huffman code

where:

P = probability of occurrence of the letter

$f(x)$ = the closest integer greater than or equal to x.

Since the probability of E is 0.13 and $-\log_2 0.13$ is 2.94, then the integer greater than or equal to 2.94 is 3. Thus, 3 bits are required to encode the letter E (Peterson, Bitner and Howard, 1978).

Code construction considerations

When several characters have the same frequency of occurrence, you may be able to develop two or more codes to represent the Huffman code for a character set. This situation, which commonly occurs, provides us with an interesting problem—deciding which code to use. To illustrate how this problem can arise, as well as to provide a foundation for discussing the solution to the problem, let us first consider a five-character character set whose frequencies of occurrence are as indicated in Table 4.3.

Figure 4.7 illustrates the development of two tree structures, each resulting in the construction of different Huffman codes for the characters listed in Table 4.3. In the top portion of Figure 4.7 note that X_4 and X_5 are combined exactly in the same manner as X_4 and X_5 are combined in the second tree structure illustrated in the lower portion of Figure 4.7. However, after X_4 and X_5 are combined, the resulting frequency of occurrence of the two characters is 0.2 and you have several options available in developing the Huffman code.

In the top portion of Figure 4.7, X_3 was next combined with the merger of X_3 and X_4 as X_3 had a probability of occurrence of 0.2, which exactly equalled the combined probability of occurrence of X_4 and X_5. In the lower portion of Figure 4.7, X_2 and X_3 were merged, resulting in a combined probability of occurrence of 0.4. Since we

Table 4.3 Sample five-character character set

Character	Frequency of occurrence
X_1	0.4
X_2	0.2
X_3	0.2
X_4	0.1
X_5	0.1

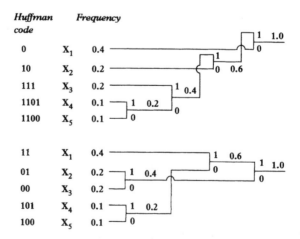

Figure 4.7 Developing different trees and codes

must merge lower frequencies of occurrences with higher frequencies of occurrence to form a Huffman code we combined the results of merging X_4 and X_5 with X_1 to produce a combined frequency of occurrence of 0.6. This was followed by merging the results of combining X_2 and X_3 with the results obtained from merging X_1 with the combination of X_4 and X_5.

In examining the resulting Huffman code at the top of Figure 4.7, how does it compare to the code at the bottom of that illustration? One way we can compare codes is by computing their average bit length, so let's do so. For the Huffman code at the top of Figure 4.7, its average length in bits is:

$$L = 0.4*1 + 0.2*2 + 0.2*3 + 0.1*4 + 0.1*4 = 2.2 \text{ bits.}$$

For the Huffman code at the bottom of Figure 4.7, its average length in bits is:

$$L = 0.4*2 + 0.2*2 + 0.2*2 + 0.1*3 + 0.1*3 = 2.2 \text{ bits.}$$

Note that both codes have the same efficiency expressed in terms of the average number of bits required to encode a character. However, the two codes do not have the same set of lengths for the characters in the character set. Although over a long period of time the use of either code should produce the same result in terms of efficiency expresssed as the sum of encoded bits, in actuality, a more appropriate choice is to select the code whose average length varies the least. To illustrate this, let us compute the variance for each Huffman code. The variance of the Huffman code at the top of Figure 4.7 is:

$$\text{Var}(1) = 0.4(1{-}2.2)^2 + 0.2(2{-}2.2)^2 + 0.2(3{-}2.2)^2 + 0.1(4{-}2.2)^2$$
$$+ \ 0.1(4{-}2.2)^2 = 1.36.$$

The variance of the Huffman code developed in the lower portion of Figure 4.7 is:

$$\text{Var}(2) \ = \ 0.4(2{-}2.2)^2 \ + \ 0.2(2{-}2.2)^2 \ + \ 0.2(2{-}2.22)^2 \ + \ 0.1 \ (3{-}2.2)^2 \ +$$
$$0.1(3{-}2.2)^2 = 1.6.$$

As indicated, the first Huffman code has less variability and should be selected.

Returning to Figure 4.7, if we examine the construction of each tree, we note that the combined X_4 and X_5 frequencies of occurrence in the lower portion of the illustration were directly moved to the top. In comparison, the combination of X_4 and X_5 frequencies of occurrence in the upper portion of Figure 4.7 were gradually merged with other states to reach the top. When you move a combined state as high as possible it reduces the variability of the resulting Huffman code and can be considered as the key to construction of a minimum variability Huffman code.

Information requirements

To develop a Huffman code whose average code length will approach its entropy requires the frequency distribution of the characters or symbols to be encoded to be known in advance. For file compression this is a relatively easy process since a program can first count the frequency of occurrence of characters. Then the program can develop a Huffman code and compress the contents of the file based upon an analysis of the contents of the file. In a transmission environment the construction of a Huffman code based upon a predefined character frequency of occurrence must consider several factors and can result in a far from optimum desired result.

Since the frequency distribution of a data stream is proportional to the end use of the stream, this factor can result in a preselected frequency distribution used to develop a Huffman code resulting in a code far from optimum during certain data transmission sequences. As an example, the frequency distribution of English text, such as that resulting from a data file used for computerized typesetting, may be quite different from the data file containing the results of a FOR-TRAN program compilation. In the first instance, the distribution of characters should follow the distribution of normal English, with E the most frequently occurring character while Z is one of the least frequent characters. For the FORTRAN compilation, special characters such as parenthesis, + for addition, – for substraction, ∗ for multiplication and / for division have a high degree of occurrence not normally encountered in English text.

To compensate for frequency distribution differences, several encoding schemes can be considered. First, the analysis of mixed data files can be conducted employing the computer programs listed in Appendix A (p. 389). This will enable you to ascertain the appropriate relationship between the frequency of occurrence of characters of different types of data.

A second method to consider is an adaptive Huffman encoding technique. Such a technique might first require a frequency analysis of a large block of data which would then be encoded based upon that distribution. Prior to the transmission of the encoded data, a table of the symbols and Huffman codes developed for each symbol must be transmitted to enable the encoded data to be successfully decoded. With a little imagination, you can visualize that frequently changing data streams would result in numerous tables as well as encoded data being transmitted. These tables can be considered as overhead, resulting in the compression frequency decreasing as the number of data-stream frequency distributions change per unit time. Another problem encountered with some adaptive Huffman coding techniques is determining the size of the data stream to sample and the sample intervals. The larger the sample, the greater the processing requirement becomes. If the data is to be transmitted, a buffer area is required to place the sample in while the frequency analysis is conducted. Concerning the sample interval, if three FORTRAN jobs are followed by an English text job, all of equal size, T, sampling at T, $T+2$ and $T+4$ would result in the English text job being excluded from the sample. Since a remote batch terminal operator submits jobs and pulls system output, he or she knows ahead of time the type of job that will be transmitted to or received from the computer. For this type of operating situation, predefined frequency distributions can be selected and conveyed to the opposite end of the transmission link by the transmission of a special code.

To eliminate the previously described problems resulting from the generation of frequency tables, you can construct a truly adaptive or self-adapting Huffman encoding technique. This technique builds frequency tables at both ends of a transmission link as data transmission occurs and adaptively adjusts those tables during transmission. The reader is referred to Section 4.4 which discusses this technique in detail.

A third method of compensating for frequency distribution differences is by the use of a plain text code, which is used to indicate that the character following it should be reproduced exactly as received. This permits characters that rarely occur in the source data to be excluded from the encoding process and results in the development of one type of modified Huffman code. Here, you could group all characters of low frequency of occurrence into one probability of occurrence and assign a Huffman code to represent that summed probability. This would be the plain text code and would indicate that the next eight bits represent an actual non-encoded data character. Without the use of a plain text code, large strings of, say, 20 or more bits might result in the representation of low frequency of occurrence characters. If the plain text code were four bits in length, then a maximum of 12 bits would be required to represent any low frequency of occurrence character. The pre-emption of a 4-bit code to signify that the next eight bits are the plain text representation of an 8-bit character means that some relatively high frequency of occurrence character which would have had a 4-bit code as its Huffman representation will be represented by some longer code. Thus, although there will be no very long codes present, the mean number of bits per character will increase when a plain text code is employed.

C language programs

To provide readers with a tutorial to the construction of Huffman compression, two programs are presented in this section. Both programs were developed by Bill Demas and were placed by Mr Demas into the public domain. The first program, HUFF1.C, performs file compression while the second program, DHUFF.C, reverses the process to restore the file to its original contents. Although Mr Demas granted permission to modify the programs as users see fit, the modularity and internal documentation of each program through the use of comments make both programs good learning tools. In keeping with my desire to present the programs in their original form, their names were retained as created and are exceptions to the general naming conventions used for most files in this book. Thus, readers will find the files on the convenience disk under their original filenames of HUFF1.C and DHUFF.C.

HUFF1.C

Figure 4.8 lists the statements in the program HUFF1.C coded by Bill Demas. Although the comments in MAIN indicate that the Huffman encoding procedure is performed in five separate steps, in actuality there are seven separate functions as one function invokes a second, while a seventh displays the results of the compression operation.

In examining the statements in Figure 4.8 note that Mr Demas placed most of his defined functions in a reverse location in the program with respect to their invocation in MAIN. That is, the function GET_FREQUENCY_COUNT, which is invoked first in MAIN, was placed at the end of the program. After MAIN uses the built-in functions 'freek' and 'ftell' to determine the number of bytes in the file, the function GET_FREQUENCY_COUNT is invoked. Readers examining that function will note that the 'for' loop simply cycles through the input file one character at a time using getc(ifile) as the index to the array frequency_count to increment the value for the appropriate character. That is, getc(ifile) retrieves one character from the input file. The value of the retrieved character is then used as the index to the array frequency_count, whose value is increment by 1. Thus, once this function is completed the array frequency_count contains the number of occurrences of each byte using the index in the array as the byte identifier. That is, frequency_count[13] would contain the number of Carriage Returns (ASCII 13) encountered in the file.

The function BUILD_INITIAL_HEAP first cycles through the array frequency_count. For each non-zero element in the array it sets the same element number of the array heap to the value of the index of frequency_count. The next portion of the function cycles from the highest index value of the array heap to invoke the function RHEAP.

That function creates a new tree structure based upon the previously described BUILD_INITIAL_HEAP function. Once that is accomplished, the functions BUILD_CODE_TREE and GENERATE_CODE_ TABLE perform functions related to their names. Note that throughout the program names representative of the activity being performed were used to facilitate readers in understanding its code construction. Thus code[loop]=bitcode indicates the assignment of bit codes to each element in the array code.

One of the limitations of HUFF1.C is the overhead associated with the construction of the header placed on each output file through the use of three 'fwrite' statements in MAIN. After writing the file size the following statements output two byte codes and one byte code length. In doing so the program always writes a minimum of 768 bytes plus the number of bytes necessary to record the file size. This is a substantial overhead, especially for compressing small files, and could be substantially reduced under certain situations by recording only the code and code length for non-zero characters encountered in the source file. To accomplish this, a character could be added to the compressed file after the first 'fwrite' operation whose first bit position would serve as a flag to indicate whether or not zero count characters were encountered. If they were, the remaining bits could indicate the number of bit map characters that follow. Thirty-two bit map characters would define the locations of up to 256 two-byte codes and one byte code length. Thus, the addition of 33 bytes would begin to reduce the header when more than 11 codes are not encountered in a file. If every character was encountered this technique would add one byte to the header due to the previously described flag.

Operation of HUFF1.C

To illustrate the operation of HUFF1.C, the program was compiled using Borland's Turbo C++ and executed several times using different types of files. Figure 4.9 illustrates the execution of HUFF1 using a small source C file named ARITH.C as input. Note that the requirement to use a fixed header of 772 characters resulted in the compressed output file containing more characters than the input file. Thus, the percentage savings is negative.

To obtain an indicator of the efficiency of the program when the character set was not limited to standard ASCII, an executable (binary) file created using Turbo C++ was used as input. Figure 4.10 illustrates the execution of HUFF1 using the file ARITH.EXE as input to the program. Note that the program reduced the size of the output file by approximately 13% from the size of the input file.

The program HUFF1 becomes more efficient as the size of standard ASCII files used as input increases. This is illustrated in Figure 4.11 in which the file PACKC.C was used as input to the program. In this

```
/*********************************************************************
 *                                                                   *
 *  HUFF1.C:    Huffman Compression Program.                         *
 *              14-August-1990    Bill Demas       Version 1.0       *
 *                                                                   *
 *     This program compresses a file using the Huffman codes.       *
 *                                                                   *
 *            USAGE:    HUFF1 <input file> <output file>             *
 *                                                                   *
 *  (DISK to DISK:  Input direct from disk, output direct to disk)   *
 *********************************************************************/

#include <stdio.h>
#include <stdlib.h>

#define    VERBOSE                     /* If defined, prints verbose
                                          program progress when it's
                                          running...                  *

short          father[512];
unsigned short code[256], heap_length;
unsigned long  compress_charcount, file_size, heap[257];
unsigned char  code_length[256];
long           frequency_count[512];

FILE           *ifile, *ofile;



/*********************************************************************

 MAIN ()

 This is the main program. It performs the Huffman encoding procedure in
 5 separate steps.

 I know that this program can be made more compact & faster, but I was more
 interested in UNDERSTANDABILITY !!!

 *********************************************************************/

void main (argc, argv)
int  argc;
char *argv[];
{
   unsigned short generate_code_table ();
   void           build_code_tree (), build_initial_heap ();
   void           compress_image (), compression_report ();
   void           get_frequency_count ();

   if (argc == 3)
   {
      printf ("\nHUFF1:  Huffman Code Compression Program.");
      printf ("\n          14-Aug-90  Bill Demas.  Version 1.0\n\n");


      if ((ifile = fopen (argv[1], "rb")) != NULL)
      {
```

Figure 4.8 Huffman compression program HUFF1.C

```
            fseek (ifile, 0L, 2);
            file_size = (unsigned long) ftell (ifile);

            #ifdef VERBOSE
               printf ("(1) Getting Frequency Counts.\n");
            #endif

            fseek (ifile, 0L, 0);
            get_frequency_count ();

            #ifdef VERBOSE
               printf ("(2) Building Initial Heap.\n");
            #endif

            build_initial_heap ();

            #ifdef VERBOSE
               printf ("(3) Building the Code Tree.\n");
            #endif

            build_code_tree ();

            #ifdef VERBOSE
               printf ("(4) Generating the Code Table.\n");
            #endif

            if (!generate_code_table ())
               printf ("ERROR!  Code Value Out of Range. Cannot Compress.\n");
            else
            {
               #ifdef VERBOSE
                  printf ("(5) Compressing & Creating the Output File.\n");
               #endif

               if ((ofile = fopen (argv[2], "wb")) != NULL)
               {
                  fwrite (&file_size, sizeof (file_size), 1, ofile);
                  fwrite (code, 2, 256, ofile);
                  fwrite (code_length, 1, 256, ofile);

                  fseek (ifile, 0L, 0);
                  compress_image ();

                  fclose (ofile);
               }
               else
                  printf("\nERROR: Couldn't create output file %s\n", argv[2]);

               #ifdef VERBOSE
                  compression_report ();
               #endif
            }
            fclose (ifile);
         }
      else
         printf ("\nERROR:  %s -- File not found!\n", argv[1]);
   }
   else
      printf ("Usage:  HUFF1 <input filename> <output filename>\n\n");
}
```

Figure 4.8 Continued

```
/********************************************************************

  COMPRESS_IMAGE ()

  This function performs the actual data compression.
  ********************************************************************/

void compress_image ()
{
    register unsigned int    thebyte = 0;
    register short           loop1;
    register unsigned short  current_code;
    register unsigned long   loop;

    unsigned short  current_length, dvalue;
    unsigned long   curbyte = 0;
    short           curbit = 7;


    for (loop = 0L; loop < file_size; loop++)
    {
        dvalue         = (unsigned short) getc (ifile);
        current_code   = code[dvalue];
        current_length = (unsigned short) code_length[dvalue];

        for (loop1 = current_length-1; loop1 >= 0; --loop1)
        {
            if ((current_code >> loop1) & 1)
                thebyte |= (char) (1 << curbit);

            if (--curbit < 0)
            {
                putc (thebyte, ofile);
                thebyte = 0;
                curbyte++;
                curbit = 7;
            }
        }
    }
    putc (thebyte, ofile);
    compress_charcount = ++curbyte;
}


/********************************************************************

  COMPRESSION_REPORT ()

  This function displays the results of the compression sequence.
  ********************************************************************/

void  compression_report ()
{
    float           savings;
    unsigned short  header_charcount;
    unsigned long   output_characters;
```

Figure 4.8 *Continued*

```
      header_charcount    = 768 + sizeof (file_size);
      output_characters   = (unsigned long) header_charcount +
                            compress_charcount;

      printf ("\nRaw characters            : %ld\n", file_size);
      printf ("Header characters          : %d\n", header_charcount);
      printf ("Compressed characters      : %ld\n", compress_charcount);
      printf ("Total output characters    : %ld\n", output_characters);

      savings = 100 - ((float) output_characters / (float) file_size) * 100;
      printf ("\nPercentage savings         : %3.2f%%\n", savings);
}


/********************************************************************

  GENERATE_CODE_TABLE ()

  This function generates the compression code table.
  *******************************************************************/

unsigned short  generate_code_table ()
{
    register unsigned short  loop;
    register unsigned short  current_length;
    register unsigned short  current_bit;

    unsigned short  bitcode;
    short           parent;


    for (loop = 0; loop < 256; loop++)
        if (frequency_count[loop])
        {
            current_length = bitcode = 0;
            current_bit = 1;
            parent = father[loop];

            while (parent)
            {
                if (parent < 0)
                {
                    bitcode += current_bit;
                    parent = -parent;
                }
                parent = father[parent];
                current_bit <<= 1;
                current_length++;
            }

            code[loop] = bitcode;

            if (current_length > 16)
                return (0);
            else
                code_length[loop] = (unsigned char) current_length;
        }
        else
            code[loop] = code_length[loop] = 0;
```

Figure 4.8 *Continued*

```
      return (1);
}


/*****************************************************************************

  BUILD_CODE_TREE ()

  This function builds the compression code tree.
  *****************************************************************************/

void build_code_tree ()
{
   void     reheap ();

   register unsigned short   findex;
   register unsigned long    heap_value;


   while (heap_length != 1)
   {
      heap_value = heap[1];
      heap[1]    = heap[heap_length--];

      reheap (1);
      findex = heap_length + 255;

      frequency_count[findex] = frequency_count[heap[1]] +
                                 frequency_count[heap_value];
      father[heap_value] =  findex;
      father[heap[1]]    = -findex;
      heap[1]            =  findex;

      reheap (1);
   }

   father[256] = 0;
}


/*****************************************************************************

  REHEAP ()

  This function creates a "legal" heap from the current heap tree structure.
  *****************************************************************************/

void reheap (heap_entry)
unsigned short   heap_entry;
{
   register unsigned short   index;
   register unsigned short   flag = 1;

   unsigned long    heap_value;

   heap_value = heap[heap_entry];

   while ((heap_entry <= (heap_length >> 1)) && (flag))
   {
```

Figure 4.8 *Continued*

```
        index = heap_entry << 1;

        if (index < heap_length)
            if (frequency_count[heap[index]] >= frequency_count[heap[index+1]])
                index++;

        if (frequency_count[heap_value] < frequency_count[heap[index]])
            flag--;
        else
        {
            heap[heap_entry] = heap[index];
            heap_entry       = index;
        }
    }

    heap[heap_entry] = heap_value;
}


/*********************************************************************

 BUILD_INITIAL_HEAP ()

 This function builds a heap from the initial frequency count data.
 *********************************************************************/

void build_initial_heap ()
{
    void     reheap ();

    register unsigned short  loop;


    heap_length = 0;

    for (loop = 0; loop < 256; loop++)
        if (frequency_count[loop])
            heap[++heap_length] = (unsigned long) loop;

    for (loop = heap_length; loop > 0; loop--)
        reheap (loop);
}


/*********************************************************************

 GET_FREQUENCY_COUNT ()

 This function counts the number of occurrences of each byte in the data
 that are to be compressed.
 *********************************************************************/

void get_frequency_count ()
{
    register unsigned long  loop;


    for (loop = 0; loop < file_size; loop++)
        frequency_count[getc (ifile)]++;
}
```

Figure 4.8 *Continued*

```
C:\TC\EXAMPLES>HUFF1 ARITH.C XXX

HUFF1:        Huffman Code Compression Program.
              14-Aug-90  Bill Demas.  Version 1.0

(1) Getting Frequency Counts.
(2) Building Initial Heap.
(3) Building the Code Tree.
(4) Generating the Code Table.
(5) Compressing & Creating the Output File.

Raw characters            :    744
Header characters         :    772
Compressed characters     :    484
Total output characters   :    1256

Percentage savings        :    -68.82%
```

Figure 4.9 Executing HUFF1 using a short C language source program as input

```
C:\TC\EXAMPLES>HUFF1 ARITH.EXE COMP.TST

HUFF1:        Huffman Code Compression Program.
              14-Aug-90  Bill Demas.  Version 1.0

(1) Getting Frequency Counts.
(2) Building Initial Heap.
(3) Building the Code Tree.
(4) Generating the Code Table.
(5) Compressing & Creating the Output File.

Raw characters            :    26870
Header characters         :    772
Compressed characters     :    22632
Total output characters   :    23404

Percentage savings        :    12.90%
```

Figure 4.10 Executing HUFF1 using an executable (binary) file as input

example, the use of the program resulted in a saving of approximately 40%.

DHUFF.C

The program DHUFF.C was written by Bill Demas to decompress a file previously compressed using his HUFF1 program. Both the C language version of his program as well as an executable version compiled using Borland's Turbo C++ are contained on the diskette

```
C:\TC\EXAMPLES>HUFF1 PACKC.C XXX

HUFF1:        Huffman Code Compression Program.
              14-Aug-90  Bill Dermas.  Version 1.0

(1) Getting Frequency Counts.
(2) Building Initial Heap.
(3) Building the Code Tree.
(4) Generating the Code Table.
(5) Compressing & Creating the Output File.

Raw characters               :    7369
Header characters            :    772
Compressed characters        :    3635
Total output characters      :    4407

Percentage savings           :    40.20%
```

Figure 4.11   PACKC.C used as input to the HUFF1 program

accompanying this book. The executable version of the program is stored on the diskette under the filename DHUFF.EXE.

The statements in the program DHUFF.C are listed in Figure 4.12. In Figure 4.12 you will note that the MAIN program performs Huffman decoding through a two-step procedure. First, the program calls the user-defined function BUILD_DECOMP_TREE which, as its name implies, builds the decompression tree. To do so, MAIN first reads the previously stored header which includes the size of the original file and the binary code and code length of each of the possible 256 characters. This is accomplished through the use of three 'fread' statements in MAIN. Once the decompression tree is constructed, the user-defined function DECOMPRESS_IMAGE cycles through the remainder of the compressed file, first retrieving a character and then shifting its bit positions to obtain a match in the decomp-tree array which enables the program to output the correct decompressed character. An example of the execution of DHUFF is illustrated in Figure 4.13. As indicated, DHUFF requires you to enter the compressed file name followed by the name you want to use to store the decompressed file. This is followed by the display of a two-line status or program progress indicator as the program recreates the original file.

4.3 SHANNON–FANO CODING

Similar to Huffman coding, Shannon–Fano coding results in a variable length code that is instantly decodable (Fano, 1949, 1961). Just like the development of a Huffman code, prior to developing the Shannon–Fano code for each character in your character set you must

```
/********************************************************************
 *                                                                  *
 *  DHUFF.C:     Huffman Decompression Program.                     *
 *               14-August-1990    Bill Demas       Version 1.0     *
 *                                                                  *
 *   This program decompresses a file previously compressed with the *
 *   HUFF1 program.                                                 *
 *                                                                  *
 *          USAGE:    DHUFF <input file> <output file>              *
 *                                                                  *
 *   (DISK to DISK:  Input direct from disk, output direct to disk) *
 ********************************************************************/

#include <stdio.h>
#include <stdlib.h>

#define     VERBOSE                      /* If defined, prints verbose
                                            program progress when it's
                                            running...              */

short           decomp_tree[512];
unsigned short  code[256];
unsigned long   file_size;
unsigned char   code_length[256];

FILE            *ifile, *ofile;



/********************************************************************

  MAIN ()

  This is the main program. It performs the Huffman decoding procedure in
  2 separate steps.

  I know that this program can be made more compact & faster, but I was more
  interested in UNDERSTANDABILITY !!!

 ********************************************************************/

void main (argc, argv)
int     argc;
char    *argv[];
{
   void  build_decomp_tree (), decompress_image ();

   if (argc == 3)
   {
      printf ("\nDHUFF:  Huffman Code Decompression Program.");
      printf ("\n          14-Aug-90  Bill Demas.  Version 1.0\n\n");


      if ((ifile = fopen (argv[1], "rb")) != NULL)
      {
         fread (&file_size, sizeof (file_size), 1, ifile);
         fread (code, 2, 256, ifile);
         fread (code_length, 1, 256, ifile);
```

Figure 4.12 Huffman decompression program DHUFF.C

```
         #ifdef VERBOSE
            printf ("(1) Building the tree.\n");
         #endif

         build_decomp_tree ();

         #ifdef VERBOSE
            printf ("(2) Decompressing & Creating the Output File.\n");
         #endif

         if ((ofile = fopen (argv[2], "wb")) != NULL)
         {
            decompress_image();
            fclose (ofile);
         }
         else
            printf ("\nERROR:  Couldn't create output file %s\n", argv[2]);

         fclose (ifile);
      }
      else
         printf ("\nERROR:  %s -- File not found!\n", argv[1]);
   }
   else
      printf ("Usage:  DHUFF <input filename> <output filename>\n\n");
}


/**********************************************************************

 BUILD_DECOMP_TREE ()

 This function builds the decompression tree.
 **********************************************************************/

void  build_decomp_tree ()
{
   register unsigned short  loop1;
   register unsigned short  current_index;

   unsigned short  loop;
   unsigned short  current_node = 1;


   decomp_tree[1] = 1;

   for (loop = 0; loop < 256; loop++)
   {
      if (code_length[loop])
      {
         current_index = 1;
         for (loop1 = code_length[loop] - 1; loop1 > 0; loop1--)
         {
            current_index = (decomp_tree[current_index] << 1) +
                             ((code[loop] >> loop1) & 1);
            if (!(decomp_tree[current_index]))
               decomp_tree[current_index] = ++current_node;
         }
         decomp_tree[(decomp_tree[current_index] << 1) +
```

Figure 4.12 *Continued*

```
                 (code[loop] & 1)] = -loop;
         }
     }
}


/************************************************************************

  DECOMPRESS_IMAGE ()

  This function decompresses the compressed image.
  ***********************************************************************/
void  decompress_image ()
{
    register unsigned short  cindex = 1;
    register char            curchar;
    register short           bitshift;

    unsigned long  charcount = 0L;


    while (charcount < file_size)
    {
        curchar = (char) getc (ifile);

        for (bitshift = 7; bitshift >= 0; --bitshift)
        {
            cindex = (cindex << 1) + ((curchar >> bitshift) & 1);

            if (decomp_tree[cindex] <= 0)
            {
                putc ((int) (-decomp_tree[cindex]), ofile);

                if ((++charcount) == file_size)
                    bitshift = 0;
                else
                    cindex = 1;
            }
            else
                cindex = decomp_tree[cindex];
        }
    }
}
```

Figure 4.12 *Continued*

```
        C:\TC\EXAMPLES>DHUFF COMP.TST ARITH.EXE

        DHUFF:      Huffman Code Decompression Program.
                    14-Aug-90 Bill Demas. Version 1.0
            ,

        (1) Building the tree.
        (2) Decompressing & Creating the Output File.
```

Figure 4.13 Using DHUFF to decompress a previously compressed file.

determine the probability of occurrence of each character. Then, you will arrange your character set in descending order based upon the probability of occurrence of each character.

Once your character set is arranged in descending order of its probability of occurrence, the set must be divided into two equal or almost equal subsets based upon the probability of occurrence of the characters in each subset. The first digit in one subset which represents a group of frequencies of occurrence is assigned a binary 0 value, while a binary 1 is assigned as the first digit in the second subset. Here the second subset represents all remaining frequencies of occurrence. This process of forming subsets is repeated until the character set is completely subdivided. Then, a suffix bit is added to each character in a two-character subset as required to distinguish one character's binary composition from the other character in the subset.

To obtain an understanding of the Shannon–Fano coding procedure, let us assume our character set contains seven characters whose probabilities of occurrence are indicated in Table 4.4.

By arranging the characters in the character set in descending order based upon their probability of occurrence, we can begin to form our subsets. In our subset construction process, we will group the characters into each subset so that the probability of occurrence of the characters in each subset is equal or as nearly equal as possible. Then we will assign binary 1s to one subset and binary 0s to the other subset and continue to repeat the process until all possible subsets are constructed. Figure 4.14 illustrates this process.

Note that after the initial coding process is completed, the subsets represented by the character pairs X_6,X_3 and X_4,X_5 are not unique. Thus, a binary 1 and 0 must be added to the pairs in each subset. Doing so results in the completion of the variable-length coding process in which each character is represented by a unique bit combination that is instantaneously decodable. The completed code for each character in our character set is illustrated in Figure 4.15.

Table 4.4 Character set probability of occurrence

Character	Probability of occurrence
X_1	0.10
X_2	0.05
X_3	0.20
X_4	0.15
X_5	0.15
X_6	0.25
X_7	0.10

Character	Probability	Code		
X_6	0.25	1		
X_3	0.20	1		
X_4	0.15	0	1	
X_5	0.15	0	1	
X_1	0.10	0	0	1
X_7	0.10	0	0	0
X_2	0.05	0	0	0

Figure 4.14   Initial Shannon–Fano coding process

Character	Probability	Code				
X_6	0.25	1	1			
X_3	0.20	1	0			
X_4	0.15	0	1	1		
X_5	0.15	0	1	0		
X_1	0.10	0	0	1		
X_7	0.10	0	0	0	1	
X_2	0.05	0	0	0	0	

Figure 4.15   Completed Shannon–Fano coding process

Code ambiguity

One of the problems associated with the development of a Shannon–Fano code is the fact that the procedure to develop the code can be ambiguous. To illustrate this problem, consider the Shannon–Fano coding process illustrated in Figure 4.16. Note that the probability of occurrence of the characters in the eight-character character set is such that it is not clear whether the first division to create subsets should occur between X_1 and X_2 or between X_2 and X_3.

By making the first subdivision between X_2 and X_3 the resulting Shannon–Fano code has an average bit length of 2.43, computed as follows:

Character	Probability	Code development							Code
X_1	0.35	1	1						11
X_2	0.30	1	0						10
X_3	0.15	0	1						01
X_4	0.08	0	0	1					001
X_5	0.05	0	0	0	1				0001
X_6	0.03	0	0	0	0	1			00001
X_7	0.02	0	0	0	0	0	1		000001
X_8	0.02	0	0	0	0	0	0		000000

Figure 4.16   Ambiguous code development

$$L = 0.35*2 + 0.3*2 + 0.08*3 + 0.05*4$$
$$+ 0.03*5 + 0.02*6 + 0.02*6 = 2.43.$$

In this particular example, making the first subdivision between X_2 and X_3 results in an optimum code whose average bit length is the same as that obtained by the Huffman code development process. If a different subdivision occurs, the Shannon–Fano code's average bit length will exceed that obtained by the Huffman code development process.

Efficiency comparison

To compare the efficiency of the Shannon–Fano coding process to the previously covered Huffman coding technique, let us develop the Huffman code for the character set whose probability of occurrence was previously listed in Table 4.4. Figure 4.17 (top) illustrates the construction of a Huffman code for the seven-character character set listed in Table 4.4. The lower portion of that illustration shows the assignment of binary 1s and 0s to each path member and the resulting Huffman code for each character when the binary digits in

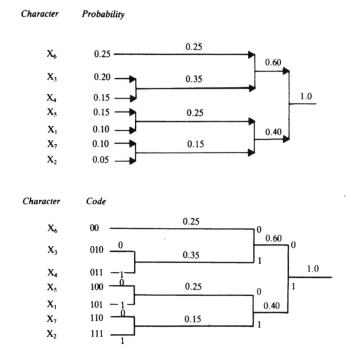

Figure 4.17 Huffman code construction

each path are recorded beginning at the unity or apex point in the coding tree.

Table 4.5 compares the codes generated by the Shannon–Fano coding procedure to the Huffman coding procedure for the seven-character character set used for each coding example. The average code length generated by each coding procedure can be computed by using the formula:

$$L = \sum_{i=1}^{7} L_i P_i$$

For the Shannon–Fano code, the average code length is:

$$L = 2*0.25 + 2*0.20 + 3*0.15 + 3*0.15$$
$$+ 3*0.10 + 4*0.10 + 4*0.05 = 2.7 \text{ bits.}$$

For the Huffman code developed in Figure 4.17, the average code length is:

$$L = 2*0.25 + 3*0.75 = 2.75 \text{ bits.}$$

Although the Shannon–Fano code appears more efficient since its average code length is less than that of the Huffman code, in actuality, it cannot be more efficient. This is because the Huffman code will result in an optimum code when the code construction is efficient, which was purposely not the case in the construction illustrated in Figure 4.17. In that illustration the author purposely did not move groups of frequencies of occurrences as high as possible in an expedient manner. This was done as, upon occasion, it is easy to construct a Huffman code that does not represent an optimum code, and the creation of a Shannon–Fano code can be used as a method for verifying whether or not the Huffman code construction resulted in an

Table 4.5 Coding comparison

Character	Probability	Shannon–Fano code	Huffman code
X_6	0.25	11	00
X_3	0.20	10	010
X_4	0.15	011	011
X_5	0.15	010	100
X_1	0.10	001	101
X_7	0.10	0001	110
X_2	0.05	0000	111

optimum code. This is accomplished by comparing the average bit length of the Shannon–Fano code to the average bit length of the Huffman code. If the average bit length of the Huffman code exceeds the average bit length of the Shannon–Fano code, this fact should be used as an indicator that you should attempt to reconstruct your Huffman code. Since the Shannon–Fano average code length was less than the Huffman average code length, let us apply the insight just gained and reconstruct the Huffman code. This reconstruction is illustrated in Figure 4.18.

Note that in reconstructing the Huffman code, more combining steps occur for groupings of lower frequency of occurrence characters than higher frequency of occurrence characters in comparison to the original code construction illustrated in Figure 4.17. Although this results in an additional code bit being required to represent characters X_7 and X_2, it also results in one less bit required to represent X_3. Since the probability of X_3 exceeds the combined probability of X_7 and X_2, the reconstructed code will have a lower average bit length. To verify this fact, let us compute the average bit length for the Huffman code developed in Figure 4.18. Doing so, we obtain:

$$L = 2*0.25 + 2*0.2 + 3*0.15 + 3*0.15 + 3*0.10 + 4*0.10$$

$$+ 4*0.05 = 2.7 \text{ bits.}$$

Note that the average code length has been reduced by 0.05 bits

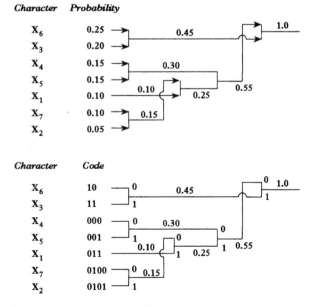

Figure 4.18 Huffman code reconstruction

and is now exactly equal to the average bit length of the Shannon–Fano code.

The previous illustrations were based upon one group of assigned probabilities of occurence to a seven-character character set. To illustrate how efficiencies between the two codes can change, let us assume that the probabilities of occurrence of the characters in the character set are now represented by the data listed in Table 4.6.

The top portion of Figure 4.19 illustrates the Shannon–Fano coding process while the lower portion of that illustration shows the Huffman coding process. Note that based upon the revisions in the probability of occurrence of the characters in the character set, the average code length for each coding technique is the same. That is, the average code length for the Shannon–Fano coding process is:

Table 4.6   Revised character set

Character	Probability of occurrence
X_1	0.0625
X_2	0.0625
X_3	0.1250
X_4	0.1250
X_5	0.0625
X_6	0.5000
X_7	0.0625

A. Shannon–Fano coding

X_6	0.50	1		
X_3	·0.125	0 1 1		
X_4	0.125	0 1 0		
X_5	0.0625	0 0	1 1	
X_1	0.0625	0 0	1 0	
X_7	0.0625	0 0	0 1	
X_2	0.0625	0 0	0 0	

B. Huffman coding

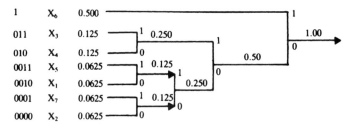

Figure 4.19   Recoding the new character set

$$L = 1*0.5 + 3*0.125 + 3*0.125 + 4*(4*0.0625) = 2.25 \text{ bits}$$

which is exactly the same code length obtained from the Huffman coding process.

Now let us assume that the probability of occurrence of each character in the character set is again altered. Suppose the new probabilities of occurrence are as indicated in Table 4.7.

The top portion of Figure 4.20 illustrates the Shannon–Fano coding process for the revised character set while the lower portion shows the Huffman coding process.

Now let us compute the average code length for each coding process. For the Shannon–Fano code, its average code length is:

Table 4.7 New character set probabilities

Character	Probability of occurrence
X_1	0.10
X_2	0.10
X_3	0.10
X_4	0.10
X_5	0.40
X_6	0.10
X_7	0.10

A. Shannon–Fano coding

X_6	0.40	1	1			
X_1	0.10	1	0			
X_2	0.10	0	1	1		
X_3	0.10	0	1	0		
X_4	0.10	0	0	1		
X_5	0.10	0	0	0	1	
X_7	0.10	0	0	0	0	

B. Huffman coding

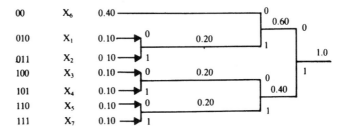

Figure 4.20 Recoding the revised character set

$$L = 2*0.4 + 2*0.1 + 3*0.1 + 3*0.1 + 3*0.1$$

$$+ 4*0.1 + 4*0.1 = 2.7 \text{ bits.}$$

For the Huffman code, its average code length is:

$$L = 2*0.4 + 3*0.6 = 2.6 \text{ bits.}$$

Thus, in this instance the Huffman code results in a more efficient bit representation of the character set than the Shannon–Fano coding method.

In general, as the probabilities of each character in the character set approach probabilities that are negative powers of 2 both codes will have their average code length approach entropy. That is, if all the probabilities of the characters in the character set were negative powers of 2 the average code length would equal entropy and the efficiency of each code would be 100%. If the probabilities of occurrence of the elements in a set have a large variance, the Shannon–Fano code will be more efficient while the Huffman code becomes more efficient as in the variance in probabilities decreases between elements in the set. However, in no event will the Shannon–Fano coding technique be more efficient than Huffman coding when the Huffman code is properly and efficiently constructed.

4.4 COMMA CODES

As previously discussed in this chapter, both Shannon–Fano and Huffman encoding methods have several advantages associated with their use. First, they produce codes whose average number of bits either represents or approaches an optimum code. Secondly, both encoding methods result in the absence of short code words that are prefixes of longer code words. This means that each code is instantaneously decodable. However, what can we say about the processing steps required to decode an instantaneously decodable code and the effect of a lost bit upon a received sequence of bits?

Concerning the decoding process, consider the Huffman code developed in Figure 4.20. The processing steps required to decode a received bit sequence into their appropriate character representations would require a significant series of bit comparisons as illustrated in Figure 4.21. Thus, we can state that Shannon–Fano and Huffman codes are processing-intensive codes.

Concerning a lost bit or sequence of lost bits, this occurrence would result in a loss of synchronization between the bit stream to be decoded and the decoding process. This situation would obviously result in decoding errors. Due to this, statistical encoding should

A. Shannon–Fano coding

X_6	0.40	1	1			
X_1	0.10	1	0			
X_2	0.10	0	1	1		
X_3	0.10	0	1	0		
X_4	0.10	0	0	1		
X_5	0.10	0	0	0	1	
X_7	0.10	0	0	0	0	

B. Huffman coding

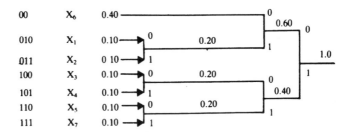

00	X_6	0.40
010	X_1	0.10
011	X_2	0 10
100	X_3	0.10
101	X_4	0.10
110	X_5	0.10
111	X_7	0.10

Figure 4.21 Decoding process for the Huffman code developed in Figure 4.20

X_6	0.4	1
X_1	0.1	01
X_2	0.1	001
X_3	0.1	0001
X_4	0.1	00001
X_5	0.1	000001
X_7	0.1	0000001

Figure 4.22 Comma code for seven-symbol character set

always be encapsulated within an error detection and correction mechanism.

To alleviate a substantial amount of processing associated with statistical encoding as well as the effect of the loss of one or more bits, you can consider the use of a comma code. This code obtains its name from the fact that it contains a terminating symbol or comma at the end of each code word.

Figure 4.22 contains a comma code developed for the seven-symbol character set for which we previously constructed a Huffman code in Figure 4.20. In constructing the comma code we used a '1' bit as the comma or terminating symbol. When that bit is encountered, it informs the receiver that a new character code will begin, providing a method of synchronization within the code.

In examining the comma code listed in Figure 4.22, let us compute its average bit length. Doing so, we obtain:

$$L = 0.4*1 + 0.1*2 + 0.1*3 + 0.1*4 + 0.1*5 + 0.1*6$$

$+ 0.1*7 = 3.1$ bits.

Here the average number of bits per character has increased from 2.6 when the Huffman code was used to 3.1 bits for a comma code. This illustrates the key trade-off between a comma code and Huffman and Shannon–Fano coding— synchronization and processing requirements are tradeoffs against efficiency.

4.5 ARITHMETIC CODING

One of the limitations associated with Huffman and Shannon–Fano coding is the fact that both methods are optimal only when the probabilities associated with each symbol are integral powers of 1/2, which is usually not a good representation of reality. For example, assume that the probability of a symbol is 1/3. Then, $-\log_2(1/3)$, i.e. $\log_2 3$ or approximately 1.59 bits would be required to represent the symbol. When Huffman or Shannon–Fano coding is used two or more bits would be assigned to the symbol. Then, each time the symbol is compressed, the compression technique would result in the addition of approximately 0.4 or more bits over the symbol's entropy. As the probability of symbols in a set increases, the coding variance between the entropy of the symbol and the number of bits required to encode it using the Huffman or Shannon–Fano technique increases. For example, assume a symbol has a 70% probability of occurrence. Then, $-\log_2 0.7$ results in an entropy of approximately 0.51 bits and a variance between the use of 1 bit to encode the symbol and its entropy of 0.49 bits. If the probability of occurrence of the symbol increases to 85%, its entropy becomes $-\log_2 0.85$ or approximately 0.2 bits, while its variance between the use of one bit to encode the symbol and its entropy increases to 0.8 bits. Here the use of a single bit to represent a symbol whose optimal code size is 0.2 bits results in the encoding of the symbol using five times the actual length necessary to represent the symbol.

Since many documents can contain a single symbol with a very high probability of occurrence, such as a typical page of correspondence which may have 70% or more of the page occupied by spaces, this tells us that we must either group sequences of bits together and use Huffman codes to represent each sequence, such as a modified Huffman code used for fax transmission which is discussed in Chapter 5, or consider a different coding technique. One such technique is arithmetic coding which does not require symbol probabilities to be integral powers of 1/2 to obtain an optimum efficiency.

Overview

Arithmetic coding is a technique in which each symbol in a symbol set is assigned an interval between 0 and 1 based upon its probability of occurrence. As a message is formed by the concatenation of separate symbols, those symbols are used to define smaller and smaller intervals between 0 and 1, resulting in a message being represented by a floating-point number. Thus, arithmetic coding results in the replacement of individual character codes used by Huffman and Shannon–Fano coding by the use of a single code for an entire sequence of symbols.

Although the concept of arithmetic compression has been known for a long time, until relatively recently it was considered more as an interesting theoretical technique than as a mechanism for implementation on computers with fixed-size words. Only in the late 1980s did arithmetic coding gain popularity due to its use in an archive program developed in Japan whose compression-performing code was distributed on bulletin boards literally around the world.

Although papers published on information theory in the early 1960s covered references to compression techniques similar to modern arithmetic coding, it wasn't until the late 1970s and early 1980s that a series of papers were published (Langdon, 1981, 1984; Rissanen and Langdon, 1979) that focused attention upon this technique. Since many of those papers were restricted to academic readerships, it required almost another decade until this compression technique was integrated into compression-performing application programs.

Operation

The process associated with arithmetic coding is relatively easy to describe; however, its actual implementation as we will note requires careful planning.

Under arithmetic coding we commence its operation by determining the probability or frequency of each symbol. Next, we list the symbols in a fixed order, usually by ascending or descending probability or by the order of the symbols in the symbol set. We can use any order as long as we use the same order for both encoding and decoding. As you should expect, the sum of the individual probabilities is always unity. The next step is to subdivide unity into ranges, giving each symbol its own range and assigning the length of the range so that it equals the probability of the symbol. Once this is accomplished, we are ready to construct a number which represents a sequence of symbols. To do so we perform the following steps:

1. Select the first symbol in the sequence and locate the ranges associated with the symbol.

2. Select the next symbol. Multiply the range associated with the previous symbol by the low and high ends of the range of the current symbol. Add the results to the low range of the prior symbol to obtain a new low range and new high range for the current symbol.
3. Continue step 2 until all of the symbols to be encoded are processed.
4. Select any number within the range of the last symbol of the sequence.

To illustrate the operation of arithmetic coding, let's first assume our alphabet is restricted to the set of eight symbols [d, e, f, g, n, r, u and space]. To simplify our illustration of the operation of arithmetic coding, our alphabet was also restricted to lowercase symbols. Since arithmetic coding uses the probability of occurrence of each symbol in the alphabet to construct appropriate coding intervals, we will next assign a probability of occurrence to each symbol. This assignment is indicated in Table 4.8.

Once the probability of occurrence of each symbol is known, we then assign a range to each symbol based upon its probability. This range assignment subdivides the interval from zero to one [0, 1] into subintervals that correspond to the probabilities of each symbol.

Table 4.9 illustrates the assignment of subinterval ranges to each symbol in our alphabet. Symbols can be assigned to subintervals in any order, as long as the compressor and expander use the same assignment; however, for most persons subinterval range assignments are normally based upon the sequence of symbols in the alphabet. In examining Table 4.9, note that since the symbol probabilities sum to one, the subinterval ranges fill the interval between zero and one.

In examining the range intervals in Table 4.9, note that each symbol owns the entire range interval up to but not including the high number. Thus, d has the range 0.0 to 0.0999. . . and so on. As symbols

Table 4.8 Probability of occurrence of alphabet symbols

Symbol	Probability
d	0.1
e	0.3
f	0.1
g	0.05
n	0.1
r	0.2
u	0.05
space	0.1

Table 4.9 Assigning subinterval ranges

Symbol	Probability	Range
d	0.1	[0.0, 0.1]
e	0.3	[0.1, 0.4]
f	0.1	[0.4, 0.5]
g	0.05	[0.5, 0.55]
n	0.1	[0.55, 0.65]
r	0.2	[0.65, 0.85]
u	0.05	[0.85, 0.90]
space	0.1	[0.90, 1.0]

that form a message are processed, each new symbol reduces the range of the floating-point number used to represent the message.

To illustrate the operation of the encoder, assume the name 'fred unger', to include the space between the first and last name of this distinguished gentleman, is to be compressed. The first symbol, 'f', results in the encoder selecting the range [0.4, 0.5] to allocate to the symbol. The next symbol, 'r', owns the range [0.65, 0.85], which narrows the new range to 65 to 85% of the prior range. That is, the old range started at 0.4 and was 0.1 units long. Multiplying 0.65 by 0.1 and adding to 0.4 gives a new low range starting point of 0.465. Similarly, multiplying 0.85 by 0.1 and adding the result to 0.4 gives a new high range of 0.485. Thus, the new interval becomes [0.465, 0.485].

The third symbol in the message is 'e', which has the interval [0.1, 0.4]. Since the previously computed range commenced at 0.465 and was 0.02 units in length, we multiply the length of 0.02 by the starting interval of 0.1 and add the result to 0.465 to obtain the new low range starting point of 0.4652. Similarly, we multiply the prior range length of 0.02 units by 0.4 and add it to 0.465 to obtain a new high range terminating point of 0.473. Thus, encoding the third symbol results in the new interval [0.4652, 0.473]. By now the encoding pattern should be observable. That pattern can be summarized as follows:

start with the range [0, 1] where low (0) becomes 0 and high (0) becomes 1

Then for i > 1

low(i) = low(i)+[high(i−1)−low(i−1)]∗low(i)

high(i) = low(i)+[high(i−1)−low(i−1)]∗high(i)

where low(i) represents the low interval of the ith symbol encoded while high(i) represents the high interval of the ith symbol.

Coding problems

Figure 4.23 lists the statements in a small BASIC language program module created to demonstrate arithmetic coding as well as a few of the problems associated with this compression technique. The program uses a READ DATA pair of statements to initialize the eight-symbol alphabet into the array C$ and the high and low interval boundaries to the arrays HIGH and LOW. Note that although the program reads in one character at a time a mechanism is required to denote when the message is completed. This message terminator indicator can be effected in two ways. A prefix to the message consisting of a count of the characters in the message is one common mechanism that can be employed. A second mechanism is to add a symbol to the character set which serves as a message indicator. For simplicity, Figure 4.23 used neither method and simply scanned the input for the forward slash character as a message terminator. For compressing a file using arithmetic compression, you would use one of the two previously mentioned termination indicators. Figure 4.24 illustrates the execution of the BASIC program listed in Figure 4.23.

Although the execution of the arithmetic coding program module indicates that any number in the range [0.467 592 804 43, 0.467 592 805 33] can represent our message, in actuality the precision of the program used by the author was only accurate to nine digits. Thus, the actual range of values that can uniquely represent

```
REM Arithmetic encoding example
DEFDBL H, L, R
DIM c$(8), high(8), low(8)
high = 1! : low = 0!
FOR i = 0 TO 7
READ c$(i)
DATA "d","e","f","g","n","r","u",""
NEXT i
FOR i = 0 TO 7
READ high(i), low(i)
DATA .1,.0,.4,.1,.5,.4,.55,.5,.65,.55,.85,.65,.9,.85,1,.9
NEXT i
START: INPUT A$
IF A$ = "/" THEN END
FOR I = 0 TO 7
IF A$ = c$(i) THEN p = i
NEXT i
range = high - low
high = low + range * high(p)
low = low + range * low(p)
PRINT USING "     & #.########## #.##########"; c$(p); low; high
GOTO START
```

Figure 4.23 Arithmetic encoding program module

```
? f
                    f  0.40000000000  0.50000000000
? r
                    r  0.46500000000  0.48500000000
? e
                    e  0.46700000000  0.47300000000
? d
                    d  0.46700000000  0.46760000000
?
                       0.46754000000  0.46760000000
? u
                    u  0.46759100000  0.46759400000
? n
                    n  0.46759265000  0.46759295000
? g
                    g  0.46759280000  0.46759281500
? e
                    e  0.46759280150  0.46759280600
? r
                    r  0.46759280443  0.46759280533
? /

Press any key to continue
```

Figure 4.24 Execution of arithmetic encoding module listed in Figure 4.23

our message is [0.467 592 804, 0.467 5592 805]. For example, 0.467 592 804 43, 0.467 592 804 5 or even 0.467 592 804 99 would uniquely represent our message. To compute the entropy of the message, we would use logarithms to the base 10, since the previous encoding example was performed using decimal digits. Thus, the entropy of the 10-symbol message 'fred unger' is:

$$- \log 0.1 - \log 0.2 - \log 0.3 - \log 0.1 - \log 0.1 - \log 0.05$$
$$- \log 0.1 - \log 0.05 - \log 0.3 - \log 0.2,$$

which is $-\log 0.000\,000\,000\,9$, whose decimal value is 8.045 75. This explains why nine decimal digits are required to encode our message. For readers who wish to verify the prior computations using a programming language, remember that the LOG function in most languages is the natural log. To obtain the base 10 log you must divide the natural log of the number by the natural logarithm of 10.

In examining the execution of the arithmetic encoding program module, note that as the intervals get narrower more digits are required to define the expanding message. In fact, after n symbols are coded the interval I which is the product of all the symbols' encodes becomes:

$$I = P_1 \times P_2 \times P_3 \times \ldots \times P_n$$

Thus, the precision of the arithmetic required increases with the length of the message. By itself this fact would rule out the use of computers doing the computations required for long messages; however, mathematicians discovered methods whereby the required calculations can be approximated through the use of finite-precision binary integer arithmetic which resulted in the development of programs that can perform arithmetic compression regardless of the size of the message. Prior to examining those techniques, let's focus our attention upon how the decoding of a previously encoded message occurs.

Decoding

The range associated with our previously encoded message was [0.467 592 804, 0.467 592 805]. To decode the message we first locate the symbol that owns the space the encoded message falls in. Since 0.467 592 804 falls in the range [0.4, 0.5] (see Table 4.9), then f must be the first symbol in the message. To find the next symbol in the message we remove the first symbol. To accomplish this task we subtract the low value of f or 0.4, resulting in the remaining portion of the message being 0.067 592 804. Since the range of f was 0.1, we divide the remaining portion of the message by 0.1, resulting in a value of 0.675 928 04. That value falls within the range [0.65, 0.85] and results in the second character being decoded as 'r'. Next we subtract the low range of 0.65 from the message, resulting in a value of 0.25 928 04. Again, we divide by the range, which is now 0.2. Doing so we obtain 0.129 640 22 which falls within the range [0.1, 0.4]. Thus, the third character in the message must be 'e'.

Figure 4.25 lists the statements in a BASIC program module developed to illustrate the arithmetic decoding process. Again, readers should note that the version of BASIC used by the author was only accurate to nine decimal digits as standard precision was used in the coding example. Figure 4.26 shows the execution of the program based upon the entry of the number representing the previously encoded message.

Computer models

As previously discussed, the computational task involved in encoding a long sequence of symbols grows in proportion to the number of symbols to be encoded. From a practical perspective you would not want to attempt to use decimal arithmetic to perform arithmetic codings as you would soon exceed the precision of the computer. Recognizing this problem, binary arithmetic coding was used to implement arithmetic coding. The selection of binary arithmetic permits simple

```
CLS
REM Arithmetic decoding example
DEFDBL A, H, L, R
DIM c$(8), HIGH(8), LOW(8)
FOR i = 0 TO 7
READ c$(i)
DATA "d","e","f","g","n","r","u","  "
NEXT i
FOR i = 0 TO 7
READ HIGH (i), LOW(i)
DATA .1,.0,.4,.1,.5,.4,.55,.5,.65,.55,.85,.65,.9,.85,1,.9
NEXT i
INPUT A
PRINT "Number    Symbol    Low  High  Range"
FOR J = 1 TO 10
FOR i = 0 TO 7
IF (A >= LOW(i) AND A < HIGH(i)) THEN P = i
NEXT i
RANGE = HIGH(P) - LOW(P)
PRINT USING "#.#########  &  #.## #.## #.##"; A; c$(P); LOW(P); HIGH(P);
RANGE
A = (A - LOW(P)) / RANGE
NEXT J
```

Figure 4.25 Arithmetic decoding program module

```
? .46759180443
Number           Symbol   Low    High   Range
0.4675928044     f        0.40   0.50   0.10
0.6759280443     r        0.65   0.85   0.20
0.1296402215     e        0.10   0.40   0.30
0.0988007383     d        0.00   0.10   0.10
0.9880073833              0.90   1.00   0.10
0.8800738333     u        0.85   0.90   0.05
0.6014766667     n        0.55   0.65   0.10
0.5147666667     g        0.50   0.55   0.05
0.2953333330     e        0.10   0.40   0.30
0.6511111101     r        0.65   0.85   0.20
```

Figure 4.26 Execution of the arithmetic decoding module

approximations to be used in the interval scaling process which eliminates the necessity for performing time-consuming multiplication operations.

Implementations of arithmetic coding change the initial high interval of 1 to 0.999. . . or 0.111. . . in binary while the initial low interval continues to be set to 0. Those numbers are stored in integer registers so that an implied decimal point is on the left side of the word. For example, when using 16-bit registers you would set the initial high value to hex FFFF and the initial low value to hex 0. The high value would continue with hex Fs and the low value would continue with hex 0s forever. Thus, you can shift those bits into the appropriate registers as required. You would then compute new high and new low values using the previously denoted formulas; however, in doing so computations would occur using integer arithmetic.

The implementation of arithmetic coding and decoding using integer arithmetic dates to 1987 (Witten, Neal and Cleary, 1987) and is well documented in the book co-authored by two of the originators of the 1987 paper (Bell, Cleary and Witten, 1990). In both works, a special Code Value is used to indicate the maximum value of 16- or 32-bit integers based upon the word length of a computer used to implement arithmetic coding using binary integer operations. The halfway point of the Code Value range is then used as a decision criterion when performing integer arithmetic and the quantities low and high in each range are represented by integers in binary notation. A program can be developed to ensure that as the range converges as symbols are processed, 0's are shifted into the low order bits of low and 1's are shifted into high. As the range narrows, the top bits of low and high become equivalent. Any bits that are the same can be immediately output as they cannot be affected by future range narrowing as additional symbols are processed. Since low ≤ high, the encoding module required for binary integer operations would be similar to that listed in Figure 4.27. The code module listed in Figure 4.27 results in 1's shifted in at the bottom when high is scaled, while 0's are shifted into low. This technique ensures that Low < Half ≤ High.

Since arithmetic coding works by subdividing the cumulative probabilities into intervals [low, high] for each character, what happens when low and high become so close together that scaling results in the mapping of two or more symbols to the same integer in a [low, high] interval? As might be expected, this would be disastrous. Thus, the encoder must adjust the interval [low, high] so that the interval is always large enough to prevent the mapping of two or more symbols to the same interval.

One method that can be used to prevent the previously described

```
WHILE High < Half OR Low ≥ Half DO
        IF High < Half THEN
                OutputBit(0);
                Low = 2 * Low;
                High = 2 * High + 1;
        IF Low ≥ Half THEN
                OutputBit(1);
                Low = 2 * (Low – Half);
                High = 2 * (High – Half) + 1;
```

Figure 4.27 Arithmetic coding using binary integer arithmetic

problem is to ensure that the interval required to encode a symbol is at least as large as the maximum allowed cumulative frequency count. A related problem can occur due to integer multiplications that can result in an overflow of the resulting value beyond the appropriate interval. This situation can be prevented by ensuring that the product of the interval range and the maximum allowed cumulative frequency count fits within the integer word length used. Thus, for a correct binary integer operation to perform arithmetic coding, it is necessary to ensure that:

1. Cumulative Frequency [i–1] $\geq$ Cumulative Frequency [i].
2. There are no symbols i for which Cumulative Frequency [i–1] $\equiv$ Cumulative Frequency [i].
3. Total frequency count $\leq$ maximum allowed Cumulative Frequency Count.

In step 3 the total frequency count represents the number of symbols processed at any point in time.

To illustrate the coding required to implement arithmetic compression, Figure 4.28 lists the C language code from the public domain program LZARI.C written by Haruhiko Okumura. The entire program which performs Lempel–Ziv string compression and arithmetic compression is contained on the diskette accompanying this book on the file LZARI.C and is listed in its entirety in Chapter 6.

4.6 ADAPTIVE COMPRESSION

The examples of compression techniques previously covered in Chapter 3 and this chapter were based upon the assumption of prior knowledge of the data to be compressed. Using this prior knowledge permits us to predefine compression indicating characters and the character sequences which can then be substituted for strings of data containing predefined redundancy. In addition, we can construct a fixed compression table that will enable the statistical encoding of data to occur based upon the expected frequency of occurrence of the data. Run-length and diatomic encoding are examples of character sequence and character substitution where some prior knowledge or expectation of the composition of the data resulted in the definition of a single character or short sequence of characters to replace longer sequences of characters. Huffman and arithmetic encoding are examples of data-compression techniques that would employ a fixed compression table whose construction is based upon prior knowledge or assumed knowledge of the data.

The fixed compression table

Figure 4.29 illustrates the general format of a fixed compression table. In actuality, this table can be two separate tables, with a relationship established between the elements in each table, or the

```
/********** Arithmetic Compression **********/

/*  If you are not familiar with arithmetic compression, you should read
            I. E. Witten, R. M. Neal, and J. G. Cleary,
                    Communications of the ACM, Vol. 30, pp. 520-540 (1987),
        from which much have been borrowed.  */

#define M    15

/*      Q1 (= 2 to the M) must be sufficiently large, but not so
        large as the unsigned long 4 * Q1 * (Q1 - 1) overflows.  */

#define Q1   (1UL << M)
#define Q2   (2 * Q1)
#define Q3   (3 * Q1)
#define Q4   (4 * Q1)
#define MAX_CUM (Q1 - 1)

#define N_CHAR  (256 - THRESHOLD + F)
        /* character code = 0, 1, ..., N_CHAR - 1 */

unsigned long int  low = 0, high = Q4, value = 0;
int  shifts = 0;  /* counts for magnifying low and high around Q2 */
int  char_to_sym[N_CHAR], sym_to_char[N_CHAR + 1];
unsigned int
        sym_freq[N_CHAR + 1],   /* frequency for symbols */
        sym_cum[N_CHAR + 1],    /* cumulative freq for symbols */
        position_cum[N + 1];    /* cumulative freq for positions */

void StartModel(void)  /* Initialize model */

        int ch, sym, i;

        sym_cum[N_CHAR] = 0;
        for (sym = N_CHAR; sym >= 1; sym--) {
                ch = sym - 1;
                char_to_sym[ch] = sym;  sym_to_char[sym] = ch;
                sym_freq[sym] = 1;
                sym_cum[sym - 1] = sym_cum[sym] + sym_freq[sym];
        }
        sym_freq[0] = 0;  /* sentinel (!= sym_freq[1]) */
        position_cum[N] = 0;
        for (i = N; i >= 1; i--)
                position_cum[i - 1] = position_cum[i] + 10000 / (i + 200);
                        /* empirical distribution function (quite tentative) */
                        /* Please devise a better mechanism! */
}

void UpdateModel(int sym)

        int i, c, ch_i, ch_sym;

        if (sym_cum[0] >= MAX_CUM) {
                c = 0;
```

Figure 4.28 Arithmetic compression module from the program LZARI.C

```
                        for (i = N_CHAR; i > 0; i--) {
                                sym_cum[i] = c;
                                c += (sym_freq[i] = (sym_freq[i] + 1) >> 1);
                        }
                        sym_cum[0] = c;
                }
                for (i = sym; sym_freq[i] == sym_freq[i - 1]; i--) ;
                if (i < sym) {
                        ch_i = sym_to_char[i];    ch_sym = sym_to_char[sym];
                        sym_to_char[i] = ch_sym;  sym_to_char[sym] = ch_i;
                        char_to_sym[ch_i] = sym;  char_to_sym[ch_sym] = i;
                }
                sym_freq[i]++;
                while (--i >= 0) sym_cum[i]++;
}

static void Output(int bit)   /* Output 1 bit, followed by its complements */
{
        PutBit(bit);
        for ( ; shifts > 0; shifts--) PutBit(! bit);
}

void EncodeChar(int ch)
{
        int  sym;
        unsigned long int  range;

        sym = char_to_sym[ch];
        range = high - low;
        high = low + (range * sym_cum[sym - 1]) / sym_cum[0];
        low +=       (range * sym_cum[sym    ]) / sym_cum[0];
        for ( ; ; ) {
                if (high <= Q2) Output(0);
                else if (low >= Q2) {
                        Output(1);   low -= Q2;  high -= Q2;
                } else if (low >= Q1 && high <= Q3) {
                        shifts++;   low -= Q1;  high -= Q1;
                } else break;
                low += low;  high += high;
        }
        UpdateModel(sym);
}

void EncodePosition(int position)
{
        unsigned long int  range;

        range = high - low;
        high = low + (range * position_cum[position    ]) / position_cum[0];
        low +=       (range * position_cum[position + 1]) / position_cum[0];
        for ( ; ; ) {
                if (high <= Q2) Output(0);
                else if (low >= Q2) {
                        Output(1);   low -= Q2;  high -= Q2;
                } else if (low >= Q1 && high <= Q3) {
                        shifts++;   low -= Q1;  high -= Q1;
                } else break;
                low += low;  high += high;
        }
}
```

Figure 4.28 *Continued*

```
void EncodeEnd(void)
{
        shifts++;
        if (low < Q1) Output(0);  else Output(1);
        FlushBitBuffer();  /* flush bits remaining in buffer */
}

int BinarySearchSym(unsigned int x)
        /* 1       if x >= sym_cum[1],
           N_CHAR if sym_cum[N_CHAR] > x,
           i such that sym_cum[i - 1] > x >= sym_cum[i] otherwise */
{
        int i, j, k;

        i = 1;  j = N_CHAR;
        while (i < j) {
                k = (i + j) / 2;
                if (sym_cum[k] > x) i = k + 1;  else j = k;
        }
        return i;
}

int BinarySearchPos(unsigned int x)
        /* 0 if x >= position_cum[1],
           N - 1 if position_cum[N] > x,
           i such that position_cum[i] > x >= position_cum[i + 1] otherwise */
{
        int i, j, k;

        i = 1;  j = N;
        while (i < j) {
                k = (i + j) / 2;
                if (position_cum[k] > x) i = k + 1;  else j = k;
        }
        return i - 1;
}

void StartDecode(void)
{
        int i;

        for (i = 0; i < M + 2; i++)
                value = 2 * value + GetBit();
}

int DecodeChar(void)
{
        int        sym, ch;
        unsigned long int  range;

        range = high - low;
        sym = BinarySearchSym((unsigned int)
                (((value - low + 1) * sym_cum[0] - 1) / range));
        high = low + (range * sym_cum[sym - 1]) / sym_cum[0];
        low +=        (range * sym_cum[sym    ]) / sym_cum[0];
        for ( ; ; ) {
                if (low >= Q2) {
                        value -= Q2;  low -= Q2;  high -= Q2;
                } else if (low >= Q1 && high <= Q3) {
```

Figure 4.28 *Continued*

```
                        value -= Q1;   low -= Q1;   high -= Q1;
                } else if (high > Q2) break;
                low += low;  high += high;
                value = 2 * value + GetBit();
        }
        ch = sym_to_char[sym];
        UpdateModel(sym);
        return ch;
}

int DecodePosition(void)
{
        int position;
        unsigned long int  range;

        range = high - low;
        position = BinarySearchPos((unsigned int)
                (((value - low + 1) * position_cum[0] - 1) / range));
        high = low + (range * position_cum[position     ]) / position_cum[0];
        low +=        (range * position_cum[position + 1]) / position_cum[0];
        for ( ; ; ) {
                if (low >= Q2) {
                        value -= Q2;   low -= Q2;  high -= Q2;
                } else if (low >= Q1 && high <= Q3) {
                        value -= Q1;   low -= Q1;  high -= Q1;
                } else if (high > Q2) break;
                low += low;  high += high;
                value = 2 * value + GetBit();
        }
        return position;
}
```

Figure 4.28 *Continued*

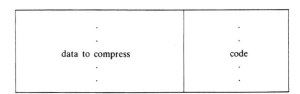

Figure 4.29 Fixed compression table format

table can consist of paired entries. Each character in the original data stream is compared to the entries in the 'data to compress' part of the compression table. When the character to be encoded matches an entry in the 'data to compress' portion of the table, the code that represents the character is extracted from the compression table. Thus, the process required to replace each character with its statistical code is reduced to a table look-up operation.

To illustrate the utilization of a fixed compression table, let us assume that as a result of an analysis of a four-character character set (X_1, X_2, X_3 and X_4) we have determined that the probability of occurrence of each character is 0.5625, 0.1875, 0.1875 and 0.0625 respectively. The Huffman code previously developed in Figure 4.3 for this character set results in the assignment of 0, 10, 110 and 111 to characters X_1 to X_4. Thus, based upon prior knowledge of the data we can develop the Huffman code for the character set which then enables us to construct the fixed compression table for this character set. This table is illustrated in Figure 4.30.

The probability of occurrence of the characters in the character set must be determined prior to constructing a fixed compression table.

The use of a fixed compression table requires each character in the original data string to be compared to the 'data to compress' entries in the table. When a match occurs, the coded entry then replaces the character in the original data string. Thus, the sequence of characters

$$X_2 X_4 X_1 X_2 X_2$$

would be replaced by the Huffman code for each character, which would result in the binary sequence:

1011101010.

Data to compress	Resulting code
X_1	0
X_2	10
X_3	110
X_4	111

Figure 4.30 Resulting fixed compression table

Efficiency

What happens to the efficiency of the predefined Huffman code when the probability of occurrence of the characters in the character set differs from the prior or expected knowledge of their frequency of occurrence? Since short codes are employed to represent frequently occurring characters while longer codes represent characters that occur less frequently, the predefined Huffman code's variance from entropy increases as the data varies from its prior or expected frequency of occurrence. One technique that can be used to maintain the efficiency of the resulting code obtained by compressing data statistically is the use of an adaptive or dynamic compression scheme, which is the main topic of this section.

Adaptive compression

When adaptive compression is performed, the data to be compressed is analysed in order to generate appropriate changes into a variable compression table.

Similar to the use of a fixed compression table, each character in the original data stream is first compared to the entries in the 'data to compress' portion of the compression table. When a match occurs, the corresponding entry in the 'resulting code' portion of the table is extracted and represents the statistically encoded character.

Where adaptive compression differs from fixed compression is in the employment of a count field in the compression table. This field is continuously updated and serves as a mechanism for the resequencing of the entries in the table. The updating of the field occurs after a character in the original data stream is matched with an entry in the 'data to compress' portion of the compression table and the 'resulting code' is extracted from the table. Then, a comparison of the entries in the count field occurs. Based upon the results of the comparison, the character and its count value may be repositioned in the compression table. This technique ensures that whenever the composition of the data changes, the compression table changes in tandem, resulting in a variable compression table that provides the most efficient statistical compression possible. Figure 4.31 illustrates how a variable compression table is resequenced based upon the composition of the data being transmitted.

Figure 4.31, part A, illustrates the initial composition of the variable compression table. Although this table was initially established based upon the frequency of occurrence of the characters in the character set, since the table is self-adjusting, we do not have to concern ourselves with the size of the sample used to initialize the entries into the table.

In Figure 4.31, part B, we assumed that the character X_2 was encountered. Since the binary code 10 is assigned to X_2 (Figure 4.31, part A), that bit string is transmitted, the count for X_2 is incremented by one and the variable compression table is resequenced. Similarly, at the receiver the bit sequence 10 is received, which is decompressed into the character X_2. The receiver then increments the count for X_2 in its compression table and its table is also resequenced.

In Figure 4.31, part C, we have assumed that the character X_4 is the next character encountered in the data to be compressed. Based upon the table then in use (Figure 4.31, part B), this character is encoded as the binary string 111. Next, the count of the frequency of occurrence of X_4 is incremented and the variable compression table is resequenced.

Figure 4.31, part D, assumes that the next character encountered in the original data string is X_2. Since the table illustrated in Figure 4.31, part C, was then in use, X_2 is encoded as the single bit 0. Then the count for X_2 is incremented by one; however, since X_2 was at the top of the compression table, the table is not resequenced.

As illustrated in Figure 4.31, adaptive compression dynamically changes the order of the entries in the compression table in tandem with the changes in the frequency of occurrence of the characters

A. Initial table   Data transmitted

Data to compress	Count	Resulting Code
X_1	0	0
X_2	0	10
X_3	0	110
X_4	0	111

B. X_2 encountered   10

Data to compress	Count	Resulting code
X_2	1	0
X_1	0	10
X_3	0	110
X_4	0	111

C. X_4 encountered   111

Data to compress	Count	Resulting code
X_2	1	0
X_4	1	10
X_1	0	110
X_3	0	111

D. X_2 encountered   0

Data to compress	Count	Resulting code
X_2	2	0
X_4	1	10
X_1	0	110
X_3	0	111

Figure 4.31 The variable compression table

in the character set. Thus, this method of implementing a statistical compression technique should always be more efficient than the utilization of fixed compression.

Coded example

Figure 4.32 contains the ADAPTC.BAS program listing. This BASIC language program was developed to illustrate many of the programming concepts involved in adaptive compression. For simplicity of illustration only four characters—E,T,I and O—are considered to be in the character set suitable for adaptive compression. All other characters encountered in the data strings the program will operate upon will be passed 'as-is' to the output buffer.

In line 115, the program branches to the subroutine commencing at line 400 which initializes the character table P$(I) to the characters E, T, I and O. Similar to the other BASIC coding examples presented in this chapter, line 130 obtains a line of up to 132 characters from a data file while line 140 obtains the length of the line.

The subroutine commencing at line 180 processes the records read from the data file. To illustrate the operation of adaptive compression, when the characters E, T, I and O are encountered they will be replaced by the characters, #, $, % and &. For simplicity, the resulting Huffman table will be displayed on a line-by-line basis instead of on an individual character basis while the code changes in the adaptive compression table will similarly occur on a line-by-line basis.

In line 230, the subroutine commencing at line 2120 is invoked. This subroutine prints the current values of the compression table. Next, lines 240 to 280 examine the extracted record from the data file on a line-by-line basis, comparing each character in the record to any of the characters in our compressible four-character character set (E, T, I, O). If a match occurs, the subroutine commencing at line 350 is invoked. Otherwise, the program simply places the character extracted from the input record into the output buffer.

The subroutine commencing at line 350 sets the character match flag to one and then adds 34 to the value of K in line 365. This action sets the ASCII value of V to either the #, $, % or & character which is used in this example to illustrate the substitution of a Huffman code for an appropriate character in the four-character character set we are using. Next, line 370 inserts the substituted character into the output buffer and the count is then incremented in line 380.

When the match flag is set, line 290 causes a branch to line 310, where the input string index is incremented by one, after which the match flag is reset to zero in line 320. If the match flag was not set, line 300 simply extracts one character from its appropriate position in the input record and places it into the output buffer.

Each time prior to a line of input being processed in this program, the subroutine call contained in line 230 will be invoked. This subroutine simply prints out the current status of the adaptive 'Huffman' compression table to include the character order and the frequency of occurrence of each character. Although this program was constructed to facilitate the visual observation of the changes in an adaptive compression table on a line-by-line basis, in developing an actual adaptive compression routine the tables would be subject to change on an individual character basis.

The actual resequencing of the adaptive compression table occurs

```
10 REM ADAPTC.BAS PROGRAM
20 DIM O$(132)
30 WIDTH 80:CLS
40 '**********MAIN ROUTINE***********************
50 '* THIS ROUTINE READS RECORDS FROM AN ASCII *
60 '* FILE INTO A STRING CALLED X$ WHICH IS     *
70 '* THEN PASSED TO SUBROUTINES FOR COMPRESSION
80 '*********************************************
90 PRINT "ENTER ASCII FILENAME. EG, ADAPT.DAT"
100 INPUT F$: OPEN F$ FOR INPUT AS #2
105 OPEN "ADAPTC.DAT" FOR OUTPUT AS #3
110 PRINT "PATIENCE - INPUT PROCESSING"
112 PRINT "SUBSTITUTION BASED ON ENTRY IN TABLE: 1=# 2=$ 3=% 4=&"
115 GOSUB 400                        PAUSE TO SET UP TABLE
120 IF EOF(2) THEN GOTO 9000
130 LINE INPUT #2, X$
140 N= LEN(X$)
150 GOSUB 180
160 GOSUB 900
170 GOTO 120
180 '*****ADAPTIVE COMPRESSION SUBROUTINE*******
190 '* THIS ROUTINE PROCESSES RECORDS FROM X$  *
200 '* AND COMPRESSES WITH HUFFMAN CODES        *
210 '* USING O$ AS THE OUTPUT BUFFER.           *
220 '*********************************************
230 GOSUB 2120                       'PRINT HUFFMAN TABLE USED
235 I=1                              'RESET INDICES
240 FOR J= 1 TO N                    'STEP THRU RECORD
250 A$= MID$(X$,J,1)                 'EXTRACT A CHARACTER
260 FOR K = 1 TO 4                   'SETUP HUFFMAN LOOP
270 IF A$=P$(K) THEN GOSUB 350       'IS INPUT CHAR IN TABLE?
280 NEXT K                           'NO - TRY NEXT
290 IF M = 1 THEN 310                'IS MATCH FLAG SET?
300 O$(I) = MID$(A$,1,1)             'NO-STUFF CHAR IN BUFFER
310 I=I+1                            'BUMP INPUT STRING INDEX
320 M=0                              'RESET MATCH FLAG
330 NEXT J                           'GO BACK FOR MORE
340 RETURN                           'DONE
350 M=1                              'SET CHAR MATCH FLAG
355 '****************************************************************
360 'INSERT COMPRESSION NOTATION IN OUTPUT BUFFER
365 V = K + 34                       'INDEX OUT TO SUBSTITUTE CHAR
370 O$(I)=CHR$(V)                    'INSERT SUBSTITUTION
380 P(K)= P(K) + 1                   'BUMP COUNT OF OCCURANCE
390 K = 4                            'FORCE END OF SEARCH
395 RETURN                           'GO BACK FOR MORE
400 DIM P$(4)                        'COMMON HUFFMAN CANDIDATES
410 DATA E,T,I,O
420 FOR I = 1 TO 4                   'SETUP CHARACTER TABLE
430 READ Z$                          'GET CHARACTER
440 P$(I) = Z$: NEXT I               'AND STUFF INTO TABLE
450 RETURN                           'DONE - TABLE COMPLETE
```

Figure 4.32 ADAPTC.BAS program listing

```
900 '*****TALLY THE COMPRESSION COUNT & WRITE BUFFER******
910 '* DISPLAY BEFORE & AFTER RESULTS OF COMPRESSION      *
920 '* AND SHOW THE NET RESULTS OBTAINED BY EACH METHOD   *
930 '********************************************************
931 N1=N1+N                         'TALLY INPUT CHAR COUNT
932 T=N-I+1                         'NET DIFFERENCE IN BUFFERS
936 T1=T1+T                         'SAVE COUNT FOR SUMMARY
940 FOR I=1 TO J-1
950 PRINT #3, O$(I);
960 NEXT I
965 PRINT #3, ""
966 GOSUB 2000                      'RESEQUENCE HUFFMAN TABLE
970 RETURN
2000 '*****RESEQUENCE & PRINT TABLE FOR ADAPTIVE COMPRESSION*****
2010 FOR J=1 TO 3                   'SETUP 1ST LOOP
2020 FOR K=J+1 TO 4                 'SETUP 2ND LOOP
2030 IF P(J) >= P(K) THEN 2100      'IS CURRENT ENTRY GREATER?
2040 TEMP= P(J)                     'NO-SAVE IN TEMP
2050 TEMP$= P$(J)                   'AND SAVE CHAR
2060 P(J)= P(K)                     'PICKUP GREATER COUNT
2070 P$(J)= P$(K)                   'AND ASSOC CHAR
2080 P(K)= TEMP                     'SWAP LESSER COUNT
2090 P$(K)= TEMP$                   'AND ASSOC CHAR
2100 NEXT K                         'FINISH 2ND LOOP
2110 NEXT J                         'FINISH 1ST LOOP
2115 RETURN                         'DONE-TABLE RESEQUENCED
2120 L= L + 1                       'REMEMBER LINE NO.
2130 PRINT "HUFFMAN TABLE USED FOR LINE";L;": ";
2140 FOR I=1 TO 4                   'SETUP PRINT TABLE LOOP
2150 PRINT P$(I);:PRINT P(I);       'PRINT CHAR AND COUNT
2160 NEXT I
2175 PRINT
2180 RETURN                         'DONE-TABLE PRINTED
9000 CLOSE: OPEN F$ FOR INPUT AS #2
9010 PRINT "FILE ";F$;" BEFORE SUBSTITUTION:"
9020 LINE INPUT #2,X$
9030 IF EOF(2) THEN 9060
9040 PRINT X$
9050 GOTO 9020
9060 PRINT X$:OPEN "ADAPTC.DAT" FOR INPUT AS #3
9070 PRINT "FILE ";F$;" AFTER SUBSTITUTION:"
9080 LINE INPUT #3,O$
9090 IF EOF(3) THEN 9998
9100 PRINT O$
9110 GOTO 9080
9998 PRINT O$
9999 CLOSE:END
```

Figure 4.32 *Continued*

in lines 2000 to 2115 of the program. This subroutine module sorts the characters in the adaptive compression table based upon their frequency of occurrence.

Figure 4.33 illustrates the sample execution of the ADAPTC.BAS program, with the status of the compression table displayed for each line of data in the file to be processed. In addition, the program displays the contents of the file prior to and after the substitution of characters from the previously defined four-character character set.

```
ENTER ASCII FILENAME. EG. ADAPT.DAT
? ADAPT.DAT
PATIENCE - INPUT PROCESSING
SUBSTITUTION BASED ON ENTRY IN TABLE: 1=# 2=$ 3=% 4=&
HUFFMAN TABLE USED FOR LINE 1 : E 0 T 0 I 0 0 0
HUFFMAN TABLE USED FOR LINE 2 : E 1 I 1 T 0 0 0
HUFFMAN TABLE USED FOR LINE 3 : I 5 0 5 T 3 E 2
HUFFMAN TABLE USED FOR LINE 4 : 0 9 E 7 T 5 I 5
HUFFMAN TABLE USED FOR LINE 5 : 0 19 I 13 T 11 E 11
FILE ADAPT.DAT BEFORE SUBSTITUTION:
1 BEGIN*********************************
2 OVATION OVATION FOR THE MUSICIAN
3 ENCORE ENCORE FOR THE ACTOR
4 OOOOOOOOOO IIIIIIII TTTTT EEEE
5 ****************************************END
FILE ADAPT.DAT AFTER SUBSTITUTION:
1 B#G%N*********************************
2 &VA%$&N &VA%$&N F&R %H# MUS$C$AN
3 &NC$R& &NC$R& F$R %H& AC%$R
4 ########## &&&&&&&& %%%%% $$$$
5 ********************************%&ND
Ok
```

Figure 4.33 Sample execution of ADAPTC.BAS program

As an example of the operation of the program, note that prior to line 1 being processed all entries in the compression table have a count of zero and the order of the entries is E, T, I and O.

The first line in the data file contains the string BEGIN, followed by many asterisks. Since the characters E and I will be replaced by the 'Huffman' codes # and %, after line 1 is processed the count for E and I should be one, while the adaptive compression table should be resequenced to account for the new frequency of occurrence.

Examining the Huffman table used for line 2 in Figure 4.33, note that the count of E and I are set to 1, while the order of the characters in the table has been rearranged to take into consideration their new frequency of occurrence. Examining line 2 in the ADAPT.DAT data file, note that OVATION contains four characters that can be substituted by the adaptive 'Huffman' code. Since the character O did not change its place in the compression table, the 'Huffman' code of & is substituted for that character. Next, the T in OVATION, which would have initially been replaced by the 'Huffman' code of $, is replaced by the 'Huffman' code of % since the adaptive table entries changed, which caused the 'Huffman' code substitutions to change. Table 4.10 summarizes the changes in the adaptive compression table prior to and after the first line of data in the input file is processed. As an exercise, you may wish to follow the code substitutions for the four-character character set for the remaining lines in the ADAPT.DAT file that are processed by the ADAPTC.BAS program.

Table 4.10 Adaptive compression table change

Initial table				
Character sequence	E	T	I	O
Code substitution	#	$	%	&
After line 1 processed				
Character sequence	E	I	T	O
Code substitution	#	$	%	&

Count field considerations

Until now, we have assumed that the count field of a variable compression table is infinite. Obviously, the count field has a finite value based upon the size of the field which must be considered when developing an adaptive compression technique.

There are two easy-to-implement methods you can consider to eliminate the possibility of a count-field overflow occurring. Each of these methods is based upon the use of integer division to reduce the values previously assigned to the count field while maintaining the relative order of the frequency of occurrence of characters transmitted.

One method to eliminate the occurrence of a count-field overflow is to examine the values of the top field entry after a field update. Then, if the bits in that field are all set, you would implement an algorithm that would perform an integer division of 2 on each field entry. Implementing this technique would result in any remainder after division being discarded. As an example, 3/2 would equal 1, with the remainder being discarded.

A second method that can be used to prevent a count-field overflow would be to simply count transmitted characters and implement integer division by 2 each time the count equalled the maximum field value. Although this method obviously results in more frequent integer division operations, it eliminates the necessity of comparing the value of the highest position count field to an all set bit condition each time a character is received.

Sixpack program

To illustrate the C language coding required to implement adaptive compression this author is again indebted to Philip Gage, who is also the developer of the BPE diatomic encoding method previously described in Chapter 3. Mr Gage developed the Sixpack program for the data compression contest announced in the February 1991 issue of *Dr Dobb's Journal.* The whimsical program name was selected to reference the fact that it combines six algorithms into a single data packing method. The source code of the C language program is listed in Figure 4.34. The algorithms used by Mr Gage illustrate such data structures as a binary tree, a hash table, doubly linked lists and a circular array. The C language source program, SIXPACK.C, which is in the public domain as well as the file NOTES.DOS, which are the original author's notes concerning the program, are included on the diskette accompanying this book. To generate an executable file from the source file, compile and link SIXPACK.C using a small memory model such as the optimize switches –G –O –Z for Turbo C++, but these are not essential.

> Turbo C++:   tcc –G –O –Z sixpack.c
>
> Power C: pc/e sixpack.c

To compress a file, enter:

> SIXPACK inputfilename outputfilename

To decompress a file, enter:

> SIXPACK inputfilename outputfilename d

(the 'd' means 'decompress', but any argument will do). In the remainder of this section information from Mr Gage's NOTES.DOC file is repeated to provide readers with a brief description of each of the algorithms included in his program.

Finite window compression

The finite window is a buffer which stores the most recent few thousand characters of input. The algorithm searches this buffer for the longest string matching the current text. If such a string is located and it meets or exceeds a minimum length, then compression can be achieved by encoding the matching section of the current text as the number of characters to copy and the distance from which to copy them. If no matching string of the minimum length or longer is found, the current character is output as a literal without compression and the algorithm proceeds to the next input character.

This finite window scheme generates two types of codes, single literal character and string copies consisting of length and distance values. To avoid useless copy length/distance pairs, the distance is measured from the last character of the string to be copied instead of the first character. Several distance formats with a different number of bits are used to minimize the distance code size. Another enhancement is not to issue a copy if a better copy exists at the next character. A final improvement is to check for an alphabetized dictionary word file and restrict copies to roughly a one-word distance on such files for greater compression.

This algorithm is more similar to the original Lempel–Ziv approach than to the later LZW implementation, and resembles methods described in 'Data compression with finite windows', *Communications of the ACM*, April 1989. The original Lempel–Ziv idea combines each copy with a literal character, while the ACM article uses blocks of literal characters. The well-known LHARC/ICE program uses a similar method to achieve impressive results.

Circular buffer array

The first problem is how to restore the buffer of recent text. To maintain a queue using a linked list would complicate searching. Shifting the entire contents of an array to add a new character would be too slow.

The buffering technique used in Sixpack is to store the text in a circular array which wraps around on itself. When the end of the array is reached, the position is reset to the beginning of the array and old text is overwritten. No additional data structures are needed and the array occupies minimum space.

Since character positions are fixed during their lifetime in the array, the linked lists described later can be allocated in parallel with the buffer array, using the character positions as the corresponding linked list node numbers. The disadvantage of this method is that all operations involving text strings in the buffer must account for the wrap at the end of the buffer.

Hash table

The fundamental problem is finding the longest string match in a large block of text. A brute force search gives very slow performance. Several search algorithms were tested, including a binary search tree, a direct lookup table and fast text search techniques. For this application, the best method seemed to be a hash table where the key is derived from the first few characters at each location in the buffer.

Each entry in the hash table is a doubly linked list containing the indices of all buffer positions with matching text. Each list requires both a head and a tail pointer. Since several string prefixes may generate the same hash key, some collisions may occur and the entire string must be checked during a search.

Doubly linked lists

Linked lists are efficient for storing string prefixes in the hash table, since the prefixes are continually being deleted when they reach the

```
/*******************************************/
/*  SIXPACK.C -- Data compression program  */
/*  Written by Philip G. Gage, April 1991   */
/*******************************************/

#include <stdio.h>
#include <alloc.h>          /* Use <malloc.c> for Power C */

#define TEXTSEARCH 1000    /* Max strings to search in text file */
#define BINSEARCH   200    /* Max strings to search in binary file */
#define TEXTNEXT     50    /* Max search at next character in text file */
#define BINNEXT      20    /* Max search at next character in binary file */
#define MAXFREQ    2000    /* Max frequency count before table reset */
#define MINCOPY       3    /* Shortest string copy length */
#define MAXCOPY      64    /* Longest string copy length */
#define SHORTRANGE    3    /* Max distance range for shortest length copy */
#define COPYRANGES    6    /* Number of string copy distance bit ranges */
short copybits[COPYRANGES] = {4,6,8,10,12,14};   /* Distance bits */

#define CODESPERRANGE (MAXCOPY - MINCOPY + 1)
int copymin[COPYRANGES], copymax[COPYRANGES];
int maxdistance, maxsize;
int distance, insert = MINCOPY, dictfile = 0, binary = 0;

#define NIL -1                  /* End of linked list marker */
#define HASHSIZE 16384          /* Number of entries in hash table */
#define HASHMASK (HASHSIZE - 1) /* Mask for hash key wrap */
short far *head, far *tail;      /* Hash table */
short far *succ, far *pred;      /* Doubly linked lists */
unsigned char *buffer;           /* Text buffer */

/* Define hash key function using MINCOPY characters of string prefix */
#define getkey(n) ((buffer[n] ^ (buffer[(n+1)%maxsize]<<4) ^ \
                   (buffer[(n+2)%maxsize]<<8)) & HASHMASK)

/* Adaptive Huffman variables */
#define TERMINATE 256           /* EOF code */
#define FIRSTCODE 257           /* First code for copy lengths */
#define MAXCHAR (FIRSTCODE+COPYRANGES*CODESPERRANGE-1)
#define SUCCMAX (MAXCHAR+1)
#define TWICEMAX (2*MAXCHAR+1)
#define ROOT 1
short left[MAXCHAR+1], right[MAXCHAR+1];  /* Huffman tree */
short up[TWICEMAX+1], freq[TWICEMAX+1];

/*** Bit packing routines ***/

int input_bit_count = 0;        /* Input bits buffered */
int input_bit_buffer = 0;       /* Input buffer */
int output_bit_count = 0;       /* Output bits buffered */
int output_bit_buffer = 0;      /* Output buffer */
long bytes_in = 0, bytes_out = 0;  /* File size counters */

/* Write one bit to output file */
output_bit(output,bit)
  FILE *output;
  int bit;
{
  output_bit_buffer <<= 1;
  if (bit) output_bit_buffer |= 1;
```

Figure 4.34 SIXPACK.C source code listing

```
  if (++output_bit_count == 8) {
    putc(output_bit_buffer,output);
    output_bit_count = 0;
    ++bytes_out;
  }

}

/* Read a bit from input file */
int input_bit(input)
  FILE *input;
{
  int bit;

  if (input_bit_count-- == 0) {
    input_bit_buffer = getc(input);
    if (input_bit_buffer == EOF) {
      printf(" UNEXPECTED END OF FILE\n");
      exit(1);
    }
    ++bytes_in;
    input_bit_count = 7;
  }
  bit = (input_bit_buffer & 0x80) != 0;
  input_bit_buffer <<= 1;
  return(bit);
}

/* Write multibit code to output file */
output_code(output,code,bits)
  FILE *output;
  int code,bits;
{
  int i;

  for (i = 0; i<bits; i++) {
    output_bit(output,code & 0x01);
    code >>= 1;
  }
}

/* Read multibit code from input file */
int input_code(input,bits)
  FILE *input;
  int bits;
{
  int i, bit = 1, code = 0;

  for (i = 0; i<bits; i++) {
    if (input_bit(input)) code |= bit;
    bit <<= 1;
  }
  return(code);
}

/* Flush any remaining bits to output file before closing file */
flush_bits(output)
  FILE *output;
{
  if (output_bit_count > 0) {
    putc((output_bit_buffer << (8-output_bit_count)),output);
```

Figure 4.34 *Continued*

```
      ++bytes_out;
  }
}

/  * Adaptive Huffman frequency compression ***/

/* Data structure based partly on "Application of Splay Trees
   to Data Compression", Communications of the ACM 8/88 */

/* Initialize data for compression or decompression */
initialize()
{
  int i, j;

  /* Initialize Huffman frequency tree */
  for (i = 2; i<=TWICEMAX; i++) {
    up[i] = i/2;
    freq[i] = 1;
  }
  for (i = 1; i<=MAXCHAR; i++) {
    left[i] = 2*i;
    right[i] = 2*i+1;
  }

  /* Initialize copy distance ranges */
  j = 0;
  for (i = 0; i<COPYRANGES; i++) {
    copymin[i] = j;
    j += 1 << copybits[i];
    copymax[i] = j - 1;
  }

  maxdistance = j - 1;
  maxsize = maxdistance + MAXCOPY;
}

/* Update frequency counts from leaf to root */
update_freq(a,b)
  int a,b;
{
  do {
    freq[up[a]] = freq[a] + freq[b];
    a = up[a];
    if (a != ROOT) {
      if (left[up[a]] == a) b = right[up[a]];
      else b = left[up[a]];
    }
  } while (a != ROOT);

  /* Periodically scale frequencies down by half to avoid overflow */
  /* This also provides some local adaption and better compression */
  if (freq[ROOT] == MAXFREQ)
    for (a = 1; a<=TWICEMAX; a++) freq[a] >>= 1;
}

/* Update Huffman model for each character code */
Update_model(code)
  int code;
{
  int a, b, c, ua, uua;
```

Figure 4.34 *Continued*

```
        a = code + SUCCMAX;
        ++freq[a];
        if (up[a] != ROOT) {
          ua = up[a];
          if (left[ua] == a) update_freq(a,right[ua]);
          else update_freq(a,left[ua]);
          do {
            uua = up[ua];
            if (left[uua] == ua) b = right[uua];
            else b = left[uua];

            /* If high freq lower in tree, swap nodes */
            if (freq[a] > freq[b]) {
              if (left[uua] == ua) right[uua] = a;
              else left[uua] = a;
              if (left[ua] == a) {
                left[ua] = b; c = right[ua];
              } else {
                right[ua] = b; c = left[ua];
              }
              up[b] = ua; up[a] = uua;
              update_freq(b,c); a = b;
            }
            a = up[a]; ua = up[a];
          } while (ua != ROOT);
        }
}

/* Compress a character code to output stream */
Compress(output,code)
  FILE *output;
  int code;
{
  int a, sp = 0;
  int stack[50];

  a = code + SUCCMAX;
  do {
    stack[sp++] = (right[up[a]] == a);
    a = up[a];
  } while (a != ROOT);
  do {
    output_bit(output,stack[--sp]);
  } while (sp);
  update_model(code);
}

/* Uncompress a character code from input stream */
int uncompress(input)
  FILE *input;
{
  int a = ROOT;

  do {
    if (input_bit(input)) a = right[a];
    else a = left[a];
  } while (a <= MAXCHAR);
  update_model(a-SUCCMAX);
  return(a-SUCCMAX);
}
```

Figure 4.34 *Continued*

```
/*** Hash table linked list string search routines ***/

/* Add node to head of list */
add_node(n)
  int n;
{
  int key;

  key = getkey(n);
  if (head[key] == NIL) {
    tail[key] = n;
    succ[n] = NIL;
  } else {
    succ[n] = head[key];
    pred[head[key]] = n;
  }
  head[key] = n;
  pred[n] = NIL;
}

/* Delete node from tail of list */
delete_node(n)
  int n;
{
  int key;

  key = getkey(n);
  if (head[key] == tail[key])
    head[key] = NIL;
  else {
    succ[pred[tail[key]]] = NIL;
    tail[key] = pred[tail[key]];
  }
}

/* Find longest string matching lookahead buffer string */
int match(n,depth)
  int n,depth;
{
  int i, j, index, key, dist, len, best = 0, count = 0;

  if (n == maxsize) n = 0;
  key = getkey(n);
  index = head[key];
  while (index != NIL) {
    if (++count > depth) break;       /* Quit if depth exceeded */
    if (buffer[(n+best)%maxsize] == buffer[(index+best)%maxsize]) {
      len = 0;  i = n;  j = index;
      while (buffer[i]==buffer[j] && len<MAXCOPY && j!=n && i!=insert) {
        ++len;
        if (++i == maxsize) i = 0;
        if (++j == maxsize) j = 0;
      }
      dist = n - index;
      if (dist < 0) dist += maxsize;
      dist -= len;
      /* If dict file, quit at shortest distance range */
      if (dictfile && dist > copymax[0]) break;
      if (len > best && dist <= maxdistance) {      /* Update best match */
```

Figure 4.34 _Continued_

```
            if (len > MINCOPY || dist <= copymax[SHORTRANGE+binary]) {
              best = len; distance = dist;
            }
          }
        }
      index = succ[index];
    }
    return(best);
}

/*** Finite Window compression routines ***/

#define IDLE 0      /* Not processing a copy */
#define COPY 1      /* Currently processing copy */

/* Check first buffer for ordered dictionary file */
/* Better compression using short distance copies */
dictionary()
{
  int i = 0, j = 0, k, count = 0;

  /* Count matching chars at start of adjacent lines */
  while (++j < MINCOPY+MAXCOPY) {
    if (buffer[j-1] == 10) {
      k = j;
      while (buffer[i++] == buffer[k++]) ++count;
      i = j;
    }
  }
  /* If matching line prefixes > 25% assume dictionary */
  If (count > (MINCOPY+MAXCOPY)/4) dictfile = 1;
}

/* Encode file from input to output */
encode(input,output)
  FILE *input, *output;
{
  int c, i, n=MINCOPY, addpos=0, len=0, full=0, state=IDLE, nextlen;

  initialize();
  head = farmalloc((unsigned long)HASHSIZE*sizeof(short));
  tail = farmalloc((unsigned long)HASHSIZE*sizeof(short));
  succ = farmalloc((unsigned long)maxsize*sizeof(short));
  pred = farmalloc((unsigned long)maxsize*sizeof(short));
  buffer = (unsigned char *) malloc(maxsize*sizeof(unsigned char));
  if (head==NULL || tail==NULL || succ==NULL || pred==NULL || buffer==NULL) {
    printf("Error allocating memory\n");
    exit(1);
  }

  /* Initialize hash table to empty */
  for (i = 0; i<HASHSIZE; i++) {
    head[i] = NIL;
  }

  /* Compress first few characters using Huffman */
  for (i = 0; i<MINCOPY; i++) {
    if ((c = getc(input)) == EOF) {
      compress(output,TERMINATE);
      flush_bits(output);
```

Figure 4.34 *Continued*

```
      return(bytes_in);
   }
   compress(output,c);   ++bytes_in;
   buffer[i] = c;


   /* Preload next few characters into lookahead buffer */
   for (i = 0; i<MAXCOPY; i++) {
     if ((c = getc(input)) == EOF) break;
     buffer[insert++] = c;   ++bytes_in;
     if (c > 127) binary = 1;      /* Binary file ? */
   }
   dictionary();   /* Check for dictionary file */

   while (n != insert) {
     /* Check compression to insure really a dictionary file */
     if (dictfile && ((bytes_in % MAXCOPY) == 0))
       if (bytes_in/bytes_out < 2)
         dictfile = 0;        /* Oops, not a dictionary file ! */

     /* Update nodes in hash table lists */
     if (full) delete_node(insert);
     add_node(addpos);

     /* If doing copy, process character, else check for new copy */
     if (state == COPY) {
       if (--len == 1) state = IDLE;
     } else {

       /* Get match length at next character and current char */
       if (binary) {
         nextlen = match(n+1,BINNEXT);
         len = match(n,BINSEARCH);
       } else {
         nextlen = match(n+1,TEXTNEXT);
         len = match(n,TEXTSEARCH);
       }

       /* If long enough and no better match at next char, start copy */
       if (len >= MINCOPY && len >= nextlen) {
         state = COPY;

         /* Look up minimum bits to encode distance */
         for (i = 0; i<COPYRANGES; i++) {
           if (distance <= copymax[i]) {
             compress(output,FIRSTCODE-MINCOPY+len+i*CODESPERRANGE);
             output_code(output,distance-copymin[i],copybits[i]);
             break;
           }
         }
       }
       else    /* Else output single literal character */
         compress(output,buffer[n]);
     }

     /* Advance buffer pointers */
     if (++n == maxsize) n = 0;
     if (++addpos == maxsize) addpos = 0;

     /* Add next input character to buffer */
```

Figure 4.34 *Continued*

```
      if (c != EOF) {
        if ((c = getc(input)) != EOF) {
          buffer[insert++] = c;   ++bytes_in;
        } else full = 0;
        if (insert == maxsize) {
          insert = 0; full = 1;
        }
      }
    }

    /* Output EOF code and free memory */
    compress(output,TERMINATE);
    flush_bits(output);
    farfree(head); farfree(tail); farfree(succ); farfree(pred);
    free(buffer);
}

/* Decode file from input to output */
decode(input,output)
  FILE *input,*output;
{
  int c, i, j, k, dist, len, n = 0, index;

  initialize();
  buffer = (unsigned char *) malloc(maxsize*sizeof(unsigned char));
  if (buffer == NULL) {
    printf("Error allocating memory\n");
    exit(1);
  }
  while ((c = uncompress(input)) != TERMINATE) {
    if (c < 256) {        /* Single literal character ? */
      putc(c,output);
      ++bytes_out;
      buffer[n++] = c;
      if (n == maxsize) n = 0;
    } else {              /* Else string copy length/distance codes */
      index = (c - FIRSTCODE)/CODESPERRANGE;
      len = c - FIRSTCODE + MINCOPY - index*CODESPERRANGE;
      dist = input_code(input,copybits[index]) + len + copymin[index];
      j = n; k = n - dist;
      if (k < 0) k += maxsize;
      for (i = 0; i<len; i++) {
        putc(buffer[k],output);   ++bytes_out;
        buffer[j++] = buffer[k++];
        if (j == maxsize) j = 0;
        if (k == maxsize) k = 0;
      }
      n += len;
      if (n >= maxsize) n -= maxsize;
    }
  }
  free(buffer);
}

/* Main program */
main(argc,argv)
  int argc;
  char *argv[];
{
  FILE *infile, *outfile;
```

Figure 4.34 *Continued*

```
  if (argc < 3 || argc > 4)
    printf("Usage: %s inputfile outputfile [decompress]\n",argv[0]);
  else if (!strcmp(argv[1],argv[2]))
    printf("File names must be different\n");
  else if ((infile = fopen(argv[1],"rb")) == NULL)
    printf("Error opening input file %s\n",argv[1]);
  else if ((outfile = fopen(argv[2],"wb")) == NULL)
    printf("Error opening output file %s\n",argv[2]);
  else {
    if (argc == 3) {
      encode(infile,outfile);
      printf("Packed from %ld bytes to %ld bytes\n",bytes_in,bytes_out);
    } else {
      decode(infile,outfile);
      printf("Unpacked from %ld bytes to %ld bytes\n",bytes_in,bytes_out);
    }
    fclose(outfile); fclose(infile);
  }
}
```

Figure 4.34 *Continued*

maximum search distance and many duplicate keys may exist in some lists. Hash table chaining techniques would be awkward in this situation.

Both successor and predecessor pointers must be kept for a doubly linked list. New nodes are added at the head of the list and old nodes are deleted at the tail of the list. A singly linked list would result in slow delete times, since the entire list must be scanned to find the last node. Searching begins at the head of the list, keeping track of the best matching string seen so far. This method has the advantage that the most recent string match is always found, resulting in shorter distance copies that can be encoded in fewer bits. No actual information needs to be stored in the lists because the node pointers also indicate the character positions in the buffer.

Adaptive Huffman coding

As a final compression stage, each literal character and copy length code is passed through an adaptive frequency filter which squeezes frequently occurring characters into short bit strings. The possible copy length codes for each distance range are added to the end of the normal character set. The copy distance values are likely to be more random and not susceptible to frequency encoding, so they are output using fixed length bit strings.

A binary prefix code tree which approximates the famous Huffman tree is maintained by counting each character and propagating the count upward through the tree. During this propagation the frequency of each node is calculated as the sum of the frequencies of both children. The new frequency of each traversed node is then compared to that of the node that is up two levels and down one. If the higher frequency is lower in the tree, the two nodes are swapped. To avoid overflow and provide local adaption to changing data, the frequencies are periodically scaled down by a factor of two.

The data structures and compress/uncompress routines are derived from Pascal versions presented in 'Application of splay trees to data compression', *Communications of the ACM*, August 1988. Mr Gage replaced the original semisplaying by frequency coding, giving better results for this application but running slower.

Bit packing

The final topic to be covered is packing and unpacking of variable length bit strings. Several different sizes of codes are used for copy distance values, while the adaptive Huffman algorithm processes individual bits. Routines to handle single bits and multibit codes are used in the program. A flush routine writes any buffered bits to the output file before it is closed.

In summary, the Sixpack program provides very good compression with low memory usage, about 200K for compression and 50K for decompression. The code is fairly simple and generates an executable file only 14K in size. It uses a one-pass method suitable for large files and redirected data streams. The main disadvantage is slow compression speed, making it more suitable for archive distribution than for routine backups. There is much room for performance improvement, making this a potentially useful method.

4.7 MNP COMPRESSION

In concluding this chapter, we will focus our attention upon two popular methods of modem data compression incorporated into the Microcom Networking Protocol (MNP). Although the CCITT (now known as the ITU) V.42 bis recommendation uses string compression described in Chapter 6, the large base of over several million installed MNP modems makes MNP a *de facto* standard. In addition, MNP compression methods include several interesting algorithms which results in their examination providing us with additional information concerning the implementation and operation of data compression.

MNP Class 5

MNP Class 5 data compression is one of two methods of data compression supported by the Microcom Networking Protocol (MNP). The MNP protocol was developed by Microcom, Inc., to enhance the communications capability between modems and currently consists of 10 classes. Each class supports lower-class operations, providing downward compatibility between different products that support different MNP classes.

Operations defined by MNP include the adaptive packetization of data, conversion of asynchronous data into a synchronous format, different modulation schemes, as well as two types of data compression. MNP Class 5 data compression was the first method of compression incorporated into the protocol. As previously explained in Chapter 3, MNP Class 5 incorporates two manipulations to a transmitted data stream. The first manipulation is a version of run-length encoding in which sequences of three or more repeating characters are compressed by the insertion of a count after the third character and the removal of all repeating characters in excess of three. Under the MNP Class 5 version of run-length encoding, if the repeated character sequence is three characters in length, the sequence is followed by a repetition count of 0. The maximum repetition count supported is 250.

Both the run-length count and the three repetitions of a character, as well as any other characters in the data stream, are next compressed through the use of an adaptive frequency encoding technique.

The adaptive frequency encoding method employed by MNP Class 5 data compression results in the substitution of a compression token for each 8-bit data character. The token used changes with the frequency of occurrence of the actual data character so that shorter tokens are substituted for more frequently occurring data characters similar to the method described in Section 4.6 which covered adaptive compression.

The formation of tokens to represent 8-bit characters was based upon the fact that there are 2^8 or 256 different characters that can occur and each character must be mapped onto 256 different codes. Since conventional ASCII and non-eighth-bit-set EBCDIC represent a majority of characters used in data transmission, Microcom considered the fact that there are 128 characters that have fewer than eight bits in developing tokens. The compression tokens were developed to recognize that fact by representing the more frequently occurring characters with codes that have leading zeros removed. To separate recoded data characters from each other, a two-part token was employed. The first part of the token, known as its header, is three bits in length as that enables a variable length of up to eight bits to be defined. Thus, the token header can be viewed as a reverse comma code. The second part of the token is referred to as the token's body and is variable in length. The length of the token's body is defined by the bit composition of the header, which can range in value from binary 000 to 111.

Table 4.11 lists the relationship between the decimal value of each

Table 4.11 Character to token initial mapping

Data character decimal value	Token composition	
	Header	Body
0	000	0
1	000	1
2	001	0
3	001	1
4	010	00
5	010	01
6	010	10
7	010	11
8	011	000
9	011	001
10	011	010
11	011	011
12	011	100
13	011	101
14	011	110
15	011	111
16	100	0000
17	100	0001
18	100	0010
19	100	0011
20	100	0100
21	100	0101
22	100	0110
23	100	0111
24	100	1000
25	100	1001
26	100	1010
27	100	1011
28	100	1100
29	100	1101
30	100	1110
31	100	1111
32	101	00000
33	101	00001
34	101	00010

(35–246 continue in the same pattern)

247	111	1110111
248	111	1111000
249	111	1111001
250	111	1111010
251	111	1111011
252	111	1111100
253	111	1111101
254	111	1111110
255	111	11111110

character and the substituted token at compression initialization. In examining the entries in Table 4.11, there are three special cases that do not follow the previously described relationships. First, there are two tokens with a header of 000 where the actual length of the body is 1. Lastly, when the header portion of the token indicates a length of 7 and the body is seven 1-bits, the actual length of the body is eight bits. Otherwise, the header portion of the token indicates the length of the variable-length token body.

When MNP Class 5 compression is initiated, the frequency of occurrence of each of the 256 possible characters is zero. For purposes of character-to-token mapping, the character whose bit value represents decimal 0 (00000000 binary) is assumed to be the most frequently occurring character and is represented by the first of the shortest tokens in Table 4.11. Characters with increasing decimal values are represented by succeeding tokens, until the character whose decimal value is 255 (11111111 binary) is assumed to be the most infrequently occurring character.

As a character is processed, the token to which its decimal value is currently mapped is substituted for the character. After this substitution occurs, the frequency of occurrence of the character is incremented by 1. If the frequency of occurrence of the character just encoded into a token is greater after incrementing than the frequency of the next most frequently occurring character, then the compression tokens of the current character and the next most frequently occurring character are exchanged. The frequency of the current character is then compared to the frequency of the character which is now the next most frequently occurring character. If the frequency of the current character is greater, then the compressed tokens are once again swapped. This process is repeated until no more swaps are required, at which time the mapping of characters and compression tokens is correctly adapted based upon the relative frequency of occurrence of the characters processed.

The frequency of occurrence for each character is an 8-bit field which has a maximum value of 255 (11111111 binary). To prevent the field from overflowing, each time the character-compression token mappings are sorted by frequency the frequency count of the current character is compared to the maximum count value of 255. If the maximum count has been reached, the frequency of occurrence of each character is scaled downward by dividing each frequency by 2 using integer division.

As previously discussed in Chapter 3, as well as briefly mentioned in this section, sequences of three or more repeating characters are first placed into a four-character run-length encoded sequence. Here the first three characters are the actual repeated characters, while the fourth character is the repetition count. Although the repeated characters and count are mapped into tokens, the count character is not used to increase the frequency of occurrence of the character to which the token is mapped.

Flushing

During the compression process, 8-bit characters can be encoded into tokens ranging from four to 11 bits in length. Since computers operate upon 8-bit characters, it may be necessary to terminate a transmission sequence by inserting a number of 'pad' bits to fill an 8-bit boundary. To accomplish this under MNP Class 5 compression, the transmitter inserts a special token into the data stream after the last user data token. The special token is known as a flush token and has the bit composition 11111111111. After the flush token, '1' bits are appended by adding them to the low-order end of the body of the flush token as required to produce an integral number of 8-bit characters.

Efficiency

Although the efficiency of MNP Class 5 data compression, like all compression methods, is dependent upon the data stream it operates upon, we can note several characteristics with respect to its efficiency. First, discounting repeating strings of characters, its maximum compression ratio is 2. This is because the smallest token an 8-bit character can be mapped into is four bits. Secondly, only 32 out of 256 characters will be mapped into tokens that are seven bits or less in length. This means that this compression method will result in either no actual compression or expansion for 224 out of 256 8-bit character representations. In spite of these limitations, the use of run-length encoding, and the fact that the 15 most frequently occurring characters in most data streams cumulatively exceed 50% of all data and can be encoded into six bits or less, can result in a compression ratio between 1.5 and 2.0. Thus, MNP Class 5 data compression can provide 2400 bps modems with an effective throughput between 3600 and 4800 bps. Similarly, 9600 bps modems that support MNP Class 5 data compression can be expected to provide an effective throughput between 14 400 and 19 200 bps when communicating with a compatible modem.

MNP Class 7 enhanced data compression

MNP Class 7 enhanced data compression is the second method of compression supported by the Microcom Networking Protocol. This method of data compression uses what is known as a first-order Markov model to predict the probability of character occurrence based upon the previous character and performs adaptive Huffman coding on the data stream. In addition, MNP Class 7 uses run-length encoding to compress streams of duplicate characters. However, the method of run-length encoding differs from that used by MNP Class 5 data compression.

MNP Class 7 enhanced data compression can be considered as a two-stage process. First, multiple consecutive copies of the same character or 8-bit pattern are compressed using run-length encoding. Under MNP Class 7, if the encoder has sent the same character three times it sends the count of the remaining identical consecutive characters as a single 4-bit nibble. Once a run-length encoding algorithm is applied to the data stream it is then compressed using a first-order Markov model for code selection.

The first-order Markov model used for code selection can be viewed as a two-dimensional 256×256 element matrix, which is illustrated in Figure 4.35. The top row of the matrix contains 256 entries that can be viewed as pointers to an appropriate column location within each row entry. Each pointer within the top row of the matrix corresponds to one of 256 8-bit character values while the column below the pointer can be considered as one of 256 coding tables.

To encode a character for transmission the previous character is used to select the appropriate coding table by using the pointer entry to select the appropriate column. Next, the column is searched until the current character is encountered. In place of transmitting the 8-bit composition of the character the Huffman code developed for that character is used. In addition, the frequency of occurrence of the character is updated and the coding table may be adjusted based upon the frequency of the character transmitted being compared to the frequency of occurrence of other characters in that particular coding table. Thus, each of the coding tables previously shown in Figure 4.35 actually consists of three fields. One field contains the 8-bit composition of the character, while the other fields contain the frequency of occurrence of the character and its Huffman code. Similar to the previously described adaptive Huffman coding technique, the Huffman field is static and remains fixed. In comparison, each character's bit composition and count fields will be adjusted by location in the coding table based upon their frequency of occurrence.

Pointers	00000000	11111111
	00000000	 00000000
	.	....
Coding tables	.	....
	.	....
	11111111	 11111111

Figure 4.35 Initial Markov model

	Pointers	A	B	C	D	C ...
		T	L	H	O	D
	Character positions	H	E	O	A	R
	within coding tables	C	U	R	E	S
		.	.	.	.	.
		.	.	.	.	.

Figure 4.36   First-order Markov model processing the alphabet

The encoding process used by MNP Class 7 recognizes the fact that there is a high probability that many characters are followed by other characters. To illustrate this concept consider the English language. Here, the probability of a U following a Q is very high. Thus, the Q coding table will most likely have the character U move rapidly to the top of the table. Since Huffman coding is employed, normally U following a Q will be encoded as one bit. If we assume that only the letters of the alphabet are being encoded, a portion of the Markov model might appear as indicated in Figure 4.36.

Note that the letter T is in the top position in the coding table under the A pointer as the letter T frequently follows the letter A. Similarly, the letter L is illustrated at the top of the coding table under the B pointer as the letter L frequently follows the letter B.

Efficiency

Unlike MNP Class 5 compression in which a minimum of four bits is required to represent a character, MNP Class 7 can represent a character with only one bit. In fact, the top character in each of 256 coding tables can be represented by one bit. Due to the use of adaptive Huffman coding, MNP Class 7 enhanced data compression can be expected to provide, as a minimum, a compression ratio between 15 and 25% above that obtainable using MNP Class 5 compression.

5

FACSIMILE COMPRESSION

Facsimile, or its popular abbreviation 'fax', has become an almost indispensable method for conveying information. While you can correctly consider a fax to represent an image, separate chapters in this book were used to subdivide fax from image compression. The rationale for this subdivision is based upon the fact that fax compression significantly differs from most other methods of image compression. A second reason for this subdivision relates to the fact that whereas specific file formats have been standardized to store different types of images, fax presently lacks a computer file storage standard. Although the development of facsimile dates from the late nineteenth century, its popular use is a relatively recent phenomenon due to the development of a compression standard and the use of an integrated modem on a chip reducing the cost of fax machines to under a few hundred dollars. Today most businesses and many home owners routinely send correspondence via fax. Not only does it arrive at its destination near instantaneously compared with the days or even weeks required for postal communications, but in many instances it is less expensive. The key to the ability to transmit a page in under a minute is through the use of compression, the subject of this chapter.

In this chapter we will focus our attention upon the evolution of different compression techniques used to compress fax, ranging from *ad hoc* developed methods to modern fax compression standards. However, prior to doing so we will obtain an overview of the operation of the modern Group 3 digital facsimile systems.

5.1 GROUP 3 FAX OPERATIONAL OVERVIEW

As its nomenclature implies, Group 3 fax represents a third facsimile standard. Group 1 and Group 2 fax systems were based upon analog technology. Group 1 equipment standards, which like Group 3 were actually CCITT (now ITU) Recommendations, used one frequency to represent a black picture element (pel) and another frequency to represent a white picture element. By switching frequencies, one fax system informed the other of the composition of scanner pels. Group 1 equipment was capable of scanning 180 lines per minute and required approximately six minutes to transmit a standard page and four minutes to transmit a page in a lower resolution. Group 2 equipment used a different type of modulation technique which reduced the time required to transmit a standard page to three minutes; however, it was still based upon the use of analog technology.

The CCITT Group 3 digital facsimile standard was promulgated in 1980. However, the high cost of implementing the technology to include its then high-speed 9600 bps modem delayed its wide-scale adoption by businesses and home users to the late 1980s and early 1990s. By 1995 you could purchase a Group 3 fax that cost several thousand dollars in 1985 for under $300.

Figure 5.1 illustrates in block diagram form the architecture of a Group 3 fax system. The upper portion of Figure 5.1 illustrates the transmitter of a Group 3 fax, while the lower portion of Figure 5.1 illustrates the receiver of a second Group 3 fax.

The scanner illuminates each line of the page being read. A charge coupled device (CCD) measures the intensity of the reflection of the light and generates pulses whose amplitude varies according to the brightness of the reflection of each picture element. The Analog to Digital (A/D) converter converts the CCD analog pulses to digital, generating one bit per picture element. Next, the compressor reduces the digital data stream which is then modulated into an analog signal for transmission over the analog public switched telephone network (PSTN). At the receiver the modem demodulates the received analog signal back into its compressed digital form. The expander reverses the compression process, which enables the printer to convert the bit stream into a copy of the original scanned page.

The compression method employed by modern Group 3 digital fax is well defined and based upon a modification of the Huffman coding technique previously described in Chapter 4. An option left for further

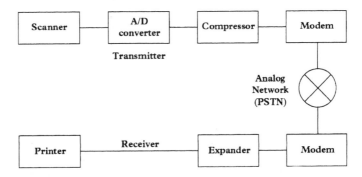

Figure 5.1　Group 3 fax architecture

study under Group 3 is a two-dimensional compression scheme whose original development dated to telemetry operations. Although two-dimensional coding is not defined for Group 3 fax machines, several vendors have included it in their top-of-the-line models. In addition, Group 4 fax systems use a modified form of the two-dimensional encoding used in some Group 3 systems. Group 4 fax standards were developed in 1984; however, such equipment is based upon the use of Integrated Services Digital Network (ISDN) transmission facilities operating at 64 kbps. Due to the limited development of communications carrier ISDN transmission facilities, Group 4 equipment represents less than 1% of the fax market, although it provides a mechanism to transmit a page in under six seconds.

Since both Group 3 and Group 4 two-dimensional compression methods can be traced to telemetry applications that are based upon relative encoding, we will first examine that compression method. This will be followed by the one-dimensional modified Huffman technique employed in fax systems and the two-dimensional compression technique referenced in the Group 3 standard and included in the Group 4 standard.

5.2 RELATIVE ENCODING

Relative encoding is a compression technique that is not normally applicable to the transmission of conventional data files. This type of compression is effectively employed when there are sequences of runs in the original data stream that vary only slightly from each other or the run sequences can be broken into patterns relative to each other. An example of the former is telemetry data, while the bit patterns of digital facsimile machines represent a version of the latter.

Telemetry compression

In telemetry data generation, a sensing device is used to record measurements at predefined intervals. These measurements are then transmitted to a central location for additional processing. One example of telemetry signals is the numerous space probes which transmit temperature readings, colour spectrum analysis and other data, either upon command from earth stations or at predefined time intervals. Normally, telemetry signals contain a sequence of numeric fields consisting of subsequences or runs of numerics that vary only slightly from each other as illustrated in the top portion of Figure 5.2.

Prior to actual data transmission, compression occurs to reduce the total amount of data necessary to represent the recorded measurements. Each measurement other than the first is coded with the relative difference between it and the preceding measurement, as

Original telemetry measurements

46 46 46.1 46.1 46.1 46 46 46 46.1 46.1 46.1 46.2

Relative encoding

46 0 .1 0 0 —.1 0 0 .1 0 0 .1

Figure 5.2 Relative encoding process. Telemetry signals often consist of a sequence of numerics that vary only slightly from each other during a certain time interval

long as the absolute value of the increment is less than some predetermined value. This is shown in the lower portion of Figure 5.2. If the increment should exceed this value, a special character is inserted to denote that the particular value at that location is not available or the special character could be followed by the measurement that is out of the boundary range for the relative encoding process. This limits wide fluctuations and is one disadvantage associated with the utilization of this technique. Another disadvantage is that if data values consistently vary both within and outside the relative encoding boundary range and a combination of a special character and actual value is transmitted, this will cause an expansion instead of a compression of the data stream.

Additional techniques may be employed to obtain a higher degree of compression depending upon the original telemetry measurements and the resultant data due to the relative encoding process. In the top portion of Figure 5.2, the original telemetry measurements illustrated consist of 38 characters to include numerics and decimal points. As a result of the relative encoding process, the number of numerics and decimal point characters has been reduced to 18. By the incorporation of a second compression technique, the number of characters used to represent the relative encoding process may be further reduced. One method that could be used is the half-byte packing process where each numeric digit is stripped of its first four bits and packed two per character. If we use a 4-bit representation for the decimal point and minus sign, half-byte packing will result in the transmission of nine 8-bit bytes of data. Thus, while the relative encoding process resulted in a 2.24 (38/17) compression ratio, recompressing the relative encoding results employing the half-byte packing technique approximately doubles the compression ratio to 4.223 (38/9).

While the half-byte packing process was illustrated as the combining or second compression technique, other techniques may be employed with results dependent upon the variability of the original telemetry measurements. If the original telemetry measurements indicated a stable 46 for the time interval sampled, the relative encoding process would result in a long string of zeros after the value indicator of 46. For this situation, run-length encoding would be more effective as the second compression technique.

Digital facsimile

Several relative encoding techniques have been developed to compress digital facsimile data. In the remainder of this section we will obtain an overview of several relative encoding techniques that were implemented in non-standardized fax machines prior to 1980, deferring a discussion of the vertical correlation method used in Group 3 and Group 4 equipment to the last section in this chapter as they are used in tandem with a modified Huffman coding technique which is described in the next section in this chapter. However, prior to obtaining an overview of relative encoding techniques, a review of the elements of facsimile technology is warranted.

Facsimile systems use the basic concept of scanning—normally on a line by line basis—to create a stream of information concerning the lightness or darkness of the small area being scanned at any given point in time. The resulting stream of information is then transmitted and used to drive an image-reproducing device at a facsimile receiver where the original information is reproduced. In general, the operation of a facsimile device is quite similar to the technology employed in television, where 525 lines on the US domestic television system are used to reproduce images. For facsimile systems, the clarity depends upon the fineness of the scan. Normally, approximately 100 scan lines per inch are required to successfully reproduce a page of typewritten material. Thus, a normal $8\frac{1}{2}$ inch × 11 inch sheet of paper, scanned longitudinally, would require approximately 1100 scan lines. Each scan line in turn consists of approximately 1730 picture elements (pixels, or pels), resulting in approximately 1.9 million bits for an $8\frac{1}{2}$ × 11 sheet of paper. To transmit this data at 9600 bps without compression, 178 s or approximately 3.3 min are required.

For facsimile systems, the degree of compression theoretically obtainable is normally very large for the typical facsimile message. As an example, consider a typewritten memorandum containing 500 characters. In conventional data transmission, each character can be represented and transmitted by eight bits. Thus, the entire message could be transmitted, ignoring control characters, by 500 × 8 or 4000 bits of information. If transmitted at 9600 bps, the total transmission time would be less than 0.5 s. In comparison, the same message sent by conventional facsimile requires the transmission of almost 1.9 million bits and takes about 3.3 min without data compression, a difference of approximately 396 to 1 between conventional facsimile code and character transmission.

Facsimile techniques

One of the earliest facsimile compression techniques was run-length encoding. Here, the transmission of the digital line scan is replaced by the transmission of a quantity count of each of the successive runs of black or white scanned pels.

Since the vast majority of documents to be scanned contains a much higher quantity of white pels than black ones, transmitting the difference between scans may significantly reduce the quantity of data to be transmitted. In this method of compression, one complete scan is held in a memory area of the device and compared with the subsequent scan. Transmitting only changes relative to the preceding scan results in the relative compression process. Once the differences between the first and second scans are transmitted, the first scan is removed from memory and replaced by the second scan. Next, a third scan is compared with the second scan now located in memory. A flow chart showing the required steps for this type of relative encoding process is illustrated in Figure 5.3.

In Figure 5.4, a portion of the relative changes resulting from the comparison of two scan lines is shown. Several methods can be used to denote the relative changes between the Nth and $(N+1)$th scan lines. One method is to denote the position of the change by what is normally called a positional identification. Here, the position of each relative change is denoted with respect to the first pel of the line. If there are many consecutive changes, the transmission of each individual position could require more data bits than the transmission of the original line prior to comparison with the preceding line. To take advantage of successive relative changes between scanned lines, the

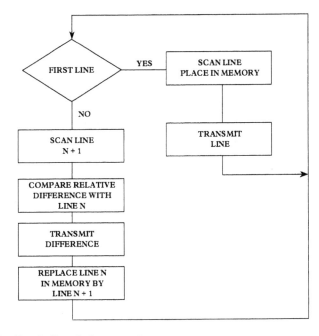

Figure 5.3 Facsimile relative encoding process

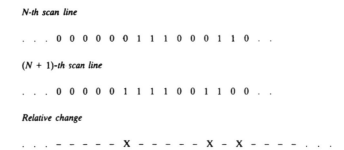

N-th scan line

. . . 0 0 0 0 0 0 1 1 1 0 0 0 1 1 0 . .

(N + 1)-th scan line

. . . 0 0 0 0 0 1 1 1 1 0 0 1 1 0 0 . .

Relative change

. . . – – – – – X – – – – – X – X – – – – . . .

Figure 5.4 Relative change. To denote the relative changes between scan lines several methods can be employed to include identification by position and displacement

Initial position of relative change	Number of successive relative changes
40	6
80	20
175	4
350	31
480	8
930	14
1250	16
1310	5
1340	4

Transmitted data

40	6	80	20	175	4	350	31	480	8	930	14	1250	16	1310	5	1340

Figure 5.5 Transmitting positional information. Using the positional relative process, the initial position of each relative change is followed by the number of successive relative changes

position indicator can be followed by a quantity count which contains the number of successive relative changes. This is illustrated in Figure 5.5 where the table at the top of the figure tabulates the initial position of the relative change between line scans and the number of succeeding relative changes, while the transmission sequence is indicated at the lower portion of that figure. Under the Consultative Committee for International Telephone and Telegraph (CCITT) digital facsimile standards, there are 1728 picture elements or points to be read by the scanner along the width of a document 215 mm wide. Due to the large number of positions, the transmission of positional information can rapidly increase in duration, especially when a number of relative changes occur at the far end of the scan line. One method used to alleviate this 'end of the line' increase is by the use of displacement notation. As with positional notation, the relative changes between scan lines are first computed. Then, instead of transmitting all of the initial positions of the relative changes and the number of successive changes as illustrated in Figure 5.5, only the first initial position is transmitted. Thereafter, the displacement between relative changes is transmitted. This displacement method can include the transmission of successive relative change information and is illustrated in Figure 5.6. This figure is based upon the data provided in the tabular portion of Figure 5.5. In comparing the illustrated positional and displacement methods, the positional method requires 41 numeric characters while the displacement method can be accomplished by the use of 35 such characters. If numerics are packed two per byte, then the displacement technique will result in 140 bits being required to represent the 1728 points in the example while the positional method would require 164 bits.

The transmission of both positional and displacement information has been applied to various techniques used to compress facsimile information. Another fax compression technique is obtained by assigning predefined codes to different runs of black and white pixels. The codes are then substituted for each sequence of pixel runs encountered in each scanned line. This technique is based upon a modified version of Huffman encoding and is discussed in detail in the next section in this chapter.

5.3 MODIFIED HUFFMAN CODES

The representation of characters and symbols by an appropriate Huffman code is excellent in theory if you desire to have the average number of bits per symbol approach entropy. In practice, however, a number of difficulties can arise when Huffman coding is applied to certain applications, most particularly in the area of facsimile transmission.

When applying Huffman coding to facsimile transmission, each facsimile line can be viewed as consisting of a series of black or white 'runs', each run consisting of a series of similar picture elements. If

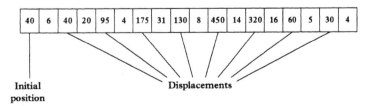

Figure 5.6 Transmitting displacement information. Another relative encoding technique results in the transmission of displacement information

the type of the first run is known, then the type of all successive runs will be known, as black and white runs must alternate.

The probability of occurrence of each run of a given length of pels can be calculated and short code words can be used to represent runs that have a high frequency of occurrence, while longer code words can be used to represent runs that have a low probability of occurrence. In a way similar to the changing of data processing jobs, statistics for the run-length probabilities associated with line scans change on a line-to-line and document-to-document basis. Thus, an optimum or near optimum code for a particular line or document may be far from optimum for a different line or document. A second major problem is the fact that the creation of the Huffman code on a real-time basis requires a large degree of processing power, normally in excess of the capabilities of facsimile machines developed in the early and mid 1980s where the cost of the scanner, transmitter/receiver, central logic and power supply resulted in a machine cost under $1000 to remain competitive.

To reduce some real-time processing requirements, a table look-up approach can be employed. Since CCITT standards require 1728 pels per line, the use of a table look-up technique would require storage for 1728 variable-length locations for each facsimile machine, each location containing a binary code word corresponding to a particular run length. The implementation problems associated with applying the full Huffman coding technique to facsimile applications resulted in the development of one modified Huffman coding scheme more suitable to the hardware cost constraints of the competitive facsimile marketplace.

In the development of a modified Huffman coding scheme for facsimile applications, a change was made which, while only rarely permitting the average symbol length to approach entropy, does permit significant compression while minimizing hardware and processing requirements. Here, the probability of occurrences of different run lengths of picture elements (pels) was calculated for all lengths of white and black runs based upon statistics obtained from the analysis of a group of 11 documents recommended by the CCITT as being typical. To reduce table look-up storage requirements, the Huffman code set was truncated by the creation of a base 64 representation of each run length and the utilization of two code tables to reduce the overall table size in comparison with the table size that would be required if only one table were used (McCullough, 1977).

Based upon the run-length probabilities of 11 typical documents, code tables were developed for run lengths ranging from 1 to 63 pels. Since the probability of occurrence of white runs differs from the frequency of occurrence of black runs, a table must be developed for both runs. This dual table set is listed in Table 5.1 for run lengths ranging from 0 to 63 pels. The codes in this table set represent the least significant digit (LSD) of the code word and are often referred to as the termination code. In order to permit the encoding of runs in excess of 63 pels, a second set of code tables must be employed to handle runs ranging in size from 64 pels to the maximum line scan length of 1728 pels. These codes are listed in Table 5.2. These represent the most significant digit of the code word and are known as the master or make up code. Recognizing that fax systems exist which accommodate larger paper widths, Group 3 fax standards support an option for the addition of 'Extended Make Up Codes'. Those Extended Make Up Codes are listed in Table 5.3 and are equivalent for both black and white runs.

When a run of 63 pels or less is encountered, the appropriate type of LSD code set is accessed to obtain a single base 64 code word. To encode a run of 64 pels or more, two base 64 code words must be used. First, the most significant digit code word is obtained from the MSD code table such that $N \times 64$, $1 \leq N \leq 27$, does not exceed the run length. Here N cannot exceed 27 as when $N = 27$ the maximum of 1728 pels per line occurs for a standard $8\frac{1}{2}$ inch $\times$ 11 inch document. If the fax system has the capability to use a larger page width, it communicates this fact during the call setup process in its Digital Identification Signal (DIS). Next, the difference between the run length and $N \times 64$ is obtained and the least significant digit is accessed from the appropriate LSD code table.

Figure 5.7 shows an example of the table look-up operations for a sample sequence of black and white runs of various pel sizes. In the upper portion of this illustration, the relationship between a series of original video data and its representation in the modified Huffman code is tabulated. Based upon the preceding you can consider the application of the described modified Huffman coding technique as a one-dimensional encoding scheme. This encoding scheme is applied on a line by line basis to each scanned line and represents the compressed horizontal correlation of pels within the same line.

To employ the modified Huffman coding scheme successfully, some rules must be developed and followed to alleviate a number of deficiencies inherent from employing a statistical encoding technique. In such techniques, code words do not contain any inherent positional information which is necessary for synchronization. This can be compensated for by making it a rule that the first run of each line must be a white run, even if it results in a run length of zero. Thereafter, runs must alternate between black runs and white runs. To denote the end of each scan line and the beginning of the next line, a unique line delineation code, called an end-of-line code (EOL), is employed. The EOL code is a unique code word that can never be encountered within a valid line of data. Once each line is encoded, fill bits of 0s may be employed as pad bits prior to transmitting the EOL for timing purposes. Fill bits are variable length strings of 0s inserted between a line of data and an EOL code word. Fill bits are added to insure each line of data, to include Fill and EOL, exceeds

Table 5.1 Least significant digit (terminating) codes for the modified Huffman
process

White run length	Code word	Base 64 representation	Black run length	Code word
0	00110101	0	0	0000110111
1	000111	1	1	010
2	0111	2	2	11
3	1000	3	3	10
4	1011	4	4	011
5	1100	5	5	0011
6	1110	6	6	0010
7	1111	7	7	00011
8	10011	8	8	000101
9	10100	9	9	000100
10	00111	a	10	0000100
11	01000	b	11	0000101
12	001000	c	12	0000111
13	000011	d	13	00000100
14	110100	e	14	00000111
15	110101	f	15	000011000
16	101010	g	16	0000010111
17	101011	h	17	0000011000
18	0100111	i	18	0000001000
19	0001100	j	19	00001100111
20	0001000	k	20	00001101000
21	0010111	l	21	00001101100
22	0000011	m	22	00000110111
23	0000100	n	23	00000101000
24	0101000	o	24	00000010111
25	0101011	p	25	00000011000
26	0010011	q	26	000011001010
27	0100100	r	27	000011001011
28	0011000	s	28	000011001100
29	00000010	t	29	000011001101
30	00000011	u	30	000001101000
31	00011010	v	31	000001101001
32	00011011	w	32	000001101010
33	00010010	x	33	000001101011
34	00010011	y	34	000011010010
35	00010100	z	35	000011010011
36	00010101	A	36	000011010100
37	00010110	B	37	000011010101
38	00010111	C	38	000011010110
39	00101000	D	39	000011010111
40	00101001	E	40	000001101100
41	00101010	F	41	000001101101
42	00101011	G	42	000011011010
43	00101100	H	43	000011011011

Table 5.1 Continued

White run length	Code word	Base 64 representation	Black run length	Code word
44	00101101	I	44	000001010100
45	00000100	J	45	000001010101
46	00000101	K	46	000001010110
47	00001010	L	47	000001010111
48	00001011	M	48	000001100100
49	01010010	N	49	000001100101
50	01010011	O	50	000001010010
51	01010100	P	51	000001010011
52	01010101	Q	52	000000100100
53	00100100	R	53	000000110111
54	00100101	S	54	000000111000
55	01011000	T	55	000000100111
56	01011001	U	56	000000101000
57	01011010	V	57	000001011000
58	01011011	W	58	000001011001
59	01001010	X	59	000000101011
60	01001011	Y	60	000000101100
61	00110010	Z	61	000001011010
62	00110011	*	62	000001100110
63	00110100	#	63	000001100111

the minimum transmission time of a total scanning line established during the initial portion of communications between fax machines. The end result of the incorporation of these rules permits a line format to be defined as shown in Figure 5.8. Once the transmission of a document is completed, this fact is indicated by the originating fax machine sending six EOL code words. The sequence of six EOL code words is referred to as a Return to Control (RTC). Through the incorporation of the modified Huffman coding technique, the transmission time of a typical business document has been reduced to under 20 s at a transmission rate of 9600 bps.

The significance of the reduction becomes apparent when you consider that the resolution of 1780 pels per line and 96 horizontal lines per inch results in a total of 1 410 048 pels for an $8\frac{1}{2} \times 11$ document. Without compression, a transmission time of approximately 2.5 minutes would be required for the data without considering the transmission of the end-of-line codes.

5.4 MODIFIED READ CODING

The modified Huffman code described in the previous section operates on a scanned line-by-line basis. Thus, it does not consider the

Table 5.2 Most significant digit (make up) codes for the modified Huffman process

White run length	Code word	Base 64 representation	Black run length	Code word
64	11011	1	64	0000001111
128	10010	2	128	000011001000
192	010111	3	192	000011001001
256	0110111	4	256	000001011011
320	00110110	5	320	000000110011
384	00110111	6	384	000000110010
448	01100100	7	448	000000110101
512	01100101	8	512	0000001101100
576	01101000	9	576	0000001101101
640	01100111	a	640	0000001001010
704	011001100	b	704	0000001001011
768	011001101	c	768	0000001001100
832	011010010	d	832	0000001001101
836	011010011	e	836	0000001110010
960	011010100	f	960	0000001110011
1024	011010101	g	1024	0000001110100
1088	011010110	h	1088	0000001110101
1152	011010111	i	1152	0000001110110
1216	011011000	j	1216	0000001110111
1280	011011001	k	1280	0000001010010
1344	011011010	l	1344	0000001010011
1408	011011011	m	1408	0000001010100
1472	010011000	n	1472	0000001010101
1536	010011001	o	1536	0000001011010
1600	010011010	p	1600	0000001011011
1664	011000	q	1664	0000001100100
1728	010011011	r	1728	0000001100101
EOL	0000000000001		EOL	000000000001

fact that most fax documents have a high degree of vertical correlation between scanned lines. Recognizing this fact, the Group 3 fax Recommendation permits the one-dimensional modified Huffman code to be extended as an option to a two-dimensional scheme, while the Group 4 Recommendation formally defines the use of a two-dimensional scheme.

The compression scheme optionally used in Group 3 fax is formally referred to as Modified Read Coding (MRC) while the compression scheme used in Group 4 fax is referred to as Modified Modified Read Coding (MMRC). Both compression schemes are relative addressing methods which use the vertical correlation between lines to obtain a higher compression ratio than is normally obtainable through the exclusive use of modified Huffman coding.

Vertical correlation is based upon the fact that a fax system stores

Table 5.3 Extended Make Up codes

White or black run length	Code word
00000001000	1792
00000001100	1856
00000001101	1920
000000010010	1984
000000010011	2048
000000010100	2112
000000010101	2176
000000010110	2240
000000010111	2304
000000011100	2368
000000011101	2432
000000011110	2496
000000011111	2560

Original video data	Modified Huffman code base 64 representation	Modified Huffman code base 64 representation	
		MSD	LSD
5 black pels	5 (black)	NA	0011
17 white pels	h (white)	NA	101011
32 black pels	w (black)	NA	000001101010
32 white pels	w (white)	NA	00011011
728 black pels	bφ (black)	0000001001011	00000010111
1728 white pels	rφ (white)	010011011	00110101
64 black pels	1φ (black)	0000001111	0000110111
55 white pels	T (white)	NA	01011000
1028 white pels	g2 (black)	011010101	0111

Figure 5.7 Encoding using the modified Huffman code. By a sequence of tabular references for black and white runs the modified Huffman code is constructed

each pel of two or more scanning lines. The first scanning line in memory is known as the reference line and the location of each black pel of this line functions as a reference for the next scanned line. That line is known as the coding line.

Figure 5.9 illustrates an example of changing pels in a reference

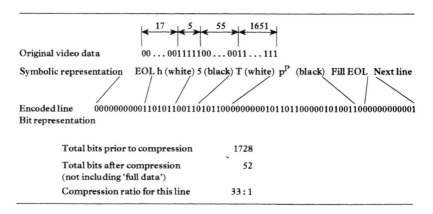

Figure 5.8 Rules define line format. To denote the beginning and end of each scan line an end-of-line code (EOL) is employed

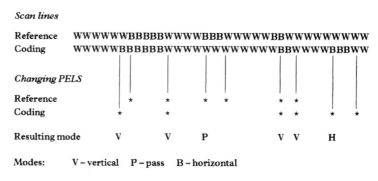

Figure 5.9 Forming a modified read code by recognizing changing pels

and coding line as well as when vertical (V), pass (P) and horizontal (H) compression is used to encode the data in the coding line. In examining Figure 5.9, note that a changing pel on the reference and coding lines represents a position where a pel changes its color, either from white (W) to black (B) or from black to white.

If a pair of changing pels on both the reference and coding lines match or are within three pels of one another, vertical mode coding is used. This is indicated by a 'V' on the 'Resulting Mode' line at the bottom of Figure 5.9. Seven codes are used to indicate modified read vertical coding since the pel pairs can vary from +3 to −3 positions between adjacent lines to include being directly under one another. Table 5.4 lists the modified read vertical codes.

When the distance between changing pels exceeds three, one of two additional coding modes are used. If changing pels are limited to the reference line pass, mode coding indicated by a 'P' in the 'Resulting Mode' line in Figure 5.9 is used. A code of 0001 is used to indicate the skipping of two changing pels on the reference line. If two chang-

Table 5.4 Modified read vertical codes

'Changing' pel position between reference and coding lines	Vertical code
−3	0000010
−2	000010
−1	010
0	1
+1	001
+2	000011
+3	0000011

ing pels are on the coding line, horizontal mode coding indicated by an 'H' in the 'Resulting Mode' line in Figure 5.9 is used. Here a code of 001 indicates horizontal mode coding as well as the occurrence of modified Huffman coding for the next two runs.

Since a received fax signal can be distorted by the generation of noise resulting from lightning, telephone line switches and machinery, errors will result in the pattern of black and white dots on the received document. To prevent this incorrect pattern from propagating vertically down the page the use of modified read requires the periodic use of modified Huffman coding. At normal resolution this occurs every second line and is referred to as having a K factor of 2. At fine resolution the use of modified Huffman occurs every fourth line, resulting in a K factor of 4. Since the Group 4 standard blocks data and includes error detection and correction via retransmission, the Modified Modified Read (MMR) read technique used by Group 4 fax does not use a periodic modified Huffman code. Thus, MMR has a K factor of infinity. Figure 5.10 illustrates Group 3 and Group 4 K factors.

The actual time required to transmit a page using modified read coding has the same dependencies as the use of modified Huffman coding. That is, it depends upon the composition of black and white information on a page as well as the operating rate of the modem used by the fax system.

The previously described modified read code was designed for use by fax systems that are mechanically limited in their ability to rapidly advance paper when a redundant all white or all black portion of a page is transmitted. In fact, under the Group 3 Recommendation a fax machine is required to transmit Fill bits when a large portion of a page is white and the transmission of a compressed modified Huffman code to represent the line would require less time than the minimum transmission time of a total scanning line established when one fax machine calls another and exchanges pre-message control procedures. Under Group 4 fax a modified modified read code (MMRC)

```
K = 2 (Standard Resolution Group 3)
            MH coding          ...
            2-D coding         ...
            MH coding          ...
            2-D coding         ...
                  ⋮

K = 4 (Fine Resolution Group 3)
            MH coding          ...
            2-D coding         ...
            2-D coding         ...
            2-D coding         ...
            MH coded all white ...

K = ∞ (Modified Modified Read Group 4)
            First line all white...
            2-D coding         ...
            2-D coding         ...
                  ⋮
            2-D coding         ...
```

Figure 5.10   K factors

is used. This code represents an adaptation of the MR code, with a K factor of infinity; however, the International Standards Organization (ISO) and the CCITT (now the ITU) formed a committee known as the Joint Bi-level Image Group (JBIG) to develop a new standard for black and white transmission.

DICTIONARY BASED STRING COMPRESSION

Dictionary based string compression has its foundation originating in work on information storage and retrieval during the 1960s. At that time a data structure call the TRIE, a name derived from the word 'reTRIEval', was developed as a mechanism to represent learned strings. In 1975, two professors in the United Kingdom, Mayne and James, published a paper on string learning based upon the use of an internal dictionary, with an intended application to data compression.

In 1977 and 1978, two professors at the Technion in Haifa, Israel, Jacob Ziv and Abraham Lempel, published what are now considered landmark papers which established the mathematical basis for the development of a series of string learning algorithms. The paper published in 1977 (Ziv and Lempel, 1977) resulted in a dictionary based string compression technique now commonly referred to as LZ77. The paper published in 1978 (Ziv and Lempel, 1978) resulted in a dictionary based string compression technique now commonly referred to as LZ78. Although both papers are now considered landmarks similar to Huffman's 1972 paper on character redundancy, both papers were conspicuous by their initial lack of attention until 1984. In that year, the Englishman Terry Welch published a detailed algorithmic description of the application of Lempel and Ziv's prior publications to disk data compression (Welch, 1984). The effort of Welch in modifying the Lempel and Ziv algorithm resulted in what is now referred to as the LZW algorithm which has become a widely used data compression technique employed in several file archiving and compression performing programs, used to compress images by CompuServe's GIF file format, and many other applications. In fact, the V.42 bis modem data compression is based upon Welch's LZW technique.

In an attempt to improve upon dictionary based string compression, several multilevel techniques were developed which use different dictionary based string compression techniques as a base. In this chapter we will first examine mathematically why string compression can provide a higher level of efficiency than single character based compression methods. Once this is accomplished, we will turn our attention to the Lempel and Ziv 1977 and 1978 dictionary string compression techniques and the work of Welch which resulted in V.42 bis modem compression. In doing so, when appropriate, we will examine several multilevel compression techniques which use dictionary based string compression as a foundation.

6.1 STRING COMPRESSION EFFICIENCIES

To understand why string compression provides more efficiency than character-based compression methods, let us first examine the entropy of a two-character character set as the probability of occurrence of each character changes. If X_1 and X_2 represent our two-character character set, let us assume X_1 occurs 90% of the time and X_2 occurs 10% of the time. Then, the entropy is:

$$H = -[-0.9 \log_2 0.9 + 0.1 \log_2 0.1]$$

$$H = 0.9*(0.15) + 0.1*(3.3) = 0.47 \text{ bits.}$$

Now suppose X_1 occurs 80% of the time, while X_2 occurs 20% of the time. The entropy then becomes:

$$H = -[-0.8 \log_2 0.8 + 0.2 \log_2 0.8]$$

$$H = 0.8*(0.32) + 0.2*(2.31) = 0.72 \text{ bits.}$$

We can continue varying the frequency of occurrence of each character in our character set. Table 6.1 lists the resulting entropy for the two-character character set as X_1 decreased in frequency of occurrence from 90% to 50%, while X_2 increased from 10% to 50%.

Table 6.1 Entropy versus frequency of occurrence

Frequency of occurrence		
X_1	X_2	Entropy in bits
90	10	0.47
80	20	0.72
70	30	0.88
60	40	0.97
50	50	1.00

In examining the entropy column contained in Table 6.1, note that the greater the inequality in probability of occurrence the lower the theoretical average number of bits required to represent each character. This indicates that a compression method that recognizes common strings, which on a character basis constitute a large percentage of characters, offers the potential for representation by a low number of bits. Let us verify this observation by examining the effect of encoding groups of characters in place of single characters.

Suppose the frequency of occurrence of X_1 is 80% and X_2 is 20%. As previously calculated, the entropy for this character set is 0.72 bits. Using either Huffman or Shannon–Fano encoding would result in assigning a 0 to one character and a 1 to the other character. Here the average length is one bit per character, which is 28% greater than the entropy. If we combine the characters in pairs we obtain four possible character sets whose resulting probabilities are indicated in Table 6.2.

Figure 6.1 illustrates the construction of a Huffman code for the character pairs listed in Table 6.2 based upon their probabilities of occurrence. Note that their average bit length is:

$$L = 1*0.64 + 2*0.16 + 3*0.16 + 3*0.04 = 1.56 \text{ bits.}$$

Since each character pair represents two characters, the average bit length per pair must be divided by 2 to find the average bit length per character. Doing so results in an average of 0.78 bits per character when characters are paired.

Now let us proceed one step further and encode three characters

Table 6.2 Character pair probabilities

Character pair	Probability of occurrence
X_1X_1	$0.8*0.8 = 0.64$
X_1X_2	$0.8*0.2 = 0.16$
X_2X_1	$0.2*0.8 = 0.16$
X_2X_2	$0.2*0.2 = 0.4$

Huffman code	Character pair	Frequency of occurrence			
1	X_1X_1	0.64		1 0.36	1 1.0
000	X_1X_2	0.16		0	0
011	X_2X_1	0.16	1 0.2		
010	X_2X_2	0.04	0		

Figure 6.1 Huffman encoding of character pairs listed in Table 6.2

at a time. Table 6.3 lists the eight distinct tri-character combinations and their resulting probability of occurrence.

Figure 6.2 illustrates the construction of the Huffman code for the tri-character groupings based upon their probability of occurrence listed in Table 6.3. Now let us examine the average bit length per tri-character group. The average bit length is:

$$L = 1*0.512 + 3*0.128*3 + 5*0.032*3 + 5*0.008 = 2.184 \text{ bits.}$$

Since the average number of bits per tri-character group is 2.184, dividing that number by 3 results in the average number of bits per character, which is 0.728. From the preceding computations, we note that as we combine more and more characters the average character bit length approaches entropy. In fact, the average length will equal entropy if all of the probabilities of occurrence are inverse powers of 2. As we combine additional characters into groups of strings their probabilities of occurrence approach inverse powers of 2, which is the reason why a compression technique based upon encoding strings should be more efficient than one based upon encoding characters

Table 6.3 Tri-character probabilities

Character pair	Probability of occurrence
$X_1X_1X_1$	$0.8*0.8*0.8 = 0.512$
$X_1X_1X_2$	$0.8*0.8*0.2 = 0.128$
$X_1X_2X_1$	$0.8*0.2*0.8 = 0.128$
$X_1X_2X_2$	$0.8*0.2*0.2 = 0.032$
$X_2X_1X_1$	$0.2*0.8*0.8 = 0.128$
$X_2X_1X_2$	$0.2*0.8*0.2 = 0.032$
$X_2X_2X_1$	$0.2*0.2*0.8 = 0.032$
$X_2X_2X_2$	$0.2*0.2*0.2 = 0.008$

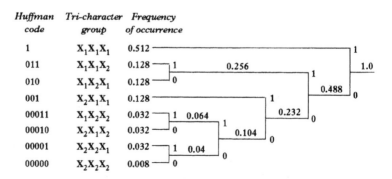

Figure 6.2 Huffman encoding of tri-character groupings listed in Table 6.3

and was one of the reasons for the selection of the Lempel–Ziv technique for use in the CCITT V.42 bis data-compression recommendation.

6.2 LZ77 COMPRESSION

The first in a pair of landmark papers covering dictionary based string compression techniques, 'A universal algorithm for sequential data compression', was written by Jacob Ziv and Abraham Lempel, members of the Department of Electrical Engineering at the Technion in Haifa, Israel, during 1975. The publication of the paper in *IEEE Transactions on Information Theory* occurred in May 1977 and although a family of adaptive dictionary encoding techniques based upon the algorithm published in 1977 were originally called Ziv–Lempel coding, for reasons unknown other than perhaps a historical mistake the reversal of the initials of the last names of the authors resulted in the techniques being abbreviated as LZ coding. Thus, the first algorithm which was defined in their 1977 paper is referred to as LZ77 compression.

The LZ77 algorithm, as well as subsequent dictionary based string compression techniques which use the initial paper published in 1977 as a foundation, assumes no prior knowledge of the source to be compressed. This resulted in the term 'universal algorithm' placed in the title of the paper.

Encoding

The basic LZ77 algorithm subdivides a data source into two parts. The first part represents previous encoded output, which becomes the dictionary. The second part represents text to be encoded.

Figure 6.3 illustrates a block diagram of the LZ77 data flow. Initially, both the lookahead buffer and dictionary are empty and represent a sliding window of n characters through which the source data flows into the lookahead buffer and from that buffer into the dictionary which represents the previously encoded part of the sliding

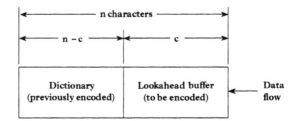

Figure 6.3 LZ77 sliding window. Under the LZ77 algorithm data entering the lookahead buffer is encoded based upon entries in a dictionary consisting of previously encoded information. Since the lookahead buffer and dictionary "slide" over the data, the algorithm functions as a sliding window

window. As originally defined in the referenced paper, the encoding algorithm checks the symbols in the lookahead buffer to determine if they match a string in the dictionary. If a match occurs, the algorithm creates a token that defines the location and length of the string in the dictionary. If no match occurs, the symbol is passed through unencoded. Thus, the output of the LZ77 algorithm consists of a series of tokens and symbols.

Under the original LZ77 algorithm the token was referred to as a codeword which was composed of three items:

1. An offset from a string in the dictionary that matches a string in the lookahead buffer.
2. The length of the string.
3. The first symbol in the lookahead buffer that follows the phrase.

The operation of the LZ77 algorithm commences with the first $n - c$ characters of the dictionary set to spaces, while the first c characters of text are loaded into the lookahead buffer. Next, the first $n - c$ characters in the dictionary are searched to locate the longest match with entries in the lookahead buffer. During this attempt to match strings between the dictionary and the lookahead buffer, the matching process can overlap from the dictionary into the buffer.

To illustrate an example of the LZ77 sliding window compression process, consider the top portion of Figure 6.4 which indicates the current state of an LZ77 sliding window of 15 character positions consisting of a dictionary of 11 positions and a lookahead buffer of four positions. The longest match for the lookahead buffer is the string sequence 'aba' which commences at character position 10. The longest match is then coded into a token of the form $\langle i, j, k \rangle$, where i is the offset of the longest match in the dictionary with respect to the beginning of the dictionary, j is the length of the match, and k is the first character in the lookahead buffer that follows the string match. Thus, in this example, the output token would be $\langle 10, 3, c \rangle$. Next, the window slides $j+1$ positions to the right to provide new entries into the lookahead buffer and the trailing dictionary. Once the sliding window operation is completed, the next coding step will occur. If there is no match between the dictionary and the contents of the lookahead buffer, the contents of the lookahead buffer can be encoded as a single character at a time using an offset location and a string length of zero. Although doing so is obviously not efficient, it ensures that the algorithm can encode any message.

Decoding

The power of the LZ77 algorithm is best illustrated by its decoding or decompression capability. Turning our attention to Figure 6.4b, let's

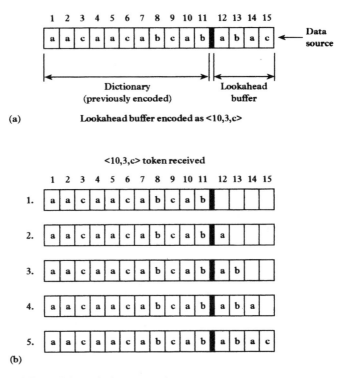

Figure 6.4 LZ77 sliding window operation

examine how the algorithm decodes or decompresses a received token.

In (1) the token ⟨10,3,c⟩ was received with the dictionary holding the previously decoded text but having no knowledge of the contents of the lookahead buffer. Since the first part of the token (10) indicates the position within the dictionary from which a match began, the decoder knows that the first character in the lookahead buffer is 'a' and places it in position 12. This is indicated in (2). Next, since the second entry in the token (3) indicates a matched string length of three, the decoder knows that position 11 in the dictionary had the character 'b' and places it in position 13 in the lookahead buffer. This is illustrated in (3). Since only two characters were placed in the lookahead buffer when a match of three characters occurred, the decoder starts at position 10 and picks the third character from that position. Note that this technique permits decoding to be based on encoding across the dictionary/lookahead buffer boundary. Since the character in position 12 was an 'a', it is entered into position 14 as illustrated in (4). Now that the three characters in the string are decoded, the third part of the token which represents the first symbol in the original lookahead buffer that followed the encoded string is placed in the decoded or received lookahead buffer. This is indicated in (5) and results in the reconstruction of the data represented by the token.

Coding problems

There are several problems associated with the implementation of the original LZ77 algorithm that precludes its use as a practical method for implementing dictionary based string compression. First, each encoding step requires a constant amount of time which can become very large when the dictionary is large. For example, if the dictionary is $n - c$ characters in length and the lookahead buffer is c characters in length then $(n - c) \times c$ character comparisons are required per encoded string. Since the typical encoded string is only a few characters in length, a substantial number of comparisons is required for each character coded. This means you must carefully consider the data structure used as some structures will permit faster searching than others.

A second problem which is related to the first is the relationship between the length of the dictionary and the length of the lookahead buffer. As you increase the length of the dictionary, there is a greater probability that a string match will occur between the dictionary and the lookahead buffer. Unfortunately, any increase in the length of the dictionary results in an increase in the number of string comparisons that have to be performed against the lookahead buffer. A third problem is the use of a triplet based token, with the character following the encoded string included in the token. The use of the following character within the token is often wasteful, since it can often be included as part of a subsequent token. Another associated problem with respect to the length of the dictionary concerns the resulting size of a token required to indicate an encoded string. For example, assume the dictionary has a length of 4096 characters and the maximum length of a substring is 16 character positions. Then a token would require 12 bits for the dictionary position (i), four bits for the length of the string (j), and eight bits to encode either the character following an encoded substring or a single eight-bit character for which there was no entry in the dictionary. While a token with a length of 24 bits is reasonable for encoding strings of three or more characters, it results in a considerable amount of overhead when a character in the lookahead buffer is not in the dictionary or when two or fewer characters in a substring are encoded via a 24-bit token.

6.3 LZSS COMPRESSION

Recognizing some of the previously mentioned problems associated with the LZ77 algorithm, a modification was proposed (Storer and Szymanski, 1982) which became known as LZSS, with the pair of Ss added to LZ in recognition of the authors of the article.

Under the LZSS encoding method, a sliding window of n characters, with c characters forming a lookahead buffer and $n - c$ characters forming the dictionary of previously encoded strings, is used in the same manner as the LZ77 algorithm. However, the composition of the token and the data structure was modified.

LZSS modifications

Concerning the use of tokens, under the LZ77 algorithm a triplet consisting of an offset, string length and character following the string was used. This resulted in LZ77 encoding having to alternate pointers with plain characters regardless of the composition of the input text. Under LZSS tokens and plaintext characters can be freely intermixed. To accomplish this, each token or plaintext character is preceded by a single bit prefix. A prefix bit set to '1' indicates that a non-coded 8-bit symbol follows. A prefix bit set to '0' indicates that a dictionary reference consisting of a two-part token follows. The first part of the token is the offset, while the second part is the string length. Figure 6.5 illustrates the format of the LZSS output data stream.

In examining the dictionary reference format illustrated in Figure 6.5, note that the number of bits used for the offset and string length depend upon the length of the dictionary and the length of the lookahead buffer. Similar to LZ77, increasing the length of the dictionary and/or the length of substrings to be matched increases compression efficiency to the detriment of processing time.

A second change between LZ77 and LZSS involves the data structure used under the latter algorithm. Although LZSS continues to store data in a contiguous sliding window, as each substring or non-encoded character is passed out of the lookahead buffer and into the dictionary LZSS adds the entries into a tree structure. By using a binary tree to store processed data, the time required to locate the longest matching string in the tree is reduced from the time required to perform the number of searches equal to the product of the dictionary length and string length to the time required to perform the number of searches equal to the base 2 logarithm of the dictionary length multiplied by the string length. This means that a doubling of the length of the dictionary under LZSS only results in a small increase in processing time. In comparison, doubling the length of the dictionary under LZ77 doubles the processing time. Readers are referred to Section 6.6 for a description of the use of the trie data structure and its use to store the contents of a dictionary of strings.

```
plaintext              ⟨1⟩⟨eight bit symbol⟩
dictionary reference   ⟨0⟩⟨offset⟩⟨string length⟩
```

Figure 6.5 LZSS output data stream format

C language program

Figure 6.6 lists the C language code of the public domain program LZSS.C written by Haruhiko Okumura. This program was developed based upon Storer and Szymanski's slightly modifed version of the LZ77 algorithm. An earlier version of this program which did not incorporate the use of a binary tree was rewritten using Turbo Pascal and assembly language by Kazuhiko Miki and formed the basis for the archiver and compression performing program he developed and named LARC. In the LZSS.C program a ring buffer of 4096 characters is used and the upper limit for a string match is set to 18 characters. The program reports its progress each time the text input exceeds multiples of 1024 characters.

6.4 LZ78 COMPRESSION

The Lempel–Ziv dictionary based string compression technique now referred to as LZ78 derives its name from the year of publication of the authors' second, now classical, work (Ziv and Lempel, 1978). The algorithm which forms the basis for the LZ78 technique represents a completely new approach to dictionary based string compression. Instead of using pointers to reference substrings in the lookahead buffer that were previously processed and are now in the dictionary, the LZ78 technique subdivides text into phrases, a process referred to as phrasing, where each phrase represents the longest matching phrase previously encountered plus one character. Next, each phrase is encoded as an index to its prefix, plus the one additional character. The new phrase is then added to the list of phrases stored in a dictionary.

The development of the LZ78 algorithm overcame one key deficiency of the LZ77 algorithm. That algorithm uses a relatively small window of previously processed data. This means that many valuable dictionary entries will be discarded as they slide out of the dictionary as new entries are processed. For example, if this book was being compressed using the LZ77 algorithm, the phrase 'Huffman encoding' which appeared a few chapters back and which will appear later in this chapter would almost certainly have been removed from the dictionary. If popular phrases are retained, the resulting efficiency of the algorithm is enhanced, an area LZ78 was developed to address.

Operation

The best way to illustrate the operation of the LZ78 algorithm is via an example, so let's do so. Since we require a string to illustrate the LZ78 process, let's use the famous saying of Fred Flintstone, 'aaa-

baaadaabaado'. Figure 6.7 illustrates the decomposition of the string
into eight phrases or substrings. Each phrase is encoded as a phrase
that previously occurred, followed by a trailing character. The two-
character token uses a phrase code of zero to indicate an empty string
when the input to the encoder is encountered for the first time and
as a result is not currently in the dictionary. Each token of the form
(i, n) can be viewed as indicating to the decoder 'copy phrase i followed
by character n'.

Initially the LZ78 dictionary consists of the null string. Thus, the
first character in the string to be processed is not in the dictionary
and is encoded as (0,a). Next, since 'a' is in the dictionary as phrase
number 1, the substring 'aa' can be encoded as a reference to phrase
or substring number 1. Doing so results in the output (1,a) to represent the substring 'aa'. Since 'b' is not in the dictionary, it is encoded as (0,b). Next, the phrase or substring 'aaa' can be encoded as a
reference to string 2. Doing so results in the output token of (2,a).
Since 'd' was not previously encoded, the token (0,d) is used to represent that substring. Next, the string 'aab' represents string 2 followed by the character 'b' and is output as the token (2,b). The following three-character string 'aad' represents the previously encoded
phrase 2 followed by the character d. Thus, its output token becomes
(2,d). Finally, the trailing 'o' which represents a new string of one
character is output as the token (0,o).

Coding tradeoffs

Under the LZ78 algorithm there was no restriction on how far back
a token's pointer can be used to reference previously processed data.
However, in reality the dictionary cannot be allowed to grow indefi-
nitely as at some point all available memory will be used. Thus, a
'clear code' is required to indicate the flushing of prior passed phrases
or infrequently encountered phrases to enable the process either to
start anew or perhaps to use memory more expediently. Another
problem associated with the implementation of the LZ78 algorithm
concerns the length of the token, which expands as more phrases are
added to the dictionary. When n phrases are passed, a pointer of log
n bits is required. For example, a dictionary containing 256 phrases
requires an 8-bit phrase pointer, while a dictionary capable of holding
8192 phrases requires a 14-bit phrase pointer. Since each phrase is
essentially unlimited in its length, unless limits are placed on the
phrase length memory allocation problems can easily occur. For
example, consider the use of a 16-bit phrase pointer capable of supporting 65 536 phrases in a dictionary. As the average length of a
phrase expands from five to ten bytes the amount of memory required
to store just the dictionary increases by 327 650 bytes.

Although the LZ78 algorithm operates reasonably well on large files

```
/*******************************************************
        LZSS.C -- A Data Compression Program
        (tab = 4 spaces)
********************************************************
        4/6/1989 Haruhiko Okumura
        Use, distribute, and modify this program freely.
        Please send me your improved versions.
                PC-VAN          SCIENCE
                NIFTY-Serve     PAF01022
                CompuServe      74050,1022
********************************************************/
#include <stdio.h>
#include <stdlib.h>
#include <string.h>
#include <ctype.h>

#define N           4096 /* size of ring buffer */
#define F           18   /* upper limit for match_length */
#define THRESHOLD   2     /* encode string into position and length if
                            match_length is greater than this */
#define NIL         N    /* index for root of binary search trees */

unsigned long int
        textsize = 0,     /* text size counter */
        codesize = 0,     /* code size counter */
        printcount = 0;   /* counter for reporting progress every 1K bytes */
unsigned char
        text_buf[N + F - 1];    /* ring buffer of size N, with extra F-1 bytes to
                                   facilitate string comparison of longest match. */
int  match_position, match_length,  /*These are set by the InsertNode() procedure.*/
        lson[N + 1], rson[N + 257], dad[N + 1];  /* left & right children & parents --
                                        These constitute binary search trees. */
FILE *infile, *outfile;       /* input & output files */

void InitTree(void)          /* initialize trees */
{
        int  i;

        /* For i = 0 to N - 1, rson[i] and lson[i] will be the right and left children of
            node i.  These nodes need not be initialized. Also, dad[i] is the parent of
            node i.  These are initialized to NIL (= N), which stands for 'not used.'  For
            i = 0 to 255, rson[N + i + 1] is the root of the tree for strings that begin
            with character i.  These are initialized to NIL. Note there are 256 trees. */

        for (i = N + 1; i <= N + 256; i++) rson[i] = NIL;
        for (i = 0; i < N; i++) dad[i] = NIL;
}

void InsertNode(int r)
        /* Inserts string of length F, text_buf[r..r+F-1], into one of the trees
            (text_buf[r]'th tree) and returns the longest-match position and length via
            the global variables match_position and match_length. If match_length = F,
            then removes the old node in favor of the new one, because the old one will be
            deleted sooner.
            Note r plays double role, as tree node and position in buffer. */
{
        int  i, p, cmp;
        unsigned char  *key;

        cmp = 1;  key = &text_buf[r];  p = N + 1 + key[0];
        rson[r] = lson[r] = NIL;  match_length = 0;
        for ( ; ; ) {
                if (cmp >= 0) {
```

Figure 6.6 Program listing of LZSS.C

```
                    if (rson[p] != NIL) p = rson[p];
                    else {  rson[p] = r;  dad[r] = p;  return;  }
            } else {
                    if (lson[p] != NIL) p = lson[p];
                    else {  lson[p] = r;  dad[r] = p;  return;  }
            }
            for (i = 1; i < F; i++)
                    if ((cmp = key[i] - text_buf[p + i]) != 0)  break;
            if (i > match_length) {
                    match_position = p;
                    if ((match_length = i) >= F)  break;
            }
        }
    dad[r] = dad[p];  lson[r] = lson[p];  rson[r] = rson[p];
    dad[lson[p]] = r;  dad[rson[p]] = r;
    if (rson[dad[p]] == p) rson[dad[p]] = r;
    else                   lson[dad[p]] = r;
    dad[p] = NIL;    /* remove p */
}

void DeleteNode(int p)     /* deletes node p from tree */
{
    int  q;

    if (dad[p] == NIL) return;  /* not in tree */
    if (rson[p] == NIL) q = lson[p];
    else if (lson[p] == NIL) q = rson[p];
    else {
        q = lson[p];
        if (rson[q] != NIL) {
            do {  q = rson[q];  } while (rson[q] != NIL);
            rson[dad[q]] = lson[q];  dad[lson[q]] = dad[q];
            lson[q] = lson[p];  dad[lson[p]] = q;
        }
        rson[q] = rson[p];  dad[rson[p]] = q;
    }
    dad[q] = dad[p];
    if (rson[dad[p]] == p) rson[dad[p]] = q;  else lson[dad[p]] = q;
    dad[p] = NIL;
}

void Encode(void)
{
    int  i, c, len, r, s, last_match_length, code_buf_ptr;
    unsigned char  code_buf[17], mask;

    InitTree();        /* initialize trees */
     code_buf[0] = 0;  /* code_buf[1..16] saves eight units of code, and
            code_buf[0] works as eight flags, "1" representing that the unit
            is an unencoded letter (1 byte), "0" a position-and-length pair
            (2 bytes).  Thus, eight units require at most 16 bytes of code. */
    code_buf_ptr = mask = 1;
        s = 0;  r = N - F;
    for (i = s; i < r; i++) text_buf[i] = ' '; ;   /* Clear the buffer with any
                                        character that will appear often. */
    for (len = 0; len < F && (c = getc(infile)) != EOF; len++)
        text_buf[r + len] = c;  /*Read F bytes into the last F bytes of the buffer */
    if ((textsize = len) == 0) return;   /* text of size zero */
     for (i = 1; i <= F; i++) InsertNode(r - I);  /* Insert the F strings, each of
                which begins with one or more 'space' characters.  Note the order in
                which these strings are inserted.  This way, degenerate trees will be
                less likely to occur. */

     InsertNode(r);     /* Finally, insert the whole string just read.  The global
                    variables match_length and match_position are set. */
        do {
```

```
            if (match_length > len) match_length = len;  /* match_length may be
                            spuriously long near the end of text. */
        if (match_length <= THRESHOLD) {
            match_length = 1;       /* Not long enough match.  Send one byte. */
            code_buf[0] |= mask; /* 'send one byte' flag */
            code_buf[code_buf_ptr++] = text_buf[r];  /* Send uncoded. */
        } else {
            code_buf[code_buf_ptr++] = (unsigned char) match_position;
            code_buf[code_buf_ptr++] = (unsigned char)
                (((match_position >> 4) & 0xf0)
              | (match_length - (THRESHOLD + 1)));  /* Send position and
                        length pair. Note match_length > THRESHOLD. */
        }
        if ((mask <<= 1) == 0) {  /* Shift mask left one bit. */
            for (i = 0; i < code_buf_ptr; i++)  /* Send at most 8 units of */
                putc(code_buf[i], outfile);     /* code together */
            codesize += code_buf_ptr;
            code_buf[0] = 0;  code_buf_ptr = mask = 1;
        }
        last_match_length = match_length;
        for (i = 0; i < last_match_length &&
                (c = getc(infile)) != EOF; i++) {
            DeleteNode(s); /* Delete old strings and */
            text_buf[s] = c;      /* read new bytes */
            if (s < F - 1) text_buf[s + N] = c;    /* If the position is near
                    the end of buffer, extend the buffer to make
        string comparison easier. */
            s = (s + 1) & (N - 1); r = (r + 1) & (N - 1);  /* Since this is a
                        ring buffer, increment the position modulo N. */
            InsertNode(r); /* Register the string in text_buf[r..r+F-1] */
        }
        if ((textsize += i) > printcount) {
            printf("%12ld\r", textsize);  printcount += 1024;   /* Reports progress
                        each time the textsize exceeds multiples of 1024. */
        }
        while (i++ < last_match_length) {              /* After the end of text, */
            DeleteNode(s);                 /* no need to read, but */
            s = (s + 1) & (N - 1);  r = (r + 1) & (N - 1);
            if (--len) InsertNode(r);            /* buffer may not be empty. */
        }
    } while (len > 0); /* until length of string to be processed is zero */
    if (code_buf_ptr > 1) {  /* Send remaining code. */
        for (i = 0; i < code_buf_ptr; i++) putc(code_buf[i], outfile);
        codesize += code_buf_ptr;
    }
    printf("In : %ld bytes\n", textsize);/* Encoding is done. */
    printf("Out: %ld bytes\n", codesize);
    printf("Out/In: %.3f\n", (double)codesize / textsize);
}

void Decode(void)        /* Just the reverse of Encode(). */
{
    int  i, j, k, r, c;
    unsigned int  flags;

    for (i ='0; i < N - F; i++) text_buf[i] = ' ';
    r = N - F;  flags = 0;
    for ( ; ; ) {
        if (((flags >>= 1) & 256) == 0) {
            if ((c = getc(infile)) == EOF) break;
            flags = c | 0xff00; /* uses higher byte cleverly to count eight */
        }
        if (flags & 1) {
            if ((c = getc(infile)) == EOF) break;
```

Figure 6.6 *Continued*

```
                    putc(c, outfile);  text_buf[r++] = c;  r &= (N - 1);
                } else {
                    if ((i = getc(infile)) == EOF) break;
                    if ((j = getc(infile)) == EOF) break;
                    i |= ((j & 0xf0) << 4);  j = (j & 0x0f) + THRESHOLD;
                    for (k = 0; k <= j; k++) {
                        c = text_buf[(i + k) & (N - 1)];
                        putc(c, outfile);  text_buf[r++] = c;  r &= (N - 1);
                    }
                }
            }
        }
    }

int main(int argc, char *argv[])
{
    char  *s;

    if (argc != 4) {
        printf("''lzss e file1 file2' encodes file1 into file2.\n"
               "'lzss d file2 file1' decodes file2 into file1.\n");
        return EXIT_FAILURE;
    }
    if ((s = argv[1], s[1] || strpbrk(s, "DEde") == NULL)
     || (s = argv[2], (infile  = fopen(s, "rb")) == NULL)
     || (s = argv[3], (outfile = fopen(s, "wb")) == NULL)) {
        printf("??? %s\n", s);  return EXIT_FAILURE;
    }
    if (toupper(*argv[1]) == 'E') Encode();  else Decode();
    fclose(infile);  fclose(outfile);
    return EXIT_SUCCESS;
}
```

Figure 6.6 *Continued*

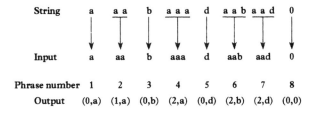

Figure 6.7 LZ78 coding example

using a phrase pointer of 12 to 14 bits, its efficiency is not the best when applied against small files. In addition, its memory requirements for a reasonably sized dictionary preclude its use in devices, such as modems, where cost is a significant manufacturing consideration and total buffer memory is usually under a few kilobytes. However, the LZ78 algorithm formed the basis for a future modification which resulted in a series of dictionary based string compression techniques that gained widespread popularity in their use in a large number of hardware and software compression performing products. The major change to the LZ78 algorithm was postulated by the Englishman Terry Welch, whose efforts are recognized by the referral of the resulting algorithm as LZW compression.

6.5 LZW COMPRESSION

The paper published by Ziv and Lempel in 1978 was similar to their paper published a year earlier in that both were presented in a very abstract mathematical form, leaving the details of their implementation and optimization to others. Among persons who made changes to the basic LZ77 and LZ78 algorithms that resulted in a significant contribution to the operation of dictionary based string compression was Terry Welch.

In 1984 Welch's article 'A technique for high-performance data compression' (Welch, 1984) was published in *IEEE Computer*. In the technique described in that article, some basic problems associated with the LZ78 algorithm were eliminated. One problem was the processing of new symbols which required the encoder to generate tokens for many symbols as the dictionary began to fill and periodically as new symbols were encountered in a data stream. Another problem was the basic format of an LZ78 token in which the inclusion of a character following a prior phrase reference was often wasteful and which LZW eliminates. This is accomplished, as we will shortly note, by the LZW algorithm initializing the list of phrases to include every character in the character set.

Operation

The LZW algorithm initially considers the character set as 256 individual string table entries whose codes range from 0 to 255. Then, the algorithm operates upon the next character in the input string of characters as follows:

1. If the character is in the table, get the next character.
2. If the character is not in the table, output the last known string's encoding and add the new string to the table.

Under the LZW algorithm, characters from the input data source are read and used to progressively form larger and larger strings until a string is formed that is not in the dictionary. When this occurs, the last known string's encoding is output and the new string is added to the table. The beauty about this technique is that the compressor and expander know that the initial string table consists of the 256 characters in the character set. Since the algorithm uses a numeric code as a token to indicate the position of a string in the dictionary's string table, this technique minimizes the token length as the dictionary begins to fill. To illustrate the simplicity of the operation of the LZW algorithm, let's view an example of its use.

Let's call the string previously read the prefix of the output string. Similarly, the last byte read becomes the suffix, where:

$$\text{prefix} + \text{suffix} = \text{new string}$$

Once a new string is formed the suffix becomes the prefix (prefix = suffix).

Initially, each character in the character set is assigned a code value equivalent to its character code. Thus, in ASCII 'a' would have the code value 97, 'b' would have the code value 98, and so on. Let's assume the input string is 'ababc. . .'.

The first operation assumes that the prefix has a value of the null string which we will indicate as the symbol 'Δ'. Thus, the first operation results in the addition of the null string to 'a' which forms the new string 'a'. Since 'a' is in the dictionary we do not output anything. However, since the suffix becomes the prefix, 'a' now becomes the prefix for the next operation. This is illustrated in the top two lines in Figure 6.8.

Processing the next character in the input string (b) results in the addition of the prefix (a) and the suffix (b) to form the new string (ab). Since this represents a new string that is not in the dictionary, we follow rule 2. That is, we output the last known string's encoding which is 97 for the character 'a' and then add the new string to the dictionary or string table. Concerning the latter, since codes 0 through 255 represent the individual characters in an 8-bit character set, we can use the code 256 to represent the string 'ab'. However, doing so obviously requires the token to be extended. Most LZW implementations use between nine and 14 bits to represent a token whose length corresponds to the size of the dictionary. Next, 'b', which was the suffix when generating the string 'ab', becomes the prefix for the next operation as indicated in line 3 in Figure 6.8. Since the next character in the string to be encoded is 'a', it functions as the suffix in creating the new string 'ba'. As that string is not presently in the string table the last known string (b) is output using its ASCII code value of 98. Next, the string 'ba' is added to the string table using the next available code which is 257.

Prefix	Suffix	New string	Output
Δ	a	a	—
a	b	ab	97
b	a	ba	98
a	b	ab	—
ab	c	abc	256
c	Δ	c	99

Figure 6.8 Encoding the string 'ababc'

Since 'a' was the suffix in creating the string 'ab', it now becomes the prefix for the next string operation as illustrated in line 4 in Figure 6.8. The fourth character in the input string (b) is processed as the suffix to forming a new string. This results in the new string 'ab' which was previously added to the string table. Since it is already in the string table, no output is generated and the string 'ab' becomes the prefix for creating the next string.

Line 5 in Figure 6.8 illustrates the creation of the next string. Here the previously created string 'ab' which was in the string table becomes the prefix for generating the next string, while the last character in the input data stream (c) becomes the suffix in forming the next string. The resulting string (abc) is not in the string table. Thus, the last known string 'ab' has its code value output. Since 'ab' was previously assigned the value 256, that value is output and the suffix (c) becomes the prefix for the creation of the next string. Since 'c' was the last character in the input string, we simply output its code value of 99.

Decoding

You would probably expect the decoding process performed by LZW compression to be the inverse of the encoding process. Unfortunately, your expectation is only about 99% accurate as there is one situation during the encoding process that does not permit accurate decoding to occur. That situation occurs when a substring is encoded using the substring immediately preceding it. To understand why this situation can cause a problem, note that, under the LZW encoding process, each time a string is added to the dictionary the update operation precedes the actual transmittal of the output of the dictionary entry. Thus, an input data string consisting of a substring encoded using the substring immediately preceding it results in the encoder outputting a code prior to the code being known to the expander. Fortunately this situation can be programmed for since the first and last characters of the substring must be the same, permitting appropriate programming to enable the expander to determine the unknown code.

Figure 6.9 lists the pseudocode required for performing LZH encoding. Figure 6.10 lists the contents of a small BASIC language program developed to illustrate both LZH encoding as well as what is happening internally in the program as the string operations are performed.

One topic not covered so far by the pseudocode and BASIC program concerns the method by which the decoder knows the operation is completed. Thus, let's discuss that and other implementation issues.

```
Prefix = read first data byte
While
    Suffix = next data byte
IF ("prefix to suffix" in string table)

    Prefix = prefix to suffix

ELSE
    output code for prefix
    add "prefix + suffix" to string table
    prefix = suffix
Output code for prefix
```

Figure 6.9 LZH pseudocode

Implementation issues

Although we can count the characters in a file and either place the count in a file header or transmit it via a communications system, doing so requires a series of extra operations. An alternative is to assign a code or token value to indicate the end of file. Another consideration important in implementing LZH concerns the number of entries in the dictionary's string table which governs the number of bits required to denote table entries. Most implementations of LZH use dictionaries that have between 1024 and 16384 string entries, requiring codes 10 to 14 bits in length. Since any string table may not accurately reflect the contents of a dynamically changing input data stream, a mechanism is required to denote when the efficiency of the algorithm falls below a certain threshold which can function as a decision criterion to flush all entries or a portion of the entries in the string table. To accomplish this, you can monitor the compression ratio for a substring of string input. If the compression ratio falls below a certain value and stays below that value for n processed characters, this condition could be used as a threshold to flush all or a portion of the string table. Concerning string table flushing, an indicator is required. That indicator can be a code not used for string table entries, similar to a code used to indicate the end of a file. One common practice in many implementations of LZH is to use a small block of codes commencing with 256 as logical compression-indicating characters. Another method is to selectively prune dictionary entries. Readers are referred to the next section in this chapter which describes the use of a trie data structure and how dictionary pruning is accomplished in a modified version of LZW used in V.42 bis modem data compression.

```
CLS
DIM code$(4095)
table = 255
FOR i = 0 TO 255                                'initialize first 256 positions
        code$(i) = CHR$(i)
        NEXT i
LINE INPUT X$                                    'get line of input
i = 0
WHILE i < LEN(X$)
        IF i <> 0 THEN GOTO nxt:
                prefix$ = ""                     'special case for 1st character
nxt:    suffix$ = MID$(X$, i + 1, 1)             'next character
        PRINT "i="; i, "prefix$+suffix$=", (prefix$ + suffix$)
        FOR j = 0 TO table                       'search table
                IF (prefix$ + suffix$) = code$(j) THEN GOTO MATCH
                NEXT j
                PRINT "not in table thru code position="; j 'not in table
                code$(table + 1) = prefix$ + suffix$
                table = table + 1       'increment table pointer
                FOR k = 1 TO table
                IF code$(k) <> prefix$ THEN GOTO skip
                PRINT "output code= ", k
skip:
                NEXT k
                prefix$ = suffix$
                i = i + 1
                GOTO nxt:
MATCH:
        prefix$ = prefix$ + suffix$
        i = i + 1                                'increment pointer in string
WEND
        FOR k = 1 TO table
        IF code$(k) = prefix$ THEN GOTO skip1
        NEXT k
skip1: PRINT "last output code= ", k
```

```
i= 0           prefix$+suffix$=                 a
i= 1           prefix$+suffix$=                 ab
not in table thru code position= 256
output code=   97
i= 2           prefix$+suffix$=                 ba
not in table thru code position= 257
output code=   98
i= 3           prefix$+suffix$=                 ab
i= 4           prefix$+suffix$=                 abc
not in table thru code position= 258
output code=   256
i= 5           prefix$+suffix$=                 c
last output code=              99
```

Figure 6.10 BASIC program which illustrates LZH encoding

6.6 BTLZ (V.42 BIS) COMPRESSION

A modification to LZW compression was developed by British Telecom and presented to the CCITT (now known as the ITU) in October 1988 in a meeting held in Edinburgh, Scotland. This meeting was called by the CCITT for the presentation of compression algorithms for use with modems as well as to develop a common set of test data for evaluating the efficiency of different algorithms. Over a four-month period technical experts evaluated five proposed compression methods to include British Telecom's modified Lempel–Ziv algorithm. That algorithm was found to be more effective for modem applications than competitive techniques due to its ability to be implemented on inexpensive 8-bit and 16-bit microprocessors, and its lower memory requirements and process requirements than those required by other algorithms. In September 1989, a final draft recommendation was approved by the CCITT Study Group XVIII and that draft was confirmed by a unanimous ballot of CCITT membership in January 1990 as V.42 bis. The result of the V.42 bis recommendation was the manufacture of millions of compression-performing modems whose use significantly reduces transmission duration and cost.

The British Telecom proposal addressed several areas that were not addressed by the two previously referenced Ziv and Lempel papers. Those areas included the dictionary size to include its representation and codeword notation and dictionary 'pruning', a method used to remove strings from the dictionary.

Under V.42 bis, codes 0, 1 and 2 are reserved as control codes, resulting in the initial 256 one-character strings occupying indices 3 through 258. This results in the string structure of new entries in the dictionary beginning with index 259.

To understand the V.42 bis trie data structure, let's begin by examining the logical tree formed by the LZ learning process. Figure 6.11 illustrates a logical tree formed at a certain point in time after nine learned strings were added to a dictionary. Note that strings representing the initial 256 characters in the character set are already in the dictionary. Thus, Figure 6.11 represents a logical tree of added dictionary entries at a certain point in time.

The physical trie structure used by V.42 bis is illustrated in Figure 6.12. Each node has an address indicated by the node index number which represents the LZH dictionary location. The contents of each node consist of four elements—the character, an index to its parent

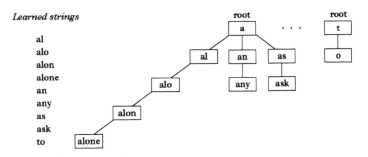

Figure 6.11 Logical tree containing nine learned strings

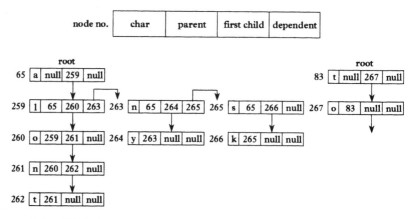

Figure 6.12 TRIE data structure implementation example

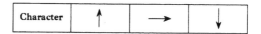

Figure 6.13 TRIE data structure representation

node, an index to its first successor or 'child' node, and an index to its first dependent in the list of successors for its parent. Each entry can contain the node index number or a physical pointer based upon whatever mechanism provides a more convenient implementation method. However, the V.42 bis standard requires the node number to be transmitted in the compressed data stream.

In examining the trie structure, note that this data structure is essentially a sequence of three pointers. The first pointer to the parent represents an upward arrow. The second pointer to the first child can be represented by a right arrow, while the pointer to the dependent can be represented by a downward pointing arrow. Thus, a more familiar trie data structure representation for some readers is illustrated in Figure 6.13.

Figure 6.12 illustrates the physical trie data structure for two root entries, 'a' at index 65 and 't' at index 83. Note that each root entry always has a null parent index and a null sibling index, indicated by the entry 'NULL'. Each node under a root in Figure 6.11 represents a learned string. The string is located by following a path from the appropriate root node. For example, the string 'ask' is represented by the node or code 266 and is located by traversing indices 65, 259, 263, 265 and 266.

To illustrate the operation of a trie data structure, assume the dictionary contents at a point in time are as depicted in Figure 6.11. Now assume the current string input is 'tom'. Having matched 't' in the sequence, the character 'o' is read and the list of strings beginning with 't' is searched. The first child pointer in node 83 points to node

267, which contains the character 'o', so another match occurs and 'm' is read. From node 267, the first child pointer is a NULL, which indicates the prior string ended. Since 'm' is the first unmatched character, it is used to update the dictionary. This is indicated in Figure 6.14 by the creation of a new node as the only descendant of 'to' with the character 'm' and a parent pointer to 'to' (node 267). If the input string had been 'action', the absence of an 'ac' entry would be discovered at the dependent traversal at node 263, since the dependent list is maintained in ascending order and the binary value of 'c' is less than the binary value of 'n'.

Dictionary pruning

Once a dictionary is filled, a mechanism is required to remove dictionary entries to enable new entries that may represent more frequently encountered strings. To accomplish this, the V.42 bis algorithm uses a pruning scheme which closely approximates a least recently used (LRU) strategy commonly used in computer systems for data storage recovery operations.

To illustrate the dictionary pruning process, let's set the variable 'NEXT' to indicate the dictionary node used next to add a new node in the string table. Let's also set the maximum number of entries in the dictionary to the constant FULL. When the algorithm commences operation, NEXT is initialized to the first free entry in the dictionary, which is node 259. As each node is added to the dictionary, NEXT is incremented until the dictionary is full and NEXT = FULL. When this situation occurs, the algorithm returns to the beginning of the string area (node number 259) and deletes the first leaf node found which is a node without descendants. The deletion process requires an update to the trie data structure so that the node is removed from its

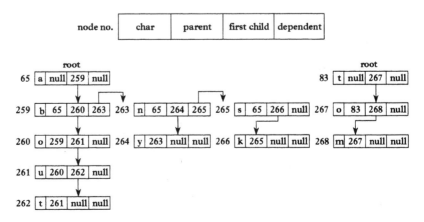

Figure 6.14   TRIE data structure after learning the string 'tom'

parent's descendant list. To illustrate this process, let's return to Figure 6.12. Assume NEXT was equal to FULL and the algorithm returned to node number 259. Here the first leaf node with no descendants would be 262. When node 262 is pruned, node 261 must then be updated. Thus, the first child entry of 262 in node 261 would be changed to a null. If the dictionary is full, on the next pass node 264 would be removed, and so on. Note that this technique maintains strings as long as they have been recently matched. This is because they will then have acquired at least one descendant.

Decoding algorithm

The decoding process of V.42 bis is similar to other LZ dictionary based string compression techniques in that it is the inverse of the encoding algorithm. Codes are received, used to reference a dictionary location for the extraction of the stored string, and the dictionary is updated in exactly the same manner as performed by the encoder. Under the V.42 bis algorithm codewords 0, 1 and 2 are command codewords, while codes 3 through 258 represent the single character strings represented by the binary values 0 through 255. Thus, newly created strings begin with 259.

To decode a received code into its original non-compressed character, the receiver follows the parent pointers from each node up to the root node, a node within the range 3 through 258 or to a NULL parent pointer. Since the code points to the last character of the string, this upward traversal results in the generation of the string in its reversed order. For example, from Figure 6.9 the receipt of the code 266 represents the string 'ask'; however, the upward traversal generates the characters 'k', 's', and 'a'. Thus, the sequence must be reversed.

Configuration and negotiation parameters

The encoding of data into strings can result in an infinite number of strings and corresponding codewords. In addition, as the number of dictionary entries increases, the number of bits required to denote a codeword increases.

To place limits upon the number of dictionary entries and resulting codeword lengths, the V.42 bis recommendation specifies a range of permitted string lengths for which a value is negotiated between the transmitting and receiving modems. This negotiation process uses three parameters, P0, P1 and P2, to tailor V.42 bis functions during an initial modem call. P0 is used to indicate if compression is used and, if so, the direction in which compression is performed. Compression can occur in both directions, the transmit direction only or the receive direction only. By specifying compression in only one direction, a double-sized dictionary can be used, a technique many communications sessions can benefit from when transmission of long data blocks occurs in one direction and short acknowledgements occur in the reverse direction.

P1 is the parameter used for negotiating the number of nodes in the dictionary. Any value from 512 to 65 535 can be used.

P2 is the parameter which defines the maximum string length. This parameter has a default value of 6 and a permitted range of 6 through 250.

Efficiency limitations

V.42 bis is similar to other LZ based compression algorithms in that the dictionary can be rapidly filled and the removal of entries either via selective pruning or flushing is not the most effective mechanism for its reconstruction. Take, for example, the occurrence of the strings 'there', 'them', and 'they'. The dictionary would have entries reflecting the substrings 'th', 'the', 'them', 'they', 'ther', and 'there' as well as the root entry 't'. Thus, a total of seven dictionary entries are required for three common words. In addition, those entries would not support the compression of 'here' even though its composition is located under the root entry 't'. Thus, not only can a Lempel–Ziv table fill rapidly but its entries can easily have built-in redundancies.

Concerning the size of the string table, when software cannot find a location for a string number it wants to assign, the table must be either pruned or erased. When either situation occurs, highly efficient string number assignments are temporarily lost, significantly reducing the algorithm's compression ratio.

Some implementations of LZ algorithms freeze the table once it is built and simply reuse the same assigned string numbers on new data. Doing so represents a new problem, since the freezing of the table assumes that new data will resemble the same or similar frequencies of characters in the previously processed data stream. In the concluding section of this chapter we will discuss methods that should be considered to alleviate or minimize the previously described efficiency limitations.

LZ variations

Following the original papers published by Ziv and Lempel, a large number of modifications to their two dictionary based string compression methods have been either implemented or proposed. Some methods, such as LZH and LZARI representing the addition of Huffman and arithmetic coding as a second stage to a method of LZ coding, can more formally be considered as multilevel compression techniques and are discussed in Section 6.7. Other LZ variations, such as LZC, LZJ, LZT and other LZ variants, primarily change the manner by which a dictionary is pruned or the number of entries permitted to be included in a dictionary.

LZC, a scheme employed by the UNIX program COMPRESS, monitors the compression ratio. As soon as that ratio falls below a present threshold, the dictionary is cleared and rebuilt from scratch. the LZJ LZ variation proposed by Jakobsson (Jakobsson, 1985) uses a dictionary that includes every unique string in previously processed text up to a maximum length. Each unique string is assigned a fixed length identification number and pruning occurs by the removal of substrings that only occurred once in the input. LZT, proposed by Tischer (Tischer, 1987), is based upon LZC but performs pruning based upon discarding the least recently used phrase. LZT also uses a different method for coding phrase numbers that slightly increases its efficiency; however, its extra processing requirement slightly reduces the speed of the algorithm.

In general, each LZ variation represents a tradeoff between memory utilization, processing time, and compression efficiency. Techniques that reduce dictionary size usually require additional processing time to achieve a compression ratio equivalent to techniques that require more memory but do not require sophisticated search routines.

Of the large number of LZ variants, the ones offering the most potential normally are those that use a multilevel compression scheme which is the subject of the next section in this chapter.

6.7 MULTILEVEL CODING AND OTHER EFFICIENCY IMPROVEMENTS

In concluding this chapter on dictionary based string compression we will focus our attention upon techniques to improve upon the building blocks provided by LZ coding. As you might expect, one method to improve upon dictionary based string compression is obtained by adding one or more compression methods, resulting in a multilevel compression process. Another method that can be utilized to improve upon the efficiency of dictionary based string compression techniques is to modify the method used to prune the dictionary. Doing so can result in the retention of more frequently encountered substrings which is not always true when the V.42 bis dictionary and similar pruning methods are used. Thus, the focus of this chapter is upon the use of multilevel compression and altering the dictionary pruning process.

LZH compression

The combination of a Ziv–Lempel type of dictionary based string compression method and Huffman coding is commonly referred to as LZH coding. Here the actual coding requires a two-pass operation. In the first pass one of the previously described LZ coding methods is employed. During the second pass the token or code pointer and, if included, the following character used in the original LZ algorithms are encoded using either a static or an adaptive Huffman algorithm.

One of the more popular LZH coding schemes uses the LZSS dictionary based string compression as a base for Huffman coding. A key problem associated with Huffman coding of previously performed LZSS coding concerns the potential number of Huffman codes that can be required. Since the dictionary can have a length of several thousand characters, the creation of a Huffman code table of that magnitude would result in a considerable amount of expansion for some table entries. To reduce the number of entries in the Huffman code table, the pointer can be split into two numbers. Doing so permits the development of code table entries consisting of a few hundred Huffman codes that correspond to those entries, permitted string lengths, and, in some cases, character codes.

A technique to accomplish the splitting of the pointer can be achieved by a process similar to the creation of modified Huffman code tables previously described in Chapter 5. That is, you can have one table of mod 64 values and another table of entries from 0 through 63. Using the two tables would enable Huffman codes to represent pointers to an extended dictionary. A second technique that can be employed to reduce the size of the Huffman coding table is to group the output of the prior encoding stage into blocks. Then, the second coding stage can result in the application of Huffman coding to the previously created blocks.

Although the initial setup of Huffman tables would be based upon a 'best guess' estimate, its efficiency can be significantly improved by making the resulting table adapt to the frequency of occurrence entries in the Huffman table. Thus, adapting arbitrarily developed tables can result in a degree of improvement in the performance of the multilevel compression used for LZH encoding.

The AR archive program

Similar to the manner in which the papers of Ziv and Lempel provided a foundation for the development of dictionary based string compression techniques, modern archiving and compression performing programs can trace their roots to the program AR, written by Haruhiko Okumura. Although the program only supported six archiving operations (adding, extracting, replacing, deleting and printing files, and listing the contents of the archive), it includes a two-stage or multilevel compression engine which was used 'as is', or in slightly modified form, by many successors' archiving and compression-performing programs. Mr Okumura's placement of the C language programs that form the basis for AR in the public domain can be considered as representing a milestone in the advancement of the application of compression technology.

The two-stage compression process begins with the use of LZSS to locate strings in a sliding dictionary. Figure 6.15 lists the statements in the program ENCODE.C. During the second stage of compression, the output of the first stage is divided into blocks and encoded using a static Huffman encoding process. The application of Huffman coding upon blocks of output from the first stage eliminates a requirement to generate an extremely large Huffman table. For example, the use of a dictionary of 32 768 positions and a lookahead buffer of 512 positions would require $32\,768 \times 512$ different match descriptions. Thus, the application of a translation table applied on a block basis significantly reduces the required table size as well as processing time required to perform Huffman encoding.

Figure 6.16 lists the contents of the file HUF.C which performs the static Huffman coding of the output of the first phase on a block basis. Readers are referred to the file AR002.EXE on diskette number 2 accompanying this book to obtain a complete set of C language files used by the archive program. The file AR002.EXE is a self-extracting archive, requiring you to simply type the file name to 'explode' the archive and to obtain its individual files. Among those files are MAKE-TREE.C which constructs the Huffman tree, AR.C which is the main program file, MAKETBL.C which is the make table for decoding, as well as a linked object program, AR.EXE, which provides you with a simple but highly efficient archiving and compression performing program.

LZARI compression

Similar to the manner in which LZ string compression techniques can be enhanced via the addition of Huffman coding, those techniques can also be enhanced through the use of another statistical compression technique. One of the more popular techniques used as a second level of operation to an LZ encoding process is arithmetic coding, resulting in the term LZARI used to represent this multilevel compression process.

LZARI compression represents a variation of the LZSS algorithm which in turn was based upon LZ77 compression. Under the LZSS algorithm, a token consisting of either a character prefixed by a '1' bit or a character pair consisting of an offset and string length prefixed by a '0' bit is used to encode the data stream. The LZSS output data stream format was previously illustrated in Figure 6.5.

To illustrate the potential improvement resulting from a second level of compression, assume the character 'e', which is the most frequently occurring character in the English language, as expected occurs more often than many, if not all, characters in the input data stream. In addition, there is a high probability that some lengths may be more commonly used to represent encoded strings than other lengths. Thus, we can anticipate that the output of the LZSS algorithm will generate characters and tokens that will vary in their frequency of occurrence. This tells us that a mechanism used to encode more frequently occurring characters using short codes and less frequently occurring characters using longer codes can enhance the LZSS encoding process. Another method that can be used to accomplish this is the use of arithmetic coding as a second stage in a dictionary based string encoding method.

Figure 6.17 lists the contents of the C language public domain program LZARI.C written by Haruhiko Okumura. In developing LZARI.C, Mr Okumura extended the character set from the normal 256 positions. The characters 0 through 255 represent the conventional 8-bit characters while positions $253 + n$ represent the position of string lengths n, where $n = 3$, 4 and so on. That extended character set is then encoded using adaptive arithmetic compression. A second technique employed by Mr Okumura concerned the match length of the LZSS token. Noting that the longest match strings tend to be the ones that were relatively recently encoded, it followed that recent positions should be encoded into fewer bits. However, since 4096 positions which represented the sizes of the buffer used in his program were too many to encode adaptively, he fixed the probability distribution of the positions by hand. The resulting program is not optimal since LZARI.C treats the positions essentially as separate entities. In addition, the string length threshold above which strings are coded into the ⟨position,length⟩ token pairs is fixed; however, a more efficient technique would change its value according to the composition of the token. In spite of these limitations the program produces a high level of compression which typically exceeds the compression ratio of normal LZ based encoders by 10 to 15%.

Improvements to consider

One common problem associated with many LZ based compression techniques is their effect upon runs of characters. For example, consider a sequence of 10 asterisks (*) which occur quite frequently in financial reports although the actual sequence length may be longer or shorter. Under LZ77 compression, the first asterisk entering the dictionary would be encoded as the triplet (0,0,*). The next nine asterisks would be encoded in the triplet (offset,9,*) where offset represents the offset location of the first asterisk placed into the dictionary with respect to either the beginning of the dictionary or its boundary with the lookahead buffer. Thus, a minimum of six characters would be required to encode a string of 10 repeating characters, although in

```
/**********************************************************
           encode.c -- sliding dictionary with percolating update
 **********************************************************/
#include "ar.h"
#include <stdlib.h>
#include <string.h>   /* memmove() */

#define PERCOLATE  1
#define NIL        0
#define MAX_HASH_VAL (3 * DICSIZ + (DICSIZ / 512 + 1) * UCHAR_MAX)

typedef short node;

static uchar *text, *childcount;
static node pos, matchpos, avail,
        *position, *parent, *prev, *next = NULL;
static int remainder, matchlen;

#if MAXMATCH <= (UCHAR_MAX + 1)
        static uchar *level;
#else
        static ushort *level;
#endif

static void allocate_memory(void)
{
        if (next != NULL) return;
    text = malloc(DICSIZ * 2 + MAXMATCH);
        level      = malloc((DICSIZ + UCHAR_MAX + 1) * sizeof(*level));
        childcount = malloc((DICSIZ + UCHAR_MAX + 1) * sizeof(*childcount));
        #if PERCOLATE
          position = malloc((DICSIZ + UCHAR_MAX + 1) * sizeof(*position));
        #else
          position = malloc(DICSIZ * sizeof(*position));
        #endif
        parent     = malloc(DICSIZ * 2 * sizeof(*parent));
        prev       = malloc(DICSIZ * 2 * sizeof(*prev));
        next       = malloc((MAX_HASH_VAL + 1) * sizeof(*next));
        if (next == NULL) error("Out of memory.");
}

static void init_slide(void)
{
        node i;

        for (i = DICSIZ; i <= DICSIZ + UCHAR_MAX; i++) {
                level[i] = 1;
                #if PERCOLATE
                        position[i] = NIL;  /* sentinel */
                #endif
        }
        for (i = DICSIZ; i < DICSIZ * 2; i++) parent[i] = NIL;
        avail = 1;
        for (i = 1; i < DICSIZ - 1; i++) next[i] = i + 1;
        next[DICSIZ - 1] = NIL;
        for (i = DICSIZ * 2; i <= MAX_HASH_VAL; i++) next[i] = NIL;
}

#define HASH(p, c) ((p) + ((c) << (DICBIT - 9)) + DICSIZ * 2)

static node child(node q, uchar c)
        /* q's child for character c (NIL if not found) */
{
        node r;

        r = next[HASH(q, c)];
```

Figure 6.15 Listing of ENCODE.C used in the AR program

```
                parent[NIL] = q;   /* sentinel */
                while (parent[r] != q) r = next[r];
                return r;
        }

        static void makechild(node q, uchar c, node r)
                /* Let r be q's child for character c. */
        {
                node h, t;

                h = HASH(q, c);
                t = next[h];   next[h] = r;   next[r] = t;
                prev[t] = r;   prev[r] = h;
                parent[r] = q;   childcount[q]++;
        }

        void split(node old)
        {
                node new, t;

                new = avail;   avail = next[new];   childcount[new] = 0;
                t = prev[old];   prev[new] = t;   next[t] = new;
                t = next[old];   next[new] = t;   prev[t] = new;
                parent[new] = parent[old];
                level[new] = matchlen;
                position[new] = pos;
                makechild(new, text[matchpos + matchlen], old);
                makechild(new, text[pos + matchlen], pos);
        }

        static void insert_node(void)
        {
                node q, r, j, t;
                uchar c, *t1, *t2;

                if (matchlen >= 4) {
                        matchlen--;
                        r = (matchpos + 1) | DICSIZ;
                        while ((q = parent[r]) == NIL) r = next[r];
                        while (level[q] >= matchlen) {
                                r = q;   q = parent[q];
                        }
                #if PERCOLATE
                        t = q;
                        while (position[t] < 0) {
                                position[t] = pos;   t = parent[t];
                        }
                        if (t < DICSIZ) position[t] = pos | PERC_FLAG;
                #else
                        t = q;
                        while (t < DICSIZ) {
                                position[t] = pos;   t = parent[t];
                        }
                #endif
                } else {
                        q = text[pos] + DICSIZ;   c = text[pos + 1];
                        if ((r = child(q, c)) == NIL) {
                                makechild(q, c, pos);   matchlen = 1;
                                return;
                        }
                        matchlen = 2;
                }
                for ( ; ; ) {
                        if (r >= DICSIZ) {
                                j = MAXMATCH;   matchpos = r;
                        } else {
```

Figure 6.15 *Continued*

```
                                    j = level[r];
                                    matchpos = position[r] & ~PERC_FLAG;
                            }
                            if (matchpos >= pos) matchpos -= DICSIZ;
                            t1 = &text[pos + matchlen];   t2 = &text[matchpos + matchlen];
                            while (matchlen < j) {
                                    if (*t1 != *t2) {  split(r);   return;   }
                                    matchlen++;  t1++;  t2++;
                            }
                            if (matchlen >= MAXMATCH) break;
                            position[r] = pos;
                            q = r;
                            if ((r = child(q, *t1)) == NIL) {
                                    makechild(q, *t1, pos);   return;
                            }
                            matchlen++;
                    }
                    t = prev[r];   prev[pos] = t;   next[t] = pos;
                    t = next[r];   next[pos] = t;   prev[t] = pos;
                    parent[pos] = q;   parent[r] = NIL;
                    next[r] = pos;   /* special use of next[] */
}

static void delete_node(void)
{
        #if PERCOLATE
                node q, r, s, t, u;
        #else
                node r, s, t, u;
        #endif

        if (parent[pos] == NIL) return;
        r = prev[pos];   s = next[pos];
        next[r] = s;   prev[s] = r;
        r = parent[pos];   parent[pos] = NIL;
        if (r >= DICSIZ || --childcount[r] > 1) return;
        #if PERCOLATE
                t = position[r] & ~PERC_FLAG;
        #else
                t = position[r];
        #endif
        if (t >= pos) t -= DICSIZ;
        #if PERCOLATE
                s = t;   q = parent[r];
                while ((u = position[q]) & PERC_FLAG) {
                        u &= ~PERC_FLAG;   if (u >= pos) u -= DICSIZ;
                        if (u > s) s = u;
                        position[q] = (s | DICSIZ);   q = parent[q];
                }
                if (q < DICSIZ) {
                        if (u >= pos) u -= DICSIZ;
                        if (u > s) s = u;
                        position[q] = s | DICSIZ | PERC_FLAG;
                }
        #endif
        s = child(r, text[t + level[r]]);
        t = prev[s];   u = next[s];
        next[t] = u;   prev[u] = t;
        t = prev[r];   next[t] = s;   prev[s] = t;
        t = next[r];   prev[t] = s;   next[s] = t;
        parent[s] = parent[r];   parent[r] = NIL;
        next[r] = avail;   avail = r;
}

static void get_next_match(void)
{
```

Figure 6.15 *Continued*

```
            int n;

            remainder--;
            if (++pos == DICSIZ * 2) {
                    memmove(&text[0], &text[DICSIZ], DICSIZ + MAXMATCH);
                    n = fread_crc(&text[DICSIZ + MAXMATCH], DICSIZ, infile);
                    remainder += n;  pos = DICSIZ;  putc('.', stderr);
            }
            delete_node();  insert_node();
    }

void encode(void)
{
        int lastmatchlen;
        node lastmatchpos;

        allocate_memory();  init_slide();  huf_encode_start();
        remainder = fread_crc(&text[DICSIZ], DICSIZ + MAXMATCH, infile);
        putc('.', stderr);
        matchlen = 0;
        pos = DICSIZ;  insert_node();
        if (matchlen > remainder) matchlen = remainder;
        while (remainder > 0 && ! unpackable) {
                lastmatchlen = matchlen;  lastmatchpos = matchpos;
                get_next_match();
                if (matchlen > remainder) matchlen = remainder;
                if (matchlen > lastmatchlen || lastmatchlen < THRESHOLD)
                        output(text[pos - 1], 0);
                else {
                        output(lastmatchlen + (UCHAR_MAX + 1 - THRESHOLD),
                                    (pos - lastmatchpos - 2) & (DICSIZ - 1));
                        while (--lastmatchlen > 0) get_next_match();
                        if (matchlen > remainder) matchlen = remainder;
                }
        }
        huf_encode_end();
}
```

Figure 6.15 *Continued*

actuality it could be more since the offset would expand in bit length as the length of the dictionary increases. Under LZ78, the sequence '**********' would be parsed as illustrated in Figure 6.18. Note that the two-character sequence of phrase and character would result in a minimum of eight characters required to encode a sequence of 10 repeating characters.

An exception to the previously described LZ77 and LZ78 compression of a sequence of repeated characters is if the sequence is already in the dictionary. Then, either a triplet (LZ77) or pair (LZ78) of characters could be used to encode a given sequence.

Suppose we preprocess an input data string to compress repeated sequences by replacing three or more repeating characters by two characters and a count character. For example, the sequence '**********' could be replaced by the sequence **8, using a modified form of run length compression similar to that employed by MNP Class 5 and Class 7 data compression. Since the third character represents the number of repeated characters following a sequence of three to include the third repeating character, this technique would reduce repeated sequences up to 258 characters in length to a three-

```c
/**********************************************************
          huf.c -- static Huffman
**********************************************************/
#include <stdlib.h>
#include "ar.h"

#define NP (DICBIT + 1)
#define NT (CODE_BIT + 3)
#define PBIT 4   /* smallest integer such that (1U << PBIT) > NP */
#define TBIT 5   /* smallest integer such that (1U << TBIT) > NT */
#if NT > NP
        #define NPT NT
#else
        #define NPT NP
#endif

ushort left[2 * NC - 1], right[2 * NC - 1];
static uchar *buf, c_len[NC], pt_len[NPT];
static uint   bufsiz = 0, blocksize;
static ushort c_freq[2 * NC - 1], c_table[4096], c_code[NC],
              p_freq[2 * NP - 1], pt_table[256], pt_code[NPT],
              t_freq[2 * NT - 1];

/***** encoding *****/

static void count_t_freq(void)
{
        int i, k, n, count;

        for (i = 0; i < NT; i++) t_freq[i] = 0;
        n = NC;
        while (n > 0 && c_len[n - 1] == 0) n--;
        i = 0;
        while (i < n) {
                k = c_len[i++];
                if (k == 0) {
                        count = 1;
                        while (i < n && c_len[i] == 0) { i++; count++; }
                        if (count <= 2) t_freq[0] += count;
                        else if (count <= 18) t_freq[1]++;
                        else if (count == 19) { t_freq[0]++; t_freq[1]++; }
                        else t_freq[2]++;
                } else t_freq[k + 2]++;
        }
}

static void write_pt_len(int n, int nbit, int i_special)
{
        int i, k;

        while (n > 0 && pt_len[n - 1] == 0) n--;
        putbits(nbit, n);
        i = 0;
        while (i < n) {
                k = pt_len[i++];
                if (k <= 6) putbits(3, k);
                else putbits(k - 3, (1U << (k - 3)) - 2);
                if (i == i_special) {
                        while (i < 6 && pt_len[i] == 0) i++;
                        putbits(2, (i - 3) & 3);
                }
        }
}

static void write_c_len(void)
{
```

Figure 6.16 Listing of HUF.C which encodes blocks of previously compressed data

```
        int i, k, n, count;

        n = NC;
        while (n > 0 && c_len[n - 1] == 0) n--;
        putbits(CBIT, n);
        i = 0;
        while (i < n) {
                k = c_len[i++];
                if (k == 0) {
                        count = 1;
                        while (i < n && c_len[i] == 0) {  i++;  count++;  }
                        if (count <= 2) {
                                for (k = 0; k < count; k++)
                                                putbits(pt_len[0], pt_code[0]);
                        } else if (count <= 18) {
                                putbits(pt_len[1], pt_code[1]);
                                putbits(4, count - 3);
                        } else if (count == 19) {
                                putbits(pt_len[0], pt_code[0]);
                                putbits(pt_len[1], pt_code[1]);
                                putbits(4, 15);
                        } else {
                                putbits(pt_len[2], pt_code[2]);
                                putbits(CBIT, count - 20);
                        }
                } else putbits(pt_len[k + 2], pt_code[k + 2]);
        }
}

static void encode_c(int c)
{
        putbits(c_len[c], c_code[c]);
}

static void encode_p(uint p)
{
        uint c, q;

        c = 0;  q = p;  while (q) {  q >>= 1;  c++;  }
        putbits(pt_len[c], pt_code[c]);
        if (c > 1) putbits(c - 1, p & (0xFFFFU >> (17 - c)));
}

static void send_block(void)
{
        uint i, k, flags, root, pos, size;

        root = make_tree(NC, c_freq, c_len, c_code);
        size = c_freq[root];  putbits(16, size);
        if (root >= NC) {
                count_t_freq();
                root = make_tree(NT, t_freq, pt_len, pt_code);
                if (root >= NT) {
                        write_pt_len(NT, TBIT, 3);
                } else {
                        putbits(TBIT, 0);  putbits(TBIT, root);
                }
                write_c_len();
        } else {
                putbits(TBIT, 0);  putbits(TBIT, 0);
                putbits(CBIT, 0);  putbits(CBIT, root);
        }
        root = make_tree(NP, p_freq, pt_len, pt_code);
        if (root >= NP) {
                write_pt_len(NP, PBIT, -1);
        } else {
```

Figure 6.16 *Continued*

```
                putbits(PBIT, 0);   putbits(PBIT, root);
        }
        pos = 0;
        for (i = 0; i < size; i++) {
                if (i % CHAR_BIT == 0) flags = buf[pos++];   else flags <<= 1;
                if (flags & (1U << (CHAR_BIT - 1))) {
                        encode_c(buf[pos++] + (1U << CHAR_BIT));
                        k = buf[pos++] << CHAR_BIT;   k += buf[pos++];
                        encode_p(k);
                } else encode_c(buf[pos++]);
                if (unpackable) return;
        }
        for (i = 0; i < NC; i++) c_freq[i] = 0;
        for (i = 0; i < NP; i++) p_freq[i] = 0;
}

static uint output_pos, output_mask;

void output(uint c, uint p)
{
        static uint cpos;

        if ((output_mask >>= 1) == 0) {
                output_mask = 1U << (CHAR_BIT - 1);
                if (output_pos >= bufsiz - 3 * CHAR_BIT) {
                        send_block();
                        if (unpackable) return;
                        output_pos = 0;
                }
                cpos = output_pos++;   buf[cpos] = 0;
        }
        buf[output_pos++] = (uchar) c;   c_freq[c]++;
        if (c >= (1U << CHAR_BIT)) {
                buf[cpos] |= output_mask;
                buf[output_pos++] = (uchar)(p >> CHAR_BIT);
                buf[output_pos++] = (uchar) p;
                c = 0;   while (p) {   p >>= 1;   c++;   }
                p_freq[c]++;
        }
}

void huf_encode_start(void)
{
        int i;

        if (bufsiz == 0) {
                bufsiz = 16 * 1024U;
                while ((buf = malloc(bufsiz)) == NULL) {
                        bufsiz = (bufsiz / 10U) * 9U;
                        if (bufsiz < 4 * 1024U) error("Out of memory.");
                }

        }
        buf[0] = 0;
        for (i = 0; i < NC; i++) c_freq[i] = 0;
        for (i = 0; i < NP; i++) p_freq[i] = 0;
        output_pos = output_mask = 0;
        init_putbits();
}

void huf_encode_end(void)
{
        if (! unpackable) {
                send_block();
                putbits(CHAR_BIT - 1, 0);   /* flush remaining bits */
        }
}
```

Figure 6.16 *Continued*

```
/***** decoding *****/
static void read_pt_len(int nn, int nbit, int i_special)
{
        int i, c, n;
        uint mask;

        n = getbits(nbit);
        if (n == 0) {
                c = getbits(nbit);
                for (i = 0; i < nn; i++) pt_len[i] = 0;
                for (i = 0; i < 256; i++) pt_table[i] = c;
        } else {
                i = 0;
                while (i < n) {
                        c = bitbuf >> (BITBUFSIZ - 3);
                        if (c == 7) {
                                mask = 1U << (BITBUFSIZ - 1 - 3);
                                while (mask & bitbuf) {  mask >>= 1;  c++;  }
                        }
                        fillbuf((c < 7) ? 3 : c - 3);
                        pt_len[i++] = c;
                        if (i == i_special) {
                                c = getbits(2);
                                while (--c >= 0) pt_len[i++] = 0;
                        }
                }
                while (i < nn) pt_len[i++] = 0;
                make_table(nn, pt_len, 8, pt_table);
        }
}

static void read_c_len(void)
{
        int i, c, n;
        uint mask;

        n = getbits(CBIT);
        if (n == 0) {
                c = getbits(CBIT);
                for (i = 0; i < NC; i++) c_len[i] = 0;
                for (i = 0; i < 4096; i++) c_table[i] = c;
        } else {
                i = 0;
                while (i < n) {
                        c = pt_table[bitbuf >> (BITBUFSIZ - 8)];
                        if (c >= NT) {
                                mask = 1U << (BITBUFSIZ - 1 - 8);
                                do {
                                        if (bitbuf & mask) c = right[c];
                                        else               c = left [c];
                                        mask >>= 1;
                                } while (c >= NT);
                        }
                        fillbuf(pt_len[c]);
                        if (c <= 2) {
                                if      (c == 0) c = 1;
                                else if (c == 1) c = getbits(4) + 3;
                                else             c = getbits(CBIT) + 20;
                                while (--c >= 0) c_len[i++] = 0;
                        } else c_len[i++] = c - 2;
                }
                while (i < NC) c_len[i++] = 0;
                make_table(NC, c_len, 12, c_table);
        }
}
```

Figure 6.16 *Continued*

```
      }
uint decode_c(void)
{
      uint j, mask;

      if (blocksize == 0) {
            blocksize = getbits(16);
            read_pt_len(NT, TBIT, 3);
            read_c_len();
            read_pt_len(NP, PBIT, -1);
      }
      blocksize--;
      j = c_table[bitbuf >> (BITBUFSIZ - 12)];
      if (j >= NC) {
            mask = 1U << (BITBUFSIZ - 1 - 12);
            do {
                  if (bitbuf & mask) j = right[j];
                  else               j = left [j];
                  mask >>= 1;
            } while (j >= NC);
      }
      fillbuf(c_len[j]);
      return j;
}

uint decode_p(void)
{
      uint j, mask;

      j = pt_table[bitbuf >> (BITBUFSIZ - 8)];
      if (j >= NP) {
            mask = 1U << (BITBUFSIZ - 1 - 8);
            do {
                  if (bitbuf & mask) j = right[j];
                  else               j = left [j];
                  mask >>= 1;
            } while (j >= NP);
      }
      fillbuf(pt_len[j]);
      if (j != 0) j = (1U << (j - 1)) + getbits(j - 1);
      return j;
}

void huf_decode_start(void)
{
      init_getbits();  blocksize = 0;
}
```

Figure 6.16 *Continued*

character sequence. Thus, a preprocessing of the input data stream prior to its encoding via the use of a dictionary based string compression algorithm offers the potential to enhance the overall operational result.

A second method that can be considered to improve dictionary based string compression could be obtained from a more efficient pruning method. V.42 bis and other LZ type implementations use a method which closely approximates a least recently used strategy. However, consider the situation where leaf nodes with arbitrary numbers 320, 330 and 350 represent three leaf nodes without descendants. Assume the frequency of occurrence of the string terminating at 320 was 25, while the frequency of occurrence of the strings termi-

```
/**************************************************************
      LZARI.C -- A Data Compression Program
      (tab = 4 spaces)
 **************************************************************
      4/7/1989 Haruhiko Okumura
      Use, distribute, and modify this program freely.
      Please send me your improved versions.
          PC-VAN      SCIENCE
          NIFTY-Serve      PAF01022
          CompuServe 74050,1022
 **************************************************************/
#include <stdio.h>
#include <stdlib.h>
#include <string.h>
#include <ctype.h>

/********* Bit I/O *********/

FILE  *infile, *outfile;
unsigned long int  textsize = 0, codesize = 0, printcount = 0;

void Error(char *message)
{
      printf("\n%s\n", message);
      exit(EXIT_FAILURE);
}

void PutBit(int bit)  /* Output one bit (bit = 0,1) */
{
      static unsigned int  buffer = 0, mask = 128;

      if (bit) buffer |= mask;
      if ((mask >>= 1) == 0) {
          if (putc(buffer, outfile) == EOF) Error("Write Error");
          buffer = 0;  mask = 128;  codesize++;
      }
}

void FlushBitBuffer(void)  /* Send remaining bits */
{
      int  i;

      for (i = 0; i < 7; i++) PutBit(0);
}

int GetBit(void)  /* Get one bit (0 or 1) */
{
      static unsigned int  buffer, mask = 0;

      if ((mask >>= 1) == 0) {
          buffer = getc(infile);  mask = 128;
      }
      return ((buffer & mask) != 0);
}

/********* LZSS with multiple binary trees *********/

#define N              4096 /* size of ring buffer */
#define F              60   /* upper limit for match_length */
#define THRESHOLD      2     /* encode string into position and length if match_length is
                               greater than this */
#define NIL            N     /* index for root of binary search trees */

unsigned char  text_buf[N + F - 1];    /* ring buffer of size N, with extra F-1 bytes to
                                          facilitate string comparison of longest match. */
int  match_position, match_length,     /* These are set by the InsertNode() procedure */
```

Figure 6.17 LZARI.C program listing

```
            lson[N + 1], rson[N + 257], dad[N + 1];   /* left & right children & parents --
                                                These constitute binary search trees. */

void InitTree(void)  /* Initialize trees */
{
    int  i;

        /* For i = 0 to N - 1, rson[i] and lson[i] will be the right and left children of node i.
           These nodes need not be initialized. Also, dad[i] is the parent of node i. These are ini-
           tialized to NIL (= N), which stands for 'not used.' For i = 0 to 255, rson[N + i + 1] is
           the root of the tree for strings that begin with character i. These are initialized to
           NIL.  Note there are 256 trees. */

        for (i = N + 1; i <= N + 256; i++) rson[i] = NIL;        /* root */
        for (i = 0; i < N; i++) dad[i] = NIL;                    /* node */
}

void InsertNode(int r)
        /* Inserts string of length F, text_buf[r..r+F-1], into one of the trees (text_buf[r]'th
           tree) and returns the longest-match position and length via the global variables
           match_position and match_length. If match_length = F, then removes the old node in favor
           of the new one, because the old one will be deleted sooner. Note r plays double role, as
           tree node and position in buffer. */
{
    int  i, p, cmp, temp;
    unsigned char  *key;

    cmp = 1;  key = &text_buf[r];  p = N + 1 + key[0];
    rson[r] = lson[r] = NIL;  match_length = 0;
    for ( ; ; ) {
        if (cmp >= 0) {
            if (rson[p] != NIL) p = rson[p];
            else {  rson[p] = r;  dad[r] = p;  return;  }
        } else {
            if (lson[p] != NIL) p = lson[p];
            else {  lson[p] = r;  dad[r] = p;  return;  }
        }
        for (i = 1; i < F; i++)
            if ((cmp = key[i] - text_buf[p + i]) != 0)  break;
        if (i > THRESHOLD) {
            if (i > match_length) {
                match_position = (r - p) & (N - 1);
                if ((match_length = i) >= F) break;
            } else if (i == match_length) {
                if ((temp = (r - p) & (N - 1)) < match_position)
                    match_position = temp;
            }
        }
    }
    dad[r] = dad[p];  lson[r] = lson[p];  rson[r] = rson[p];
    dad[lson[p]] = r;  dad[rson[p]] = r;
    if (rson[dad[p]] == p) rson[dad[p]] = r;
    else                   lson[dad[p]] = r;
    dad[p] = NIL;  /* remove p */
}

void DeleteNode(int p)  /* Delete node p from tree */
{
    int  q;

    if (dad[p] == NIL) return;  /* not in tree */
    if (rson[p] == NIL) q = lson[p];
```

Figure 6.17 *Continued*

```
             else if (lson[p] == NIL) q = rson[p];
             else {
                 q = lson[p];
                 if (rson[q] != NIL) {
                     do {  q = rson[q];  } while (rson[q] != NIL);
                     rson[dad[q]] = lson[q];  dad[lson[q]] = dad[q];
                     lson[q] = lson[p];  dad[lson[p]] = q;
                 }
                 rson[q] = rson[p];  dad[rson[p]] = q;
             }
             dad[q] = dad[p];
             if (rson[dad[p]] == p) rson[dad[p]] = q;
             else                   lson[dad[p]] = q;
             dad[p] = NIL;
         }

/********** Arithmetic Compression **********/

         /*  If you are not familiar with arithmetic compression, you should read
                 I. E. Witten, R. M. Neal, and J. G. Cleary,
                     Communications of the ACM, Vol. 30, pp. 520-540 (1987),
             from which much have been borrowed.  */

#define M    15

         /* Q1 (= 2 to the M) must be sufficiently large, but not so
            large as the unsigned long 4 * Q1 * (Q1 - 1) overflows. */

#define Q1    (1UL << M)
#define Q2    (2 * Q1)
#define Q3    (3 * Q1)
#define Q4    (4 * Q1)
#define MAX_CUM (Q1 - 1)

#define N_CHAR  (256 - THRESHOLD + F)
         /* character code = 0, 1, ..., N_CHAR - 1 */

unsigned long int  low = 0, high = Q4, value = 0;
int  shifts = 0;  /* counts for magnifying low and high around Q2 */
int  char_to_sym[N_CHAR], sym_to_char[N_CHAR + 1];
unsigned int
     sym_freq[N_CHAR + 1], /* frequency for symbols */
     sym_cum[N_CHAR + 1],  /* cumulative freq for symbols */
     position_cum[N + 1];  /* cumulative freq for positions */

void StartModel(void)       /* Initialize model */
{
     int ch, sym, i;

     sym_cum[N_CHAR] = 0;
     for (sym = N_CHAR; sym >= 1; sym--) {
         ch = sym - 1;
         char_to_sym[ch] = sym;  sym_to_char[sym] = ch;
         sym_freq[sym] = 1;
         sym_cum[sym - 1] = sym_cum[sym] + sym_freq[sym];
     }
     sym_freq[0] = 0;  /* sentinel (!= sym_freq[1]) */
     position_cum[N] = 0;
     for (i = N; i >= 1; i--)
         position_cum[i - 1] = position_cum[i] + 10000 / (i + 200);
             /* empirical distribution function (quite tentative) */
             /* Please devise a better mechanism! */
}

void UpdateModel(int sym)
{
```

Figure 6.17 *Continued*

```
            int i, c, ch_i, ch_sym;

            if (sym_cum[0] >= MAX_CUM) {
                 c = 0;
                 for (i = N_CHAR; i > 0; i--) {
                      sym_cum[i] = c;
                      c += (sym_freq[i] = (sym_freq[i] + 1) >> 1);
                 }
                 sym_cum[0] = c;
            }
            for (i = sym; sym_freq[i] == sym_freq[i - 1]; i--) ;
            if (i < sym) {
                 ch_i = sym_to_char[i];       ch_sym = sym_to_char[sym];
                 sym_to_char[i] = ch_sym;   sym_to_char[sym] = ch_i;
                 char_to_sym[ch_i] = sym;   char_to_sym[ch_sym] = i;
            }
            sym_freq[i]++;
            while (--i >= 0) sym_cum[i]++;
       }

       static void Output(int bit)  /* Output 1 bit, followed by its complements */
       {
            PutBit(bit);
            for ( ; shifts > 0; shifts--) PutBit(! bit);
       }

       void EncodeChar(int ch)
       {
            int  sym;
            unsigned long int  range;

            sym = char_to_sym[ch];
            range = high - low;
            high = low + (range * sym_cum[sym - 1]) / sym_cum[0];
            low +=        (range * sym_cum[sym    ]) / sym_cum[0];
            for ( ; ; ) {
                 if (high <= Q2) Output(0);
                 else if (low >= Q2) {
                      Output(1);  low -= Q2;  high -= Q2;
                 } else if (low >= Q1 && high <= Q3) {
                      shifts++;  low -= Q1;  high -= Q1;
                 } else break;
                 low += low;  high += high;
            }
            UpdateModel(sym);
       }

       void EncodePosition(int position)
       {
            unsigned long int  range;

            range = high - low;
            high = low + (range * position_cum[position    ]) / position_cum[0];
            low +=        (range * position_cum[position + 1]) / position_cum[0];
            for ( ; ; ) {
                 if (high <= Q2) Output(0);
                 else if (low >= Q2) {
                      Output(1);  low -= Q2;  high -= Q2;
                 } else if (low >= Q1 && high <= Q3) {
                      shifts++;  low -= Q1;  high -= Q1;
                 } else break;
                 low += low;  high += high;
            }
       }

       void EncodeEnd(void)
```

Figure 6.17 *Continued*

```
{
    shifts++;
    if (low < Q1) Output(0);  else Output(1);
    FlushBitBuffer();  /* flush bits remaining in buffer */
}

int BinarySearchSym(unsigned int x)
    /* 1      if x >= sym_cum[1],
       N_CHAR if sym_cum[N_CHAR] > x,
       i such that sym_cum[i - 1] > x >= sym_cum[i] otherwise */
{
    int i, j, k;

    i = 1;  j = N_CHAR;
    while (i < j) {
        k = (i + j) / 2;
        if (sym_cum[k] > x) i = k + 1;  else j = k;
    }
    return i;
}

int BinarySearchPos(unsigned int x)
    /* 0 if x >= position_cum[1],
       N - 1 if position_cum[N] > x,
       i such that position_cum[i] > x >= position_cum[i + 1] otherwise */
{
    int i, j, k;

    i = 1;  j = N;
    while (i < j) {
        k = (i + j) / 2;
        if (position_cum[k] > x) i = k + 1;  else j = k;
    }
    return i - 1;
}

void StartDecode(void)
{
    int i;

    for (i = 0; i < M + 2; i++)
        value = 2 * value + GetBit();
}

int DecodeChar(void)
{
    int    sym, ch;
    unsigned long int  range;

    range = high - low;
    sym = BinarySearchSym((unsigned int)
        (((value - low + 1) * sym_cum[0] - 1) / range));
    high = low + (range * sym_cum[sym - 1]) / sym_cum[0];
    low +=        (range * sym_cum[sym    ]) / sym_cum[0];
    for ( ; ; ) {
        if (low >= Q2) {
            value -= Q2;  low -= Q2; high -= Q2;
        } else if (low >= Q1 && high <= Q3) {
            value -= Q1;  low -= Q1;  high -= Q1;
        } else if (high > Q2) break;
        low += low;  high += high;
        value = 2 * value + GetBit();
    }
    ch = sym_to_char[sym];
    UpdateModel(sym);
    return ch;
```

Figure 6.17 *Continued*

```
    }

int DecodePosition(void)
{
    int position;
    unsigned long int  range;

    range = high - low;
    position = BinarySearchPos((unsigned int)
        (((value - low + 1) * position_cum[0] - 1) / range));
    high = low + (range * position_cum[position    ]) / position_cum[0];
    low +=       (range * position_cum[position + 1]) / position_cum[0];
    for ( ; ; ) {
        if (low >= Q2) {
            value -= Q2;  low -= Q2;  high -= Q2;
        } else if (low >= Q1 && high <= Q3) {
            value -= Q1;  low -= Q1;  high -= Q1;
        } else if (high > Q2) break;
        low += low;  high += high;
        value = 2 * value + GetBit();
    }
    return position;
}


/********** Encode and Decode **********/

void Encode(void)
{
    int  i, c, len, r, s, last_match_length;

    fseek(infile, 0L, SEEK_END);
    textsize = ftell(infile);
    if (fwrite(&textsize, sizeof textsize, 1, outfile) < 1)
        Error("Write Error");  /* output size of text */
    codesize += sizeof textsize;
    if (textsize == 0) return;
    rewind(infile);  textsize = 0;
    StartModel();  InitTree();
    s = 0;  r = N - F;
    for (i = s; i < r; i++) text_buf[i] = ' ';
    for (len = 0; len < F && (c = getc(infile)) != EOF; len++)
        text_buf[r + len] = c;
    textsize = len;
    for (i = 1; i <= F; i++) InsertNode(r - i);
    InsertNode(r);
    do {
        if (match_length > len) match_length = len;
        if (match_length <= THRESHOLD) {
            match_length = 1;  EncodeChar(text_buf[r]);
        } else {
            EncodeChar(255 - THRESHOLD + match_length);
            EncodePosition(match_position - 1);
        }
        last_match_length = match_length;
        for (i = 0; i < last_match_length &&
                (c = getc(infile)) != EOF; i++) {
            DeleteNode(s);  text_buf[s] = c;
            if (s < F - 1) text_buf[s + N] = c;
            s = (s + 1) & (N - 1);
            r = (r + 1) & (N - 1);
            InsertNode(r);
        }
        if ((textsize += i) > printcount) {
            printf("%12ld\r", textsize);  printcount += 1024;
        }
        while (i++ < last_match_length) {
```

Figure 6.17 *Continued*

```
                                DeleteNode(s);
                                s = (s + 1) & (N - 1);
                                r = (r + 1) & (N - 1);
                                if (--len) InsertNode(r);
                        }
                } while (len > 0);
                EncodeEnd();
                printf("In : %lu bytes\n", textsize);
                printf("Out: %lu bytes\n", codesize);
                printf("Out/In: %.3f\n", (double)codesize / textsize);
        }

        void Decode(void)
        {
                int  i, j, k, r, c;
                unsigned long int  count;

                if (fread(&textsize, sizeof textsize, 1, infile) < 1)
                        Error("Read Error");   /* read size of text */
                if (textsize == 0) return;
                StartDecode();  StartModel();
                for (i = 0; i < N - F; i++) text_buf[i] = ' ';
                r = N - F;
                for (count = 0; count < textsize; ) {
                        c = DecodeChar();
                        if (c < 256) {
                                putc(c, outfile);  text_buf[r++] = c;
                                r &= (N - 1);  count++;
                        } else {
                                i = (r - DecodePosition() - 1) & (N - 1);
                                j = c - 255 + THRESHOLD;
                                for (k = 0; k < j; k++) {
                                        c = text_buf[(i + k) & (N - 1)];
                                        putc(c, outfile);  text_buf[r++] = c;
                                        r &= (N - 1);  count++;
                                }
                        }
                        if (count > printcount) {
                                printf("%12lu\r", count);  printcount += 1024;
                        }
                }
                printf("%12lu\n", count);
        }

        int main(int argc, char *argv[])
        {
                char  *s;

                if (argc != 4) {
                        printf("'lzari e file1 file2' encodes file1 into file2.\n"
                                "'lzari d file2 file1' decodes file2 into file1.\n");
                        return EXIT_FAILURE;
                }
                if ((s = argv[1], s[1] || strpbrk(s, "DEde") == NULL)
                 || (s = argv[2], (infile  = fopen(s, "rb")) == NULL)
                 || (s = argv[3], (outfile = fopen(s, "wb")) == NULL)) {
                        printf("??? %s\n", s);  return EXIT_FAILURE;
                }
                if (toupper(*argv[1]) == 'E') Encode();  else Decode();
                fclose(infile);  fclose(outfile);
                return EXIT_SUCCESS;
        }
        .
```

Figure 6.17 *Continued*

input	*	**	***	****
phrase number	1	2	3	4
output	(0,*)	(1,*)	(2,*)	(3,*)

Figure 6.18 LZ78 parsing of a repeated sequence

nating at node numbers 330 and 350 were 15 and 20, respectively. Under the pruning method used by V.42 bis the first leaf node found without descendants, node 320, would be pruned first when the dictionary is filled, even though the string terminating at that node occurs more frequently than strings terminating at the other nodes without descendants. Thus, modifying the pruning process to consider the frequency of occurrence of strings terminating at leaf nodes that have no descendants could result in the retention of dictionary entries that are more likely to occur.

7

IMAGE COMPRESSION

New versions of Word Perfect, Microsoft's Word, and a number of electronic spreadsheet and database programs routinely provide the capability to integrate images into a variety of application programs. Although this is a relatively recent phenomenon, in actuality the use of images as well as the application of compression techniques to reduce the data storage and transmission requirements of images is far from being a recent phenomenon.

In Chapter 5 we examined the use of modified Huffman coding to compress fax transmission. Since fax represents images and the use of modified Huffman coding is over 20 years old, this technique can be considered as one of the earliest applications of compression to images. If you ever used a TIFF file or downloaded a GIF file from a bulletin board, you experienced the use of image compression. Although data compression produces image compression, we will use the term image compression to denote data compression techniques specifically applied to images. Such techniques can include lossless as well as lossy compression techniques, with an example of the former being an image stored using the GIF specification, while an example of the latter can be an image stored using the JPEG specifications.

Since a discussion of image compression requires an understanding of graphic images and how they are stored, we will first focus our attention upon this area to obtain an overview of image storage requirements. Next, we will take a short tour of image compression techniques. This tour will provide us with additional information concerning the application of a variety of compression techniques to the compression of images. This will be followed by an examination of several specific graphic file formats as well as compression techniques supported by each graphic file format. In concluding this chapter, we will focus our attention upon a few tools that can assist us in either developing graphic files or converting images from one graphic format to another.

7.1 GRAPHICS OVERVIEW

Graphic file formats are based upon the method by which graphic programs create, store and display images. Graphic programs and their resulting file formats can be subdivided into one of two categories—raster and vector.

Raster programs

A raster format program works with a series of picture elements, referred to as pels or pixels, and the resulting file format is commonly referred to as bitmapped. A raster format program divides the image area into very small points, typically 1/300 of an inch square, and stores the data for each point. Thus, one square inch at a resolution of 300 dots per inch (dpi) would require 90 000 points.

Each point in a bitmapped or raster image can have two or more states. If the image is black-and-white, each point can be represented by one bit. If the image is gray scale or color, two or more bits will be required to represent each point.

Common gray scale images have either 16 or 256 shades of gray, requiring either four or eight bits to represent the possible gray shading of each point. Color images can range from 16 colors, or four bits per point, up to 16.7 million colors, requiring 24 bits per point.

The number of bits used to represent the gray scale or color of a point is commonly referred to as the color depth of an image. Although more than 24 bits can be used to represent the color of each point, that is a practical maximum as any range of colors beyond that afforded by the use of 24 bits normally cannot be detected by the human eye.

Vector programs

A second type of graphic image program and resulting file format are vector programs which produce vector files. A vector graphics program generates shapes that are made up of line segments. Examples of vector graphics programs include computer aided design (CAD) and map generation and manipulations programs.

Raster versus vector

Raster images are shape independent and permit the input, manipulation and output of images that would be difficult, if not impossible, when using a vector image program. For this reason scanners, digital cameras and digitizer pads provide bitmapped or raster image input.

The output of raster image programs can be accomplished easier and faster to a computer monitor or graphics printer than a vector image program since no vector to raster conversion is necessary. In this chapter we will focus our attention upon raster images.

Value of image compression

To obtain an appreciation for the value of image compression, consider the storage requirements of a $3'' \times 5''$ color picture scanned using a scanner capable of recognizing 256 colors at 300 dpi. The $3'' \times 5''$ color picture consists of 15 square inches, with 90 000 points per inch. Thus, $90\,000 \times 15$ or 1 350 000 bits are required to represent the picture without considering its color depth. Since eight bits (or one byte) are required to obtain a color depth of 256, this results in a minimum data storage requirement of 1.35 Mbytes for the previously mentioned $3'' \times 5''$ color image. The reason the term 'minimum' was used is that the image file must contain a heading which enables the program to denote the type of image stored as well as information about its size, resolution, and color depth. Thus, the actual file would require a few additional bytes of storage, although additional storage would be relatively small in comparison to the total amount of storage required. Without compression we are just able to store one $3'' \times 5''$ image on a 1.44 Mbyte $3\frac{1}{2}$ inch diskette. If you are fortunate enough to have a 200 Mbyte hard drive, the storage of less than 150 $3'' \times 5''$ images would require the entire storage capacity of your drive. Similarly, the transmission time required to send or receive non-compressed images can be lengthy. For example, at 9600 bps (1200 characters or bytes per second) the $3'' \times 5''$ color image would require 1125 seconds or over 18 minutes to transmit or receive. For these reasons, compression can be considered as a necessity when working with images. Fortunately, the file formats associated with many popular graphic images define the use of one or more types of compression which permit many types of images to be stored and transmitted in a compressed form.

Table 7.1 lists nine popular types of image files, their file extension and color support. Readers should note that the documentation that provides a detailed overview of each of the image file formats listed in Table 7.1 cumulatively exceeds the page count of many comprehensive dictionaries. Thus, it should come as no surprise that we will limit our examination of image file formats to those formats that support compression and are primarily used to transport images between computers. Similarly, our description and review of image compression tools will be limited to those that work with the image formats we will examine in detail.

Table 7.1 Popular image file formats

Description	File extension	Colors
OS/2 bit map	BMP	B&W/Color
Windows bit map	BMP	B&W/Color
Windows Clipboard	CLP	B&W/Color
Paint Brush	PCX/PCC	B&W/Color
GEM environment	IMG	B&W/Color
Dr. Halo	PIC	Color
Graphics Interchange Format	GIF	B&W/Grayscale/Color
Tag Image File Format	TIF	B&W/Grayscale/Color
Joint Photographic Experts Group	JPG	Grayscale/Color

Table 7.2 Image compression techniques

Lossless
  Character based
  Statistical/entropy
  Dictionary based

Lossy
  Binary graphics
  Object model

7.2 IMAGE COMPRESSION TECHNIQUES

A wide variety of compression techniques have been developed over the past 20 years to reduce the size of bitmapped image files. Some techniques represent the application of modified lossless compression algorithms, while other techniques represent the application of lossy algorithms.

We can classify image compression techniques into a minimum of five general categories which are listed in Table 7.2. Readers should note that this classification scheme represents the views of the author of this book and other classification schemes can be developed to group compression techniques applied to images. In examining the entries in Table 7.2, note that the categories are subdivided based upon whether or not they provide a fully recoverable (lossless) image.

Character based

Character based compression techniques operate upon the grouping of pixels as a byte entity. Thus, solid backgrounds in which pixel patterns repeat for several byte groupings would be suitable for reduction from a character based compression technique.

Statistical

Figure 7.1a illustrates a poorly drawn image of a house assumed to be located in an area where most of the background represents either blue sky or green earth, except for a road leading to the house. Assume zeros are used to denote the background of the uniform sky included within a block of pixels while 1s are used to represent a block of pixels that cover a uniform green lawn. Then, Figure 7.1b can be considered to represent the subdivision of the image into blocks of pixels that have the same or similar pixel characteristics. If the blocks have the same pixel characteristics, with each pixel exactly the same as the other pixels to include their color depth sequence, then the repeating sequence may not occur on a byte basis. Instead, groups of bytes to include the pixels' color depth may repeat for a sequence. Thus, a statistical or entropy based compression method, such as Huffman coding or arithmetic coding, could be applied to the image and would more than likely produce better results.

Dictionary based coding

Since most images consist of repeating sequences of pixels a dictionary based coding method, such as any of the Lempel–Ziv algorithms or derivatives, can be applied to an image file. In fact, one of the most popular image file formats, CompuServe's GIF specification, uses a modified form of LZW compression to reduce the size of the resulting

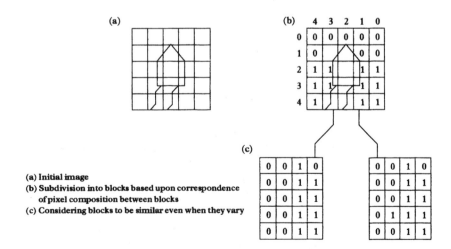

(a) Initial image
(b) Subdivision into blocks based upon correspondence
    of pixel composition between blocks
(c) Considering blocks to be similar even when they vary

Figure 7.1 Examining an image

stored image. In applying LZW compression, the GIF specification first causes the separation of the color depth from the pixel values representing the image prior to applying compression to the pixels. Later in this chapter we will examine the GIF specification in detail.

Binary graphics

There are a large number of compression techniques specifically developed to operate upon images. Some of those techniques, such as the CCITT Group 3 modified Huffman code, run-length coding and relative address coding, can be considered to represent lossless binary graphics compression techniques. Other compression techniques that fall into a binary graphics category are lossy. Examples of the latter include predictive coding, Adaptive Bilevel Image Compression (ABIC), and the Joint Photographic Experts Group (JPEG) image compression technique.

Predictive coding

Predictive coding uses previously encoded parts of an image to predict the composition of future parts, resulting in lossy compression. This scheme was first described in 1976 (Preuss, 1976) and was applied to facsimile encoding of binary data.

Under predictive coding, data is scanned in raster format and at each pixel position a prediction is made based upon the composition of neighboring previously processed pixels. If the prediction can be done 'perfectly', such as in a background area, then the pixel is coded as a zero. Otherwise, the actual value of the pixel is coded. The result of the prediction process is clusters of zero coded pixels for homogeneous areas. Those pixels can then be reduced through the use of another compression technique, such as run-length coding or relative address coding.

Adaptive Bilevel Image Compression

Adaptive Bilevel Image Compression (ABIC) refers to a class of compression techniques developed at IBM (Mitchell and Pennebaken, 1988) and is currently being standardized by the Joint Binary Image Group (JBIG). ABIC uses a two-dimensional predictive coder to determine the relative probabilities of a 0 or a 1 in the next scanned pixel. The conditional probability is then used by an arithmetic coder to minimize the size and transmission rate of an image to near the entropy of the source. The probability estimator and arithmetic coder was patented by IBM and is referred to as a Q-coder. JBIG expands upon the work of IBM by introducing adaptive templates. Here the template represents the composition of a group of pixels that can be used to define a relationship to the current pixel being modeled. For example, a template could represent the composition of three pixels on the same line preceding the pixel being modeled and five pixels centered above that pixel on the preceding line. At the time this book was written, the JBIG standard was being developed based upon IBM's original research efforts for the transmission of binary or bilevel documents between workstations.

JPEG

The JPEG compression and decompression algorithm represents a linked series of steps which are described in a comprehensive standards document available from the American National Standards Institute (ANSI). Figure 7.2 summarizes the major functions associated with the basic JPEG algorithm.

The basic JPEG algorithm can be considered as a baseline for which a number of extensions exist. Two examples of JPEG extensions include progressive coding which results in the gradual construction or decomposition of an image and the use of arithmetic coding to replace Huffman coding specified in the baseline algorithm.

Initially JPEG groups pixels into 8×8 blocks, organized as chrominance (color) and luminance (intensity or brightness) components. When applied to a standard computer monitor image composed of red, green and blue (RGB) pixels, this results in a YUV transformation, where Y represents intensity and U and V represent color values. In a YUV form the same picture image requires less storage than in RGB form; however, there is no perceptible loss of image quality.

After the pixels in an image are grouped into blocks, a discrete cosine transformation (DCT) process converts blocks of YUV pixels into sets of coefficients representing the isolated frequency components of the colors and intensities of the block. During the DCT process, each block is converted into a set of 64 coefficients—one dc coefficient and 63 ac coefficients. This action shifts the pixel block from the spatial domain to the frequency domain, resulting in most of the transformed block's energy being concentrated in the lower frequencies. Prior to the quantization step the dc coefficient is differentially encoded with respect to the prior block.

The quantization step results in the use of a table of 64 quantization

Figure 7.2 The basic JPEG compression algorithm

values which the 63 ac block coefficients are compared against. Since most matches are not exact but are approximations, this action results in the lossy component of JPEG. For example, consider the composition of the two subblocks illustrated in Figure 7.1c. During the quantization process small differences between blocks may not result in the use of a different quantization value being assigned to each block. Thus, the two subblocks could have the same coded value; however, in this example decompression would result in the loss of one pixel's correct value. In addition, since quantization is applied to the frequency matrix generated by the CDT operation, the quantization process can be varied to reduce the size of the resulting data stream. For example, since human vision is less sensitive to color than to intensity, the number of quantization values or steps used to represent color may be reduced, resulting in a higher degree of compression for color components while limiting visually percep- tible image degradation. Similarly, since human vision is less sensi- tive to high frequency details than to low frequency details, the steps used to represent high frequency values can be increased which enhances the compressibility of the image.

The use of run-length and Huffman coding further reduces the size of the image. Both a static predefined Huffman table and a two-pass Huffman table created to represent the specific image are supported by JPEG. The final step in the JPEG process is data packing, in which bit sequences produced by the Huffman coder are grouped into bytes.

Object model

The ability of an image to be modeled can significantly reduce redun- dancies. Examples of object model compression techniques include pattern match coding and fractal based coding.

Pattern matching

Pattern match coding results in the subdivision of an image into blocks that have similar patterns or shapes. Then, only a block iden- tifier and block position of blocks whose composition match a prior identified block requires encoding. For example, consider Figure 7.1b. Here the image was broken into 25 blocks and conveniently, for illustrative purposes, 19 represent sequences of pixels that were rep- licated in a similar manner in other blocks. A PMC technique would store the contents of one block numbered 0 and one block numbered 1. Then, to encode block (0,3) the sequence 0,0,3 could be used, with the first number denoting the fact that the block at position 0,3 matches the contents of block 0.

A comparison of the pixel contents between blocks results in an increase in block size, lowering the probability of an exact match of the composition of pixels. Thus, there is a tradeoff between attempting to match larger blocks and the ability to do so. One method which partially alleviates this tradeoff is to assume the pixel contents of blocks match even when they physically do not whenever the difference between blocks is under a certain threshold. For example, Figure 7.1c illustrates the composition of two adjacent image blocks. Although they do not exactly match, the PMC algorithm can be configured to consider the blocks to match. While this method results in an inability to fully replicate the pixel content of the image, it increases the ability of the algorithm to obtain a more effective compression ratio.

Fractal coding

Fractal coding of images is based upon the work of the Frenchman, Benoit Mandelbrot, whose classic book *The Fractal Geometry of Nature* (Mandelbrot, 1982) can be considered as opening a new branch of mathematics. Mandelbrot coined the word 'fractal' to describe objects that are 'fractured' for which mathematical formulas can be used to represent an image.

Based upon the work of Mandelbrot, a firm based in Atlanta, Georgia, Iterated Systems, developed software that can be used to compress images achieving compression ratios up to or higher than 10 000:1. The technique used by Iterated Systems commences with breaking a digitized image into blocks or segments. Each segment is compared to a library of iterated function system (IFS) codes that reproduce corresponding fractals, significantly reducing the size of the library. The IFS codes generate fractals that are compared to the pixel composition of a block. When the fractal provides a close approximate of the contents of the block the code replaces the block. Although this compression technique can achieve a compression ratio several orders of magnitude or greater than other techniques, the computational time required to encode an image can exceed 100 hours or more. Thus, this technique is impractical for on-the-fly compression applications; however, decompression can occur relatively fast, which makes this technique suitable for encoding and distributing large numbers of images or images associated with a large database.

7.3 GIF FILE FORMATS

There are two GIF formats presently defined—GIF87 based upon the CompuServe standard proposed in 1987, and GIF89 which represents a modification to the original standard. The latter is more formally referred to as GIF89a, referencing GIF Version 89a. Both GIF and 'Graphics Interchange Format' are trademarks of CompuServe, Inc., an H&R Block company.

The original GIF specification was developed in 1987 when the ability to display more than 16 colors on a monitor was considered to be in the distant future. Due to this, the 1987 standard, as well as its 1989 revision, are limited to 256 colors out of a palette of 16 million. While this limits the ability of GIF to represent truly stunning graphics, it also limits the size of a file required to store an image. This in turn reduces the time required to transmit GIF encoded images.

A modified form of LZW compression is built into both GIF standards. This provides a lossless or reversible compression method. In comparison, JPEG, a technique described later in this chapter, supports a lossy compression process. Although this results in degrading the image quality, it also enables file sizes to be considerably reduced. The actual image quality degradation is controlled by the user and a significant reduction in the size of a stored image can be achieved with little to no visually perceptible image distortion. However, if a user tends to be greedy and requires an additional reduction in the size of the stored image, the ability to visually recognize the image may be impaired as we will note later in this chapter. Figure 7.3 illustrates the general GIF file format essentially applicable to both GIF87a and GIF89a. As we examine each field in the file format illustrated in Figure 7.3, we will note the differences between the two standards.

GIF Signature

Under the GIF87a standard the GIF Signature field consists of the entry GIF87a in the first six bytes of the file. Under the GIF89a standard the GIF Signature field was renamed as the Header; however,

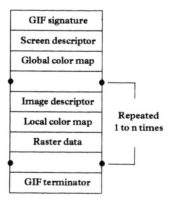

Figure 7.3 General GIF file format

six bytes are still used to identify the context of the GIF data stream. Bytes 0 to 2 continue to contain the fixed value GIF, while bytes 3 through 5 identify the version number used to format the data stream. Thus, the GIF87a standard is limited to supporting one data stream format while the GIF89a standard can support different data stream formats.

Screen Descriptor

The screen description field consists of eight bytes which define the parameters of GIF images included in the file. Those parameters include the raster width and height of the GIF image, an indicator concerning whether or not a global color map follows the screen descriptor field, the number of bits of color resolution, the number of bits per pixel in the image and the color index associated with the screen background. Figure 7.4 illustrates the format of the Screen Descriptor field.

Under the GIF87a standard the value of pixel represents the maximum number of colors used to represent an image. Since three bits are used for cr, the range of values of 0 to 7 represents one to eight bits which supports two (black and white) to 256 colors. Also under the GIF87a standard, bit 3 of word 4 as well as all bits in word 6 were reserved for future use. Under the GIF89a standard, bit 3 of word 4 becomes a sort flag. When set to 1, it indicates that a following Global Color Table is ordered by decreasing importance, with the most important color appearing first in the table. When the sort flag is set to 0, it indicates the Global Color Table is not sorted. A second change under the GIF89a standard concerns the use of word 6, which was reserved for future use under the GIF87a standard. Under the

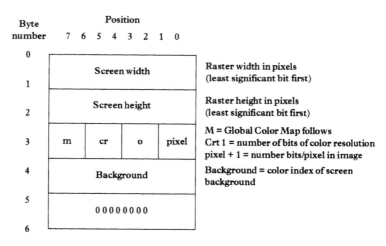

Figure 7.4 Screen Descriptor field

GIF89a standard word 6 contains the pixel aspect ratio which represents a factor used to compute an approximation of the aspect ratio of the pixel in the original image. When the value of the field is not zero the approximation of the aspect ratio is obtained from the following formula:

$$\text{Aspect ratio} = \frac{\text{Pixel aspect ratio} + 15}{64}$$

where the pixel aspect ratio is the quotient of the pixel's width over its height. The value range of this field permits specification of the widest pixel of 4:1 to the tallest pixel of 1:4 in increments of 1/64th.

Global Color Map

The Global Color Map, which was renamed the Global Color Table under the GIF89a standard, contains a sequence of bytes representing red–green–blue color triplets. Figure 7.5 illustrates the format of the Global Color Map/Table.

The presence of a Global Color Map/Table is indicated by a nonzero M flag. The number of bytes in this map/table is equal to:

$$3 \times 2^{(\text{Global Color Map/Table size} + 1)}$$

Each image pixel value received is displayed according to its closest match with an available color of the display based upon its color map.

If no Global Color Map/Table is indicated by the M bit, a default color map/table is internally developed. That map/table maps each possible incoming color index to the same hardware color index modulo $\langle n \rangle$, where $\langle n \rangle$ is the number of available hardware colors.

Byte number	Bit position 7 6 5 4 3 2 1 0
0	Red 0
1	Green 0
2	Blue 0
3	Red 1
4	Green 1
5	Blue 1
.	.
.	.
.	.
767	Blue 255

Figure 7.5 Global Color Map/Table

The color components represent a fractional intensity value ranging from none (0) to full (255). Thus, white would be represented as (255,255,255), black as (0,0,0) and medium yellow as (180,180,0).

Image Descriptor

The Image Descriptor field consists of 10 bytes which define the actual placement of an image within the space defined in the Screen Descriptor as well as how the image is formatted, whether a local or global color map is to be used, and the number of bits per pixel used for the image. Figure 7.6 illustrates the format of the Image Descriptor field.

The Image Separator functions as a synchronization character to the following image descriptor. The value of this field is fixed as hex 2C.

The Image Left field contains the column number, in pixels, of the left screen edge of the image with respect to the left edge of the logical screen. The Image Top field defines the row number, in pixels, of the top edge of the logical screen. The leftmost column and top row of the logical screen are both 0. The Image Width and Image Height fields function similarly to the Image Left and Image Top fields, defining the width and height of the image in pixels.

The packed subfields in byte 9 define whether to use a Global Color Map/Table (M = 0) and ignore the pixel subfield or use the Local Color Table (M = 1) and use the pixel subfield, whether the image is formatted in sequential order (I = 0) or the Image is formatted in interlaced order (I = 1).

Under the GIF89a format, bit 5 of byte 9 is used as a sort flag. When set to 0 this field indicates that a local color table is not ordered. When set to a value of 1, this field indicates that the Local Color Table is

Byte number	Bits 7 6 5 4 3 2 1 0
0 1	Image separator
2	Image left
3 4	Image top
5	Image width
6 7	Image height
8 9	M | I | O | O | O | Pixel

Figure 7.6 Image Descriptor Field

sorted in order of decreasing importance, usually with the most frequent color placed first in the table. The last subfield in byte 9 defines the size of the optional Local Color Map/Table and is only applicable when $M = 1$. The value of the three-bit field must be raised to the power of 2, and 1 added to the result (pixel^2 + 1), to determine the actual size of the Local Color Map/Table.

Local Color Map/Table

The optional Local Color Map/Table's presence is indicated by the setting of the M bit in byte 9 of the Image Descriptor field. This map/table contains a sequence of bytes representing red–green–blue color triplets whose format is the same as in the Global Color Map/Table previously illustrated in Figure 7.5. If the Local Color Map/Table is present, it temporarily becomes the active color table and is used to process the image that follows that map/table.

Raster Data

The Raster Data field defines the format of the actual image as a series of pixel color index values. Those pixels are stored left to right sequentially for an image row, with each image row ordered sequentially top to bottom. The exception to this ordering occurs when the I bit in byte 9 of the Image Descriptor is set to '1' to indicate an interlaced image. When this occurs, the row order of the image display follows a four-pass process in which the image is filled in by widely spaced rows. The first pass writes every eighth row, commencing with the top row of the image window. The second pass writes every fourth row, commencing at the fifth row from the top. The third pass writes every fourth row starting at the third row from the top, while the fourth pass completes the image by writing every other row, commencing at the second row from the top. Figure 7.7 illustrates the interlace process.

The image pixel values are processed as a series of color indices which map into the color map/table used. Then the resulting color values from the map/table are displayed. The series of pixel indices contained in the Raster Data field (equals the image-width × image-height) are passed to the GIF image data stream one value per pixel, compressed and packaged based upon the use of a version of LZW compression.

LZW algorithm

A variation of the LZW compression algorithm previously described in Chapter 6 is used to compress the Raster Data field. The first byte

Image Row	Pass 1	Pass 2	Pass 3	Pass 4	Result
0	**1a**				**1a**
1				**4a**	**4a**
2			**2a**		**3a**
3				**4b**	**4b**
4		**2a**			**2a**
5				**4c**	**4c**
6			**3b**		**3b**
7				**4d**	**4d**
8	**1b**				**1b**
9				**4e**	**4e**
10			**3c**		**3c**
11				**4f**	**4f**
12		**2b**			**2b**

Figure 7.7 Interlaced four-pass storage process

in the Raster Data field represents the initial LZW code size which defines the number of bits used to represent each compression code and conventional character. When the initial dictionary is filled, the algorithm increments the LZW code size by one bit. Black-and-white images that have one color bit must use a code size of 2 to obtain compressibility, representing the minimum code size. In actuality one bit is added to each code size per code, up to 12 bits per code. Thus, the maximum code size is 12 bits, which permits a dictionary of 4095 entries to be supported.

LZW compression results in variable length codes from three to 12 bits in length which are then reformatted into a series of 8-bit bytes for storage. The packing occurs right to left, with each group of 255 bytes packed into blocks prefixed by a block size header. If the last block does not contain 255 characters then the block size header indicates the size of the terminating block. Figure 7.8 illustrates the Raster Data field which contains the results of the LZW operation.

The last byte in the Raster Data field, the Block Terminator subfield, consists of a single byte with the value hex 00. As its name implies, this block terminates a sequence of data subblocks.

Although the GIF LZW algorithm matches the standard LZW algorithm, it differs from the latter in several ways. First, the GIF LZW algorithm uses a Clear Code whose value is 2 code size to reset all compression parameters and tables to the start-up state. Secondly, an End of Information Code is defined which indicates the end of the image data stream. The value of this code is the Clear Code + 1, which explains why the first available compression code value becomes ⟨clear code⟩ + 2. As previously mentioned, the output codes are variable in length commencing at ⟨code size⟩ + 1 bits per code, up to 12 bits per code. Table 7.3 summarizes the GIF LZW parameters and their values.

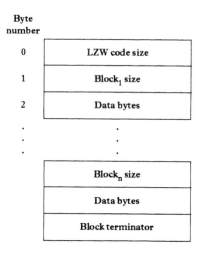

Byte
number

0	LZW code size
1	Block₁ size
2	Data bytes

Figure 7.8 Raster Data field

Table 7.3 GIF LZW parameters

LZW parameter	Value
Clear Code	$2 \times$ code size
End of Information Code	clear code + 1
First available compression code value	clear code + 2
Output codes (min, max)	code size + 1, 12 bits
Terminator block	hex 00

GIF89 extensions

Under the GIF89 standard several optional extensions can follow the GIF terminator. A graphic Control Extension field includes information on the manner in which the image is to be treated once displayed, whether or not user input is expected and a delay time for waiting prior to continuing with the processing of the data stream. Other extensions include a Comment Extension block which permits inclusion of textual information which is not past the image and a plain Text Extension which contains textual data and the parameters necessary to render the data as a graphic. Since we are primarily concerned with GIF compression, readers are referred to the CompuServe FIG89, a standard for detailed information concerning the previously mentioned extensions.

Proposed GIF specification modules

A most interesting proposal for modifying the GIF specification to improve the efficiency of its LZW compression was made public on CompuServe in June 1994 by J.F.R. 'Frank' Slinkman. Mr Slinkman, who can be reached on CompuServe at 72411,650, noted that many of the LZW codes added to the code table are never used. This causes the table to fill up sooner than necessary and degrades the efficiency of the LZW compression scheme.

If all unused codes are simply removed each time the table becomes full, a situation will develop where a few new and unreferenced codes will be placed near the top of the table and the table will fill prior to those codes having the opportunity to be referenced by higher codes. Unfortunately, removing those codes will result in their eventual replacement by additional never-to-be-referenced codes which would obstruct normal, full clearing. To rectify this situation, Mr Slinkman proposed the operation of a function that searches for unused codes in increasingly smaller and higher portions of the LZW table. Figure 7.9 lists the C language code of his proposed modification to the GIF standard. Mr Slinkman's C language function and his proposal are contained on the file PRTCLR.ARC on the convenience diskette labeled 'Graphics'.

The first time the function is called, it checks the bottom half of the table. This search is based upon the premise that a code not referenced by any of the 2048 codes in the top half of the table is unlikely to be used. Once the unused codes in the bottom half of the table are removed, the higher codes are relocated to replace the removed codes. Each subsequent function call checks the bottom half of the portion of the table commencing at the slot one above the new location of the highest code which was previously checked. If fewer than 1/32nd of the codes checked are unused during any pass, the process ends with the function resetting itself and returning zero, indicating that the entire table should be cleared in the normal manner. In addition, if the area to be checked is too small and too high in the table, the function will also reset itself and return zero since codes high in the table refer to greater lengths than codes lower in the table. This technique provides each code an opportunity to be referenced by a higher code while acknowledging the fact that if few unused codes are encountered LZW compression is being performed quite efficiently and the code table should not be modified.

The C language program listing contained in Figure 7.9 demonstrates one way to locate and remove unused LZW table codes. Readers should note that this 'partial-clear LZW' technique represents a suggestion for a change in the GIF specification and at the present time is not a part of the GIF specification.

```
/*   partclr.c
 *
 *   set tabs every 4th column to properly view this file  *
```

```
|                                                                        |
| Routine to remove unused codes from LZW table to increase compression. |
|                                                                        |
|  Author:         J.F.R. "Frank" Slinkman, 72411,650                    |
|  Date:           23-Jun-94                                             |
|  Copyright:      License hereby granted for private, personal, non-commercial |
|                  use.  Inclusion of any portion of this code, in commercial |
|                  (including shareware) or business applications/products |
|                  requires the express written consent of copyright holder. |
|  Compiler:       Pro-MC, from Misosys, Inc.                            |
|                  This is standard K&R C code.                          |
|  Libraries:      No special libraries required.                        |
|_____|
*/
```

```
/*   this code assumes the following are available globally  */

struct table {   short int        prefix;
                 unsigned char  suffix;
             } LZWtable[4096];      /* LZW code table */
int     LZW_min_code_size,          /* from data stream  */
        next_free_slot,             /* number of next unused LZW table slot */
        code_size;                  /* current LZW bitstream code size */

/*=*=*=*=*=*=*
 *
 *   Do NOT attempt to run this program!  It is only written this way to
 *   enable you to test-compile it to check for errors and determine what,
 *   if any, modifications need to be made before incorporating its code
 *   into your existing GIF encoder/decoder.
 *   *=*=*=*=*=*=*/

main()
{
 /*   partial clearing function would be called as follows:  */

     if ( next_free_slot > 4095 || code_size > 12 )   /*   choose 1 test  */
         if ( next_free_slot = part_clr( 0 ) )
           {
               int       temp;
               code_size = 0;
               temp = next_free_slot;
               while ( temp )
               {   code_size++;
                   temp >>= 1;
               }
           }
         else
             perform_normal_clear();
}
/*=*=*=*=*=*=*
 *
 *   int part_clr( int reset )
 *
 *   Arguments:
 *
```

Figure 7.9 Proposed modification to GIF standard, to remove unused codes from LZW table

```
 *  One:  TRUE (non-zero) to reset, FALSE (zero) to cause unused codes
 *  to be removed from the LZW table.  The function should be reset at
 *  the start of each new image, particularly for multi-image files.
 *
 *  Return code:
 *
 *  When called with a non-zero argument:
 *
 *       part_clr() will return zero, which should be ignored.
 *
 *  When called with a zero argument:
 *
 *       If no codes were removed from the LZW table, part_clr() returns zero,
 *       which means your program must clear the LZW table in the usual manner.
 *
 *  If unused codes have been removed from the LZW table, part_clr()
 *  returns a non-zero value: the number of the highest free slot in
 *  the table, and your program should continue (de)compression using
 *  new code-size and next-free-slot information, as illustrated above
 *  in the ersatz main() function.
 *
 *=*=*=*=*=*=*=*/

int part_clr( reset )
int  reset;
{  extern void free();
    extern char *calloc(), *malloc(), *memcpy(), *memset();

    static int    protect = 2048, first = 0;
    char       *flags;
    int        *moves, retcode = 0, ctr, j, i;

    /* reset static values and return zero if "reset" or test area too small  */
    if ( reset || protect < 200 )
    {    protect = 2048;
        first = 0;
        return retcode;
    }
    flags = calloc( 4096, sizeof(char) );/* allocate & zero flags */
    moves = malloc( 4096 * sizeof(int) );/* allocate move ctrs */

    /* protect top "protect" codes in LZW table by setting their flags  */

    memset( flags + 4096 - protect, 1, protect * sizeof(char) );

    /* if first call after a reset, calc slot number of lowest non-root code  */

    if ( !first )
        first = ( 1 << LZW_min_code_size ) + 2;
    /* set flag for each LZW code actually used  */

    for ( i = first; i < 4096; i++ )*( flags + LZWtable[i].prefix ) = 1;

    /* record how many slots to move each LZW table entry down, and count number of
       unused codes  */

    for ( ctr = 0, i = first; i < 4096; i++ )
    {    moves[i] = ctr;
        if ( !flags[i] )
            ctr++;
```

Figure 7.9 _Continued_

```
      }

      /* if too few (i.e., < 1/32nd) unused codes, reset   */

      if ( ctr < protect >> 5 )
              protect = 2048;
          first = 0;
      }

      /*   otherwise remove unused codes   */

      else
      {
      /* update prefixes in LZW table   */
          for ( i = first; i < 4096; i++ )
              LZWtable[i].prefix -= moves[i];

      /* now squeeze out unused LZW table entries */
          for ( i = first; i < 4096; i++ )
              if ( j = moves[i] )
                  memcpy( LZWtable + i - j, LZWtable + i, sizeof(struct table) );

      /* Calc starting slot number for next call.  All codes below this slot survived
         removal, and thus are referenced by higher codes. */

          first = 4096 - protect - ctr;
      /* protect top half of remaining table for next call */
          protect = ( 4096 - first ) >> 1;
          /* calc lowest free slot number for return */
          retcode = 4096 - ctr;
      }
      free ( moves );
      free( flags );
      return retcode;      /* either zero or new next_free_slot */
}
 /*=*=*=*=*=*=*/
void perform_normal_clear()
{
      /   your existing code to clear LZW table   */
}
      .
```

Figure 7.9 *Continued*

7.4 JPEG FILE FORMATS

From a strictly technical perspective, JPEG references a family of compression algorithms and does not actually reference a file format. According to an Internet newsgroup, the JPEG has been hampered in its quest to define a file format by 'turf wars' within another standards-making organization. Since the ability to exchange JPEG stored images is dependent upon the use of a common file format, C-Cube Microsystems of Milpitas, CA, picked up the gauntlet and defined two JPEG-based file formats.

The first file format defined by C-Cube Microsystems is JFIF, an acronym for JPEG File Interchange Format. This format is designed as a mechanism to transport pixels. The second file format is TIFF/JPEG, which represents an extension of the Aldus TIFF format.

The TIFF/JPEG file format is commonly referred to as TIFF 6.0 and is significantly more complex than JFIF. Since different vendors have

often implemented slightly different and resulting incompatible subsets of TIFF, this also makes this file format generally less transport-
able than JFIF. This resulted in the JFIF file specification becoming
a *de facto* standard; however, some popular software programs, such
as Handmade Software's DOS programs GIF2JPG and JPG2GIF, use
a proprietary file format. Fortunately, many of those programs are
capable of reading and writing JFIF format through the use of an
appropriate program option switch. The JPEG File Interchange Format (JFIF) can be found on the file JPEG.DOC which is contained on
the archive JPEG4.ZIP on the diskette labeled 'Graphics'
accompanying this book. In the next section in this chapter we will
discuss the programs contained on that archive.

7.5 IMAGE TOOLS

Over the past few years literally hundreds of image tools have been
developed, ranging in scope from image viewers, conversion programs
and print programs to program modules that can be incorporated into
programs you wish to develop and which facilitate the program development process. Obviously a limitation on both page and diskette
space limits the coverage of image tools in this section to but a fraction of those readily available from anonymous FTP archives on the
Internet, bulletin board systems, information utilities and commercial organizations.

In this chapter a GIF decoder program listing is presented as a
learning tool. This program was developed using QuickBasic and its
structure can be used to facilitate developing GIF programs. Because
the vast majority of computer-stored images use the GIF or JPEG
(JFIF) file formats, the data reduction obtainable from converting GIF
into JPEG has received a considerable degree of interest. Thus, this
section includes an overview of the operation of two conversion programs you can use for converting GIF stored images to JPEG and
vice versa.

GIF coding

To assist readers who want to develop programs that read GIF file
formats, such as a viewer, two program listings are presented in this
section. The first program, QBGIF.BAS, was developed by Don Babcock to operate under the QuickBASIC environment, as Mr Babcock
was doing a lot of development work using that language during 1988.
Although the program only supports the GIF87a standard, its
detailed documentation should provide you with the information
necessary to handle GIF89a extensions as well as to customize the
program to support a specific programming requirement. Figure 7.10

```
        DEFINT A-Z
        CONST Prefixes = 0
        CONST Suffixes = 1
        DIM StringTable%(4095, 1)
        DIM Stack(4095)
        False = 0
        True = NOT False

        ze$ = CHR$(0)
        cmd$ = COMMAND$
        CLS
        CLOSE 1
        OPEN cmd$ + ".gif" FOR INPUT AS 1      'make sure it exists
        CLOSE 1
        OPEN cmd$ + ".gif" FOR BINARY ACCESS READ AS 1
        'All additions to the header processing should be a part of the
        'kernel
Kernel:
        GOSUB GIFHeader
        CLS
        GOSUB ImageProcessing
        CLOSE
        BEEP
        WHILE a$ = ""     'Wait loop to keep the image on screen until a key is pressed
            a$ = INKEY$

        WEND
        END '**************** Program ENDS ******************
'***************** SUBROUTINES *****************************************
GIFHeader:        'Process the Gif Header Information
        DO               'ignore leading junk bytes
            g$ = INPUT$(1, #1)
        LOOP WHILE g$ <> "G" AND NOT EOF(1)

        if$ = INPUT$(2, #1)
        Version$ = INPUT$(3, 1)
        ScreenWidth = CVI(INPUT$(2, #1))      'Convert two bytes to an integer
        ScreenHeight = CVI(INPUT$(2, #1))     ' as above
        MapDat = ASC(INPUT$(1, #1))
        GlobalMap = (MapDat AND &H80) / &H80 'Get the Global Color Map Bit
        Cres = (MapDat AND &H70) / &H10 + 1  'Color Resolution
        R1 = (MapDat AND &H10) / &H10        'Reserved Bit
        PixelSize = (MapDat AND &H7) + 1     'Pixel Size
        BGColor = VAL(INPUT$(1, #1))         'Background Color Index
        SrtDat = ASC(INPUT$(1, #1))           'Sort and Aspect Ration information
        Sort = (SrtDat AND &H80) / &H80'Sorted Color Map Flag
        PixAspect! = ((SrtDat AND &H7F) + 31) / 64!'Pixel Aspect Ratio
        colors = 2 ^ PixelSize                'Number of Possible Colors
'**** Select overall screen mode based on screen height.
'   This select structure can is used to select the screen setup based upon
'   the parameters from the header. These were my intial choices but are by
'   no means absolute.
'
'   I use two basic methods for pixel display.
'
'   1) Select a screen mode which closely matches the global screen specs
'      and then paint actual pixels as required.
'
'   2) Select a higher resolution screen mode than required then map the
'      screen coordinates to the image screen size using the basic WINDOW
'      SCREEN function. To eliminate the gaps that would result from just
```

Figure 7.10   QBGIF.BAS program listing

```
'    plotting pixels, I draw filled boxes that are 1 logical pixel wide
'    and 1 logical pixel wide. In this way the BASIC function takes care
'    of the "gaps".
'
'    In the select structure, Pixel is set TRUE for real pixel drawing or
'    FALSE for Mapping and Box drawing.
'
'    The scale parameter is set to enable the palette selection math to
'    calculate properly.

SELECT CASE ScreenWidth
     CASE 1 TO 320
          mode = 13
          Pixels = True
          SCREEN mode
          WINDOW
          scale = 4
     CASE 640
          SELECT CASE ScreenHeight
               CASE 200
                    mode = 8
                    Pixels = True
                    SCREEN mode
                    WINDOW
                    scale = 1
               CASE 201 TO 350
                    mode = 9
                    Pixels = True
                    SCREEN mode
                    WINDOW
                    scale = 1
               CASE 351 TO 480
                    mode = 12
                    Pixels = True
                    SCREEN mode
                    WINDOW
                    scale = 4
               CASE ELSE
                    mode = 12
                    Pixels = False
                    SCREEN mode
                    WINDOW SCREEN (1, 1)-(ScreenWidth, ScreenHeight)
                    scale = 4
          END SELECT
     CASE 641 TO 700
          scale = 4
          mode = 12
          Pixels = True
          SCREEN mode
          WINDOW
     CASE 720
          scale = 4
          mode = 12
          Pixels = True
          SCREEN mode
          WINDOW
     CASE ELSE
          scale = 4
          mode = 12
          Pixels = False
          SCREEN mode
```

Figure 7.10 *Continued*

```
                    WINDOW SCREEN (1, 1)-(ScreenWidth, ScreenHeight)
               END SELECT

'****** Print some diagnostic information before image processing

     PRINT "Screen Width", ScreenWidth
     PRINT "Screen Height", ScreenHeight
     PRINT "Mode=", mode,
     PRINT "Scale=", scale
     PRINT "Press SPACE BAR to continue..."

     WHILE p$ <> " "            'clear the keyboard buffer.
            p$ = INKEY$
     WEND

     WHILE INKEY$ <> ""         'wait for a key press to proceed.
          dummy$ = INKEY$
     WEND

GlobalMap:    'Get Global Color Map information
     ColorOffset = 0  'This will be used in the future for palette fragmentation
     IF GlobalMap THEN GOSUB GetColorMap
RETURN '**************** GIFHeader Ends ************************
GetColorMap:    '********** Process Color Map Information *************

     FOR i = ColorOffset + 1 TO 2 ^ PixelSize + ColorOffset
               Red = INT(ASC(INPUT$(1, #1)) / scale)
               Green = INT(ASC(INPUT$(1, #1)) / scale)
               Blue = INT(ASC(INPUT$(1, #1)) / scale)
          SELECT CASE mode
               CASE 12, 13
                    PALETTE i - 1, 65536 * Blue + 256 * Green + Red
                    ' per QUICKBASIC manual. Colors are outstanding!
               CASE 9
                    pal = 0
                    SELECT CASE Red
                         CASE 0 TO 63                'black
                              pal = pal
                         CASE 64 TO 127              'dark red
                              pal = pal + &H20
                         CASE 128 TO 191        'medium red
                              pal = pal + &H4
                         CASE 192 TO 255        'light red
                              pal = pal + &H24
                    END SELECT
                    SELECT CASE Green                'and so forth
                         CASE 0 TO 63
                              pal = pal
                         CASE 64 TO 127
                              pal = pal + &H10
                         CASE 128 TO 191
                              pal = pal + &H2
                         CASE 192 TO 255
                              pal = pal + &H12
                    END SELECT
                    SELECT CASE Blue
                         CASE 0 TO 63
                              pal = pal
                         CASE 64 TO 127
                              pal = pal + &H8
                         CASE 128 TO 191
```

Figure 7.10 *Continued*

```
                                        pal = pal + &H1
                            CASE 192 TO 255
                                        pal = pal + &H9
                        END SELECT
                        PALETTE i - 1, pal
                CASE 8
                        pal = 0
                        SELECT CASE Red
                            CASE 0 TO 127               'red bit off
                                    pal = pal
                            CASE 128 TO 255     'red bit on
                                    pal = pal + 4
                                    intense = 8 'set intensity bit (optional)
                        END SELECT
                        SELECT CASE Green                   ' and so forth
                            CASE 0 TO 127
                                    pal = pal
                            CASE 128 TO 255
                                    pal = pal + 2
                                    intense = 8
                        END SELECT
                        SELECT CASE Blue
                            CASE 0 TO 127
                                    pal = pal
                            CASE 128 TO 255
                                    pal = pal + 1
                                    intense = 8
                        END SELECT
                    PALETTE i - 1, pal + intense
            END SELECT
    NEXT i
    'NewOffset = 2 ^ PixelSixe
                        'For future palette fragmentation (not presently used)
    RETURN  '******************** Get Color Map Ends *****************

ImageProcessing:    '****** Image processing block
    DO                  'Look for "," denoting the start of the block
        IF EOF(1) THEN RETURN
        Image$ = INPUT$(1, #1)
    LOOP UNTIL Image$ = ","    'get image block information as before
    ImageLeft = CVI(INPUT$(2, #1))
    ImageTop = CVI(INPUT$(2, #1))
    ImageWidth = CVI(INPUT$(2, #1))
    ImageHeight = CVI(INPUT$(2, #1))
    'PRINT Image$, ImageLeft, ImageTop, ImageWidth, ImageHeight
    LocalMapDat = ASC(INPUT$(1, #1))
    LocalMap = (LocalMapDat AND &H80) / &H80
    Interlace = (LocalMapDat AND &H40) / &H40
    LSrt = (LocalMapDat AND &H20) / &H20
    LR1 = (LocalMapDat AND &H10) / &H10
    LR2 = (LocalMapDat AND &H8) / &H8
    LPixelSize = (LocalMapDat AND &H7) + 1
    IF LocalMap THEN
        GOSUB GetColorMap                       'Use local Color map.
        ColorOffset = ColorOffset               ' + NewOffset (future)
    END IF
    GOSUB Decode                                'Display the GIF image
    IF NOT EOF(1) THEN GOTO ImageProcessing    'Get more images
    RETURN  '***************** ImageProcessing ENDS **************

Decode: '**************** GIF Decompression and Display *********
    x = ImageLeft
```

Figure 7.10 *Continued*

```
    y = ImageTop
    Pass = 1          'used to control interlace display
    YInterlace = 8
    FOR i = 1 TO 2 ^ PixelSize
        StringTable%(i, Suffixes) = I    'initialize the string table
    NEXT i


                        '*** read root code size
    FI$ = INPUT$(1, #1)
    PixelCodeSize = ASC(FI$ + ze$)
                        '*** initialize variables
    CodeSize = PixelCodeSize + 1
    ClearCode = 2 ^ PixelCodeSize
    EndOfImage = ClearCode + 1
    NextCode = EndOfImage + 1
    MaxCode = 2 ^ CodeSize
                        '*** read first block length
    FI$ = INPUT$(1, #1)
    BlockLength = ASC(FI$ + ze$)
    ByteCount = 1
                        '*** read first data byte
    FI$ = INPUT$(1, #1)
    Byte = ASC(FI$ + ze$)
    RightShiftCount = 0
                        '*** read code
ReadCode: '************ Image Data Read Loop Begins ********************
    Code = 0
    FOR i = 1 TO CodeSize            'peel out CodeSize bits
        IF INKEY$ <> "" THEN 'check for key press abort.
            CLOSE
            END
        END IF
        Code = Code + (Byte AND 1) * 2 ^ (i - 1)
                            ' reconstruct code - peel off the right bit
        Byte = INT(Byte / 2) ' shift right one bit
        RightShiftCount = RightShiftCount + 1
                                ' count the bits shifted out of the byte
        IF RightShiftCount < 8 THEN GOTO NextBit
                        ' Process all 8 bits of the byte before reading another
        ByteCount = ByteCount + 1 ' count the bytes processed and compare with
        IF ByteCount <= BlockLength THEN GOTO NextByte ' the block length read.
        FI$ = INPUT$(1, #1)          ' read the next block length
        BlockLength = ASC(FI$ + ze$)     ' convert to an integer
        ByteCount = 1                ' reset the byte counter
        IF BlockLength = 0 THEN RETURN  ' zero means we are done with this image
NextByte: FI$ = INPUT$(1, #1)          ' get another byte
        Byte = ASC(FI$ + ze$)        ' convert to an integer
        RightShiftCount = 0          ' reset the bit shift counter
NextBit:
    NEXT i

 'Now that we have the next compressed code, go expland it into a string if required.

    IF Code = ClearCode THEN
        CodeSize = PixelCodeSize + 1
        NextCode = ClearCode + 2
        MaxCode = 2 ^ CodeSize
        Flag = 0
        GOTO ReadCode
    END IF
    IF Code = EndOfImage THEN RETURN
    IF Flag <> 1 THEN
        Flag = 1
```

Figure 7.10 *Continued*

```
                    Value = Code
                    GOSUB Display
                    Prefix = Code
                    GOTO ReadCode
             END IF
             IF Code < NextCode THEN          '*** code exists in string table
                    Value = Code                     'set the value to the input code
                    DO WHILE Value > ClearCode 'decompose the code until it is a pixel value
                          Value = StringTable%(Value, Prefixes) 'and not another string code
                    LOOP
                    StringTable%(NextCode, Suffixes) = Value
                                                 'store the pixel value at the end of the table
                    Value = Code                     'restore the input value


             ELSE                             '*** code doesn't exist in string table
                    Value = Prefix                   'set the value to the last prefix
                    DO WHILE Value > ClearCode 'decompose the code until it is a pixel value
                          Value = StringTable%(Value, Prefixes) 'and not another string code
                    LOOP
                    StringTable%(NextCode, Suffixes) = Value
                                                 'store the pixel value at the end of the table
                    Value = NextCode                 'set value to next table index
             END IF
             StringTable%(NextCode, Prefixes) = Prefix
                                                 'store the current prefix in the table
             GOSUB Display                    'go display pixel(s)
             Prefix = Code                    'Code becomes next prefix
             NextCode = NextCode + 1          'next table index

             IF NextCode = MaxCode THEN       'increment code size if needed
                    CodeSize = CodeSize + 1
                    MaxCode = MaxCode * 2
                    IF CodeSize = 13 THEN CodeSize = 12  'codes are never larger than 12
             END IF
             GOTO ReadCode     '****************** Bottom of Image Read Loop ******

      Display:  'Display pixels by decoding from the string table while pushing them onto
                'a last in, first out stack
             StackPointer = 0
             Stack(StackPointer) = Value             'Push value onto the stack
             StackPointer = StackPointer + 1
             DO WHILE Value > ClearCode              'if value is a string code and not
                    Value = StringTable%(Value, Prefixes) 'a pixel value then
                    Stack(StackPointer) = Value      'decompose and push prefixes
                    StackPointer = StackPointer + 1 'on the stack
             LOOP
             DO UNTIL StackPointer = 0        'Now, Pop the stack and use the prefixes
                    StackPointer = StackPointer - 1 'to point to the suffixes which are the
                    Value = StringTable%(Stack(StackPointer), Suffixes) 'actual output pixel values
                    IF Pixels THEN                   'Paint pixels per mode selection
                          IF Interlace = 0 THEN 'Non interlaced image
                                IF x > (ImageLeft + ImageWidth - 1) THEN
                                       x = ImageLeft
                                       y = y + 1
                                END IF
                                x = x + 1
                                PSET (x, y), Value
                          ELSE                       'Interlaced image
                                IF x > (ImageLeft + ImageWidth - 1) THEN
                                       x = ImageLeft
                                       y = y + YInterlace
                                END IF
```

Figure 7.10 *Continued*

```
                              x = x + 1
                              IF y >= ImageTop + ImageHeight THEN
                                    SELECT CASE Pass 'y increment depends on interlace pass
                                          CASE 1
                                                y = ImageTop + 4
                                          CASE 2
                                                y = ImageTop + 2
                                                YInterlace = 4
                                          CASE 3
                                                y = ImageTop + 1
                                                YInterlace = 2
                                          CASE ELSE
                                    END SELECT
                                    Pass = Pass + 1
                              END IF
                              PSET (x, y), Value
                        END IF
                  ELSE                            'Paint Boxes instead of Pixels.
                        IF Interlace = 0 THEN 'and so forth.
                              IF x > (ImageLeft + ImageWidth - 1) THEN
                                    x = ImageLeft
                                    y = y + 1
                              END IF
                              x = x + 1
                              LINE (x, y)-(x + 1, y + 1), Value, BF
                        ELSE
                              IF x > (ImageLeft + ImageWidth - 1) THEN
                                    x = ImageLeft
                                    y = y + YInterlace
                              END IF
                              x = x + 1
                              IF y >= ImageTop + ImageHeight THEN
                                    SELECT CASE Pass
                                          CASE 1
                                                y = ImageTop + 4
                                          CASE 2
                                                y = ImageTop + 2
                                                YInterlace = 4
                                          CASE 3
                                                y = ImageTop + 1
                                                YInterlace = 2
                                          CASE ELSE
                                    END SELECT
                                    Pass = Pass + 1
                              END IF
                              LINE (x, y)-(x + 1, y + 1), Value, BF
                        END IF
                  END IF
            LOOP
      RETURN  '******************* Decode ENDS ****************
```

Figure 7.10 *Continued*

lists the statements in the program QBGIF.BAS. That program, as well as the compiled executable version of the program (QBGIF.EXE), are contained within the archive file QBGIF.ARC which is on the diskette labeled 'Graphics' accompanying this book.

JPEG4.ZIP

The file JPEG4.ZIP on the convenience diskette labeled 'Graphics' contains seven files archived using PKZIP. Thus, the extraction of files from this archive requires the use of PKUNZIP which is contained on the first convenience diskette.

The programs and documentation on the JPEG4.ZIP archive are provided by courtesy of the Independent JPEG Group. The files in the archive are summarized in Table 7.4.

The two key programs in the archive JPEG4.ZIP are CJPEG.EXE and DJPEG.EXE. The first program is used to convert an image file previously captured in PPM (PMBPLUS color format), PGM (PBMPLUS gray-scale format), and GIF and Targa formats to the JFIF format. The second program is used to decompress a JPEG (JFIF format) back into one of the previously mentioned file formats.

The field description of the operation of each program is contained on the file USAGE in the archive JPEG4.ZIP on the diskette labeled 'Graphics' accompanying this book. You can obtain a display of a help screen by simply entering the name of the program without a list of file switches and files.

The basic command line format of each program is indicated below:

> cjpeg [switches] list of image files
> djpeg [switches] list of jpeg files

Figure 7.11 illustrates the help screen display obtained by entering the CJPEG program name without any switch or file list. The -quality switch represents a scale quantization table which is used to adjust image quality. This scale varies from 0 (worst) to 100 (best). The -

Table 7.4 JPEG4.ZIP files

File	Description
CHANGELO	The change log which describes modifications made to different versions of the JPEG programs
CJPEG.EXE	An executable program which compresses an image into JPEG format
DJPEG.EXE	An executable program which decompresses a JPEG file back into a conventional image file format
JPEGDUMP.C	A C language program that dumps the header of a JPEG file
JPEGDUMP.EXE	The executable version of JPEGDUMP.C
README	Information concerning the release of the Independent JPEG Group's software and Internet archive locations where additional JPEG software can be acquired
USAGE	Describes how to use the CJPEG and DJPEG programs

```
C:\PCPLUS>cjpeg
C:\PCPLUS\CJPEG.EXE: must name one input and one output file usage:
C:\PCPLUS\CJPEG.EXE [switches] inputfile outputfile
Switches (names may be abbreviated):
 -quality N         Compression quality (0. .100; 5-95 is useful
                    range)
 -grayscale         Create monochrome JPEG file
 -optimize          Optimize Huffman table (smaller file, but slow
                    compression)
 -targa             Input file is Targa format (usually not needed)
Switches for advanced users:
 -restart N         Set restart interval in rows, or in blocks with B
 -smooth N          Smooth dithered input (N=1. .100 is strength)
 -maxmemory N       Maximum memory to use (in kbytes)
 -verbose or        -debug   Emit debut output
Switches for wizards:
 -qtables file      Use quantization tables given in file
 -sample HxV[. . .] Set JPEG sampling factors
```

Figure 7.11 CJPEG help screen

grayscale switch is used to create a monochrome JPEG file from color input. The -optimize switch results in the optimization of entropy encoding parameters instead of the use of default encoding parameters. The selection of this option causes the conversion to run slower but results in a smaller converted file. The -targa switch must be used to identify a Targa file to the program. The preceding switches as well as other program switches are fully described in the file USAGE.

The use of CJPEG is relatively straightforward. For example, to convert the file PORSCHE.GIF to a JPEG file using a scale quantization of 50, you would enter the following command line:

CJPEG -Quality 50 PORSCHE.GIF

Note that output data is written to files that have the same names except for the extension. Since CJPEG always uses .JPG for the output file name extension, the preceding command line entry results in the creation of the file PORSCHE.JPG.

Figure 7.12 illustrates the DJPEG help screen display obtained by entering the program name by itself. Since DJPEG uses a default GIF output you could enter the following command line to reconvert the previously converted file:

DJPEG PORSCHE.JPG

```
C:\PCPLUS>djpeg
C:\PCPLUS\DJPEG.EXE: must name one input and one output file
usage: C:\PCPLUS\DJPEG.EXE [switches] inputfile outputfile
Switches (names may be abbreviated):
 -colors N          Reduce image to no more than N colors
 -gif               Select GIF output format
 -pnm               Select   PBMPLUS   (PPM/PGM)   output   format
                    (default)
 -quantize N        Same as -colors N
 -targa             Select Targa output format
Switches for advanced users:
 -blocksmooth       Apply cross-block smoothing
 -grayscale         Force grayscale output
 -nodither          Don't use dithering in quantization
 -onepass           Use 1-pass quantization (fast, low quality)
 -maxmemory N       Maximum memory to use (in kbytes)
 -verbose or        -debug  Emit debug output

C:\PCPLUS>
```

Figure 7.12 DJPEG help screen

GIF2JPG/JPG2GIF

A second series of valuable GIF/JPEG conversion programs are provided in the convenience diskette labeled 'Graphics' through the courtesy of Marcos H. Woehrmann of Handmade Software, Inc. The two programs, GIF2JPG.EXE and JPG2GIF.EXE, supporting documentation in the file GIF2JPG.DOC and the sample JPG file SAMPLE.JPG are contained in the directory HANDMADE in the diskette labeled 'Graphics'.

Readers are cautioned that GIF2JPG produces a file format that is proprietary unless the -j switch is used. If you use that switch in the GIF2JPG command line, the program will produce a JPG file which is compatible with the JFIF file format.

One of the advantages obtained from the use of GIF2JPG is the fact that the Handmade Software series of programs are used on a large number of bulletin board systems to convert GIF files to JPEG in order to reduce data storage and retrieval time. Since GIF2JPG and JPG2GIF both support the proprietary Handmade Software format some readers may prefer to use this set of programs.

The command line entry format for each program is noted below:

> GIF2JPG [options][files. . .]
> JPG2GIF [options][files. . .]

There are two options which are common to both programs:

-a: automatically proceeds (overwriting any existing files and converting GIF89a files).

-k: kills (removes) the original files after they are processed.

The GIF2JPG program has two additional options:

-g[n]: specify the JPEG quality factor which will be used when com-
        pressing GIF files. The default is 55.
-j: force compatibility with the JFIF standard.

Note that you may specify wild cards as part of the file name and that
multiple file names may be specified. For example:

> GIF2JPG -q30 *.gif

will convert all .GIF files in the current directory to JPG files, using
a quality factor of 30, and

> JPG2GIF -k -a test.jpg madonna.jpg n?.jpg

will convert the specified files back to .GIF files, deleting the .JPG
files and automatically overwriting the existing GIF files. Readers are
referred to the file GIF2JPG.DOC to obtain a complete description of
the use of the two programs.

Image Alchemy

As mentioned earlier in this chapter, there are a large number of
image file formats whose descriptions would require a book as thick
as a dictionary. This can make it extremely difficult for persons with
a scanner, digital camera or other device that produces an image in
one file format to use such a device with a program that supports
different image file formats. Fortunately, one solution to this problem
can be obtained through the use of Image Alchemy, a program from
Handmade Software, Inc. which supports conversion between more
than 30 image file formats.

Ar archive named ALCHEMY.ZIP is contained on the diskette labeled 'Graphics' under the directory ALCHEMY. This archive contains
six Image Alchemy related files provided through the courtesy of
Handmade Software.

The Image Alchemy program ALCHEMY.EXE is fully functional but
cannot handle images larger than 640 by 480 pixels. By sending the
required registration fee you can obtain a registered copy of the program that can be used to convert image files representing greater res-
olutions. The file ORDER.FRM in the archive should be completed
and mailed to Handmade Software with the appropriate registration
fee to obtain a registered copy of this program.

PRINTGF

In Chapter 1 several GIF and JPEG images were displayed to illustrate the effect of lossy compression. Those images were printed on the author's laser printer using the program PRINTGF from Ravitz Software, Inc. Through the courtesy of Cary Ravitz PRINTGF is included on the diskette labeled 'Graphics' accompanying this book. The program and its documentation are included on the archive PRINTGF.ZIP.

PRINTGF is a shareware program that reads GIF, BMP, JPEG, PCX, TARGA and TIFF images and permits their display and printing on color or black-and-white monitors and printers. Like most programs included on the diskette, PRINTGF is a shareware program which you can use without charge on a trial basis to determine its suitability. If you continue to use the program after your evaluation, you should register the program.

8

COMMUNICATIONS
SOFTWARE-LINKAGE
CONSIDERATIONS

Many factors must be taken into consideration when developing software to perform compression. These factors include the type of device the software will operate on, the method used to link the compression software to other software, the transfer rate of the compressed data, either internally to or from peripheral storage units or to and from a transmission medium, and the number of instructions required for coding the appropriate software. In this chapter we will examine the software-linkage considerations applicable to the application of data compression to on-line communications systems.

8.1 COMPRESSION ROUTINE PLACEMENT

In general, compression software will be written as a modular routine whose relationship in the overall system software structure will depend upon the ultimate application—for the storage or transmission of compressed data.

The software structure for transmission compression is illustrated in general form in Figure 8.1. Here, the compression/decompression routine will be at the same level as other communications routines, such as automatic baud recognition and code conversion, with the transmission handler acting as an overall routine controller or traffic policeman, similar to the way the operating system is the controller of application programs. Upon an interrupt occurring on the line, the automatic baud detection routine might be invoked by the transmission handler to determine the operating speed of the incoming transmission. Next, assuming the baud rate was detected and proper buffering set up, the transmission control sequence routine might be invoked.

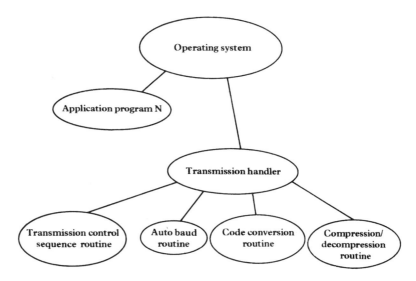

Figure 8.1 Software structure for transmission compression

Once the transmission control sequence performs its function, such as acknowledging the integrity of a BISYNC block or HDLC frame, the data contained in the block or frame information field that was previously compressed is ready to be decompressed. At this point, the decompression routine can be invoked to bring the data back to its original format. Although the preceding description makes the placement of the compression routine appear to be a simple matter, careful study of the routine must precede its placement into the system's software.

Software considerations

The compression routine, like all software routines, will require both processor memory and program execution time. The amount of memory and program execution time required by the routine or group of compression routines will determine whether or not compression can be performed in addition to the other functions on a particular machine. For buffer requirement analysis, let us assume we are considering run-length encoding. If the transmission protocol is BISYNC, we must first look at the existing data-block size handled by the protocol. If the transmission protocol is HDLC we must then consider the largest information field capable of being carried within a frame. For simplicity of analysis, if we assume that our block is 240 characters in size, compressed data will also occur in blocks of 240 characters unless we change the transmission block size. To determine the buffer area required for decompression, let us examine the worst-case requirement as illustrated in Figure 8.2. Here, the received data-block buffer is first filled with 240 characters from a transmitted data block.

If the run-length encoding is completely effective, the data block will be filled with 80 three-character sequences, consisting of a special character indicating compression followed by the repeated character of a string that was compressed and the character count that indicates the number of repeated characters. If each character count indicates that the original string consisted of 64 similar characters, a compression buffer of 5120 characters (64 × 80) becomes necessary. Although this may appear to be an excessive buffer size since the probability of encountering such a degree of compression is remote, you must consider allocating this space unless you perform decompression in stages cycling through the received data. Conversely, shortening the decompression buffer will result in additional data transfer time as we will shortly see.

If the decompression buffer is reduced to 2560 characters in size while all other conditions remain the same, two cycles will be required to complete the return of the compressed data into its original format to satisfy worst case conditions. In the first cycle, one-half of the received data-block buffer is processed, which will result in the 2560 character decompression buffer being filled. At this point, either an I/O transfer or DMA cycle transfer must be initiated to empty the buffer and transfer its contents to a peripheral storage unit. While the transfer is set up and executed, the decompression process cannot continue; therefore, reducing the buffer size results in two wait states while the smaller buffer is emptied. One way to alleviate the waiting process is to reduce the decompression buffer further in size while employing two such buffers. In this manner, double buffering can

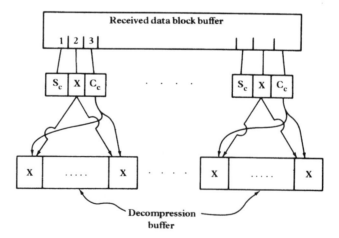

Figure 8.2 Memory requirements. When compressed data is transmitted, buffer areas must be available to restore the data to its original form. S_c = special character indication compression, C_c = compressed character count

occur which will minimize the waiting process as illustrated in Figure 8.3.

When double buffering is employed, the received data-block buffer is processed in variable-length segments. The end of the first segment occurs when the first decompression buffer is filled while the end of the second segment occurs when the second decompression buffer is filled.

After the first decompression buffer is filled and while the second buffer is being filled, a DMA transfer can occur in order to empty the first buffer. Once the second decompression buffer is filled, another DMA transfer can occur and this buffer will be emptied while the first buffer is being filled again. At this point, the end of the third segment of the received data-block buffer will occur when the first decompression buffer is again filled. This concurrent filling and emptying of the dual decompression buffers will continue until the received data-block buffer is completely processed, at which time a second transmitted block will overlay the previously processed block. The size of the decompression buffers as well as the received data-block buffer will depend upon the transmission block format as well as available processor memory. While the received data-block buffer should equal the size of the transmitted block, the decompression buffers as previously explained can be set to any size; however, the smaller the size the larger the processing and the I/O transfer time required to decompress the received data block. Thus, the programmer must also investigate the program timing prior to selecting a final decompression buffer size.

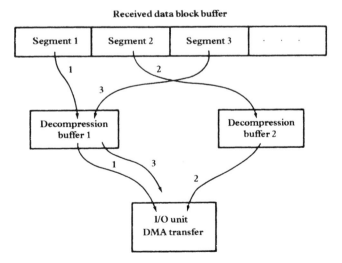

Figure 8.3 Double buffering—this can be employed to minimize transmission waiting times

8.2 TIMING CONSIDERATIONS

To estimate the amount of processor timing required for decompression, several factors require investigation. First, the programmer must estimate the number of instructions the decompression routine may require and the quantity of each type of instruction, such as single or double word, memory reference, shift instruction by number of shift positions and so on. From the computer manufacturer's programming reference manual, the timing per instruction can be obtained, either expressed by the number of machine cycles required to execute the program or denoted as a period of time and normally expressed in microseconds or 10^{-6} s. If expressed as a function of the number of machine cycles, timing per instruction can be obtained by multiplying the machine's cycle time by the number of cycles required for the instruction. Next, the sum of the product of the number of instructions of each type times the execution time per instruction is obtained. This becomes the estimated decompression time per block, less waiting time and data transfer time.

The data transfer time can be estimated by examining the coding required to set up and initiate a DMA transfer. If double buffering is employed, the actual transfer time you must consider may be zero or a very minimal value. This is because one decompression buffer is being filled while the second decompression buffer is emptied as shown in Figure 8.4. In this illustration, if one decompression buffer is emptied before the second is filled, there is no waiting time and the only extra time in addition to the normal buffer processing time is the DMA set-up time. Although this set-up time varies from computer to computer, the total time required is normally less than 20 µs. For most machines, the buffer starting and ending address are loaded

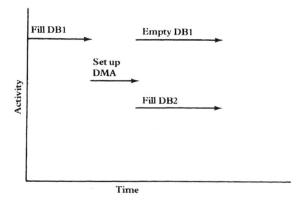

Figure 8.4 Timing considerations using double buffering. After the first buffer is filled, subsequent buffers are emptied and filled with minimal waiting time since only a DMA transfer set-up is required for each operation to occur concurrently with processing

into certain registers along with a peripheral address code and a DMA transfer request is then initiated. Since a DMA set-up is required to empty buffers, two such DMA set-up times are required to empty the two decompression buffers.

Total time

The total decompression time for the received data blocks will depend upon the data contained in the blocks. This can be estimated for the worst case condition where the most efficient type of run-length encoding has been conducted as previously explained. If double decompression buffers are employed, 40 segment cycles as listed in Table 8.1 will be required if each decompression buffer is 64 characters in length. By obtaining the time for the process cycle listed in Table 8.1 and multiplying by 40, the total decompression time can be estimated based upon the previous assumptions. If data is being received at 9600 bps, then 1200 cps (characters/s) are being received. At that data rate, ignoring control characters and retransmission time due to line errors, the received data-block buffer of 240 characters will be filled once every 0.2 (240/1200) s. This means that all of the previously discussed decompression processing as well as line handling and other communications functions must occur in that time or buffer queues must be employed to permit the processing of non-worst case conditions to be encountered which will then permit the processing to catch up with the data. By trade-offs of buffer size and timing and the employment of double buffering, most compression routines, other than some of the Huffman and modified Huffman coding techniques, can be easily adapted to on-line processing.

As an alternative to determining timing, you can compute the number of processing instructions available for operating upon the data prior to compression resulting in degraded performance. As an example, consider a data concentrator or intelligent multiplexer servicing 2400 bps terminals and connected via a 56 000 bps high-speed data

Table 8.1 Processing cycle

A.	Process received data-block buffer. Move pointer through buffer as decompression buffer 1 is filled.
B.	Set up DMA transfer. Decompression buffer 1 → storage.
C.	Process received data-block buffer. Move pointer through buffer as decompression buffer 2 is filled.
D.	Set up DMA transfer. Decompression buffer 2 → storage.

link to a host computer. If there are eight data bits per character, the 2400 bps terminals are communicating at 300 cps while the concentrator is communicating at 7000 cps.

If the terminals connected to the concentrator are used for interactive applications, their typical message length is 32 characters which takes 32/300 or 0.106 s to reach the concentrator from the terminal. To send that message from the concentrator to the host computer requires 32/7000 or 0.004 57 s. The difference between the time it takes the concentrator to receive the message and transmit the message is 0.101 43 s and can be considered a 'window of operation'. This is the amount of time available for all concentrator processing to include data compression prior to a message transmission delay resulting.

If we assume a 2 μs cycle time and four cycles per instruction, then a time of 8×10^{-6} s is required to execute an instruction. Dividing that figure into the previously computed window of operation results in 12 678 instructions which can be performed upon the data prior to a message transmission delay occurring. If you determine the current number of programmed instructions in the concentrator software, that number can be subtracted from 12 678 to determine the number of data-compression instructions that can be written prior to causing a message transmission delay.

Flow control

During periods of heavy transmission activity, buffers may become filled to capacity and subsequent data transfers will cause data to overflow such buffers, in effect becoming lost. To prevent such an occurrence, you can incorporate a number of flow-control procedures via software. These procedures are designed to selectively inhibit and enable data sources and, thus, prevent data buffers in memory from overflowing.

The most common method of flow control is obtained by the transmission of XON and XOFF characters to terminals and computers built to recognize such data characters. Transmitting an XOFF character can be used to tell a data source to inhibit all future transmission activity while one processes the data buffer contents residing in memory. Once the data buffer is emptied or has reached a certain percentage of occupancy, one can transmit an XON character to tell the data source to resume transmission. We must ensure that these two characters are not compressed if they are to be recognized and acted on by hardware which is unaware that the data is compressed.

A second method of selectively enabling and inhibiting the transmission of data can be obtained by raising and lowering the clear-to-send (CTS) signal interface lead on the RS-232 interface connection between a compression-performing device and a data source. The raising and lowering of the CTS signal would function in a way similar to the transmission of XON and XOFF characters previously discussed.

An XON, XOFF sequence and the raising and lowering of the CTS signal are normally used for asynchronous data-flow control. For synchronous data-flow control one should consider selectively altering a clock generator rate. As an example, using this method you could clock data initially at 19.2 kbps into a compression device for transmission at 9.6 kbps. As the compression buffers build up, you can lower the input clocking rate to prevent buffers from overflowing. Conversely, you can raise the clock rate if the data is susceptible to compression and buffer occupancy is minimal.

Routine linkage

When data compression is added to a front-end processor or programmable communications controller, several methods can be considered to link the compression module to existing software. The compression software can be added to an existing module or it can be coded as a separate module. For the latter situation, several methods can be used to invoke its operation. If the compression routine is to reside on a minicomputer, a 'jump and store return address' instruction or subroutine call are the two most frequently used methods for one routine to invoke another. At the completion of the routine, a return to the invoking instruction plus one location occurs and the instruction following the jump or the call instruction is executed next. If compression is to be employed with microcomputers, hardware and software differences may preclude the use of a jump or call statement. Many microprocessors are designed so that complete functions, such as a data-compression routine, may be burned into chips which are connected to the microprocessor's I/O or DMA bus. Since these coded chips are then considered part of the microprocessor's memory, linkage becomes a memory access problem. For these microprocessors that lack a 'jump and store return address' or subroutine call instruction, the address of the chip can be stored in one of the processor registers. Then an indirect address through the register will result in the processor fetching the chip address to initiate the compression routine.

9

COMPRESSION-PERFORMING HARDWARE AND SOFTWARE PRODUCT OVERVIEW

Many times you can obtain the benefits of data compression while avoiding the efforts required to analyze actual or potential data traffic and develop software to perform compression. This can be accomplished by leasing or purchasing data-compression-performing devices that are specifically designed to be used in a particular networking environment. In this chapter, we will examine the utilization of several hardware and software products to obtain a better understanding of the use of compression-performing devices. The products covered in this chapter were selected for illustrative purposes only and should not be construed as an endorsement of any hardware or software device.

The utilization of a compression-performing hardware device or the use of a software program eliminates the necessity of analyzing data transmission traffic and developing the software routines required to compress data. However, readers are cautioned that many devices and compression performing software programs are designed to be most efficient when operating upon a particular type of data traffic that may not match the data consistency you actually transmit. Thus, such devices and programs may not generate a compression ratio level equivalent to the level that could be obtained from the development of a customized system tailored to operate based upon an analysis of the consistency of your actual data traffic.

In the first section in this chapter we will focus our attention upon the operation and utilization of a few commercially available compression-performing hardware products. This will be followed by a second section focusing upon several types of compression-performing software packages. Included in this section are operational reviews of software programs used to enhance the efficiency of teleprocessing monitors operating on mainframe computers, programs that can be used to compress database files on mainframes, and programs that can be used to compress MS-DOS and PC-DOS files on IBM PC and compatible personal computers.

9.1 HARDWARE PRODUCTS

In examining hardware compression-performing products, we will distinguish such products by their functionality. First, we will examine devices that are restricted to compressing asynchronous data streams. Next, we will examine the operation and utilization of several multifunctional compression performing hardware products.

Asynchronous data compressors

Since asynchronous terminals outnumber synchronous operating terminals by a factor of 10 or more, it should be of no surprise that many vendors have developed products for use with asynchronous transmission. Such products can normally be used for transmission occurring on both leased lines and the switched telephone network; however, their primary use is for transmission on the public switched telephone network. When used on this transmission facility, the primary advantage of the compression device is its ability to reduce the duration of the transmission session. Since the cost of a long-distance call is approximately proportional to its duration, decreasing the duration of the transmission session reduces the cost of the call. Two products specifically designed to compress asynchronous data that will be examined for illustrative purposes are RAD Computers' CompressoRAD-1 and generic V.42 bis compatible modems.

CompressoRAD-1

The CompressoRAD-1 is a stand-alone compression unit that also performs asynchronous-to-synchronous/synchronous-to-asynchronous conversion and provides error detection and correction capability to data flow. Figure 9.1 illustrates how the CompressoRAD-1 could be utilized for transmitting data via the switched telephone network.

The CompressoRAD-1 accepts asynchronous data at 1200, 2400,

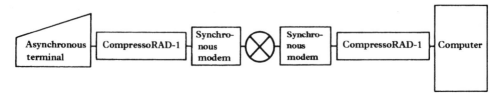

Figure 9.1 Using the CompressoRAD-1. The CompressoRAD-1 accepts asynchronous data at 1200, 2400, 4800 and 9600 bps. Data is compressed by the device, converted into a modified HDLC protocol and output to the attached modem according to the modem's clock rate setting

4800 or 9600 bps. The asynchronous data can consist of seven or eight bits per character, with either one or two stop bits and odd, even, mark, space or no parity. Utilizing an automatic adaptive algorithm, data compression ratios from 2:1 to 4:1 are obtainable according to the manufacturer.

In addition to compressing data, the device converts the asynchronous data flow into a synchronous modified higher-level data link control (HDLC) protocol. By adding a cyclic redundancy check character to each transmitted data block end-to-end error detection and correction capability is added to the data transmission.

The output data rate of the CompressoRAD-1 is determined by the clock of the attached modem, with the device capable of operating between 600 and 4800 bps. A buffer in the device is used to compensate for the differences in the compressibility of input data. Thus, when data input into the CompressoRAD is compressed, and represents a data flow greater than the rate at which data is output to the attached synchronous modem, the buffer fills, serving as a temporary storage area.

Since the buffer is finite in size, the CompressoRAD employs two methods to inhibit additional data input once the buffer is filled to a predefined level. Known as flow control, data is regulated into the device by the transmission of XOFF and XON characters and the raising and lowering of the RS-232 clear-to-send control signals.

When buffer storage fills to a predefined level, the CompressoRAD will either transmit an XOFF character or drop the CTS signal on the RS-232 interface. The choice of which method one selects is based upon the operational specifications of the terminal or computer port attached on the CompressoRAD. Some terminals and computer ports recognize XON and XOFF as flow control character signalling, enabling the CompressoRAD to be configured to use this method of flow control. For terminals and computer ports that do not recognize this method of flow control the CompressoRAD can be configured to enable and inhibit transmission to the device via the CTS control signal. Thus, after data is inhibited from transmission into the device by the CompressoRAD dropping CTS, it will raise this control signal once the buffer is emptied to another predefined level. The issuance of flow control signals based upon the quantity of data in the device's buffer is illustrated in Figure 9.2.

The use of XOFF and XON characters and the raising and lowering of the CTS control signal are used by most compression-performing products that operate upon asynchronous data. When synchronous data is compressed, flow control is normally accomplished by reducing the clocking rate signal passed from a modem attached to the compression device to a synchronous operating terminal, computer port or similar device. One common method used by hardware manufacturers is to halve the clock rate several times until the buffer empties to a predefined level. Once this occurs, the clock rate is then doubled several times until the original data rate is restored.

To understand the economics associated with the utilization of the CompressoRAD and similar compression-performing products, assume that the daily cost for communicating on the switched telephone network between a remote terminal and a central computer facility is $5, a cost representing a long-distance call duration of under 40 min. Assuming one call per day and 22 working days per month, the cost of using the switched telephone network is $110 per month or $1320 on an annual basis. Now, let us assume that the use of two asynchronous data-compression devices reduces the transmission session duration by half, resulting in the ability to reduce the cost of communications by $660 during one year of operation. To what should you then compare this potential cost saving?

In using the CompressoRAD illustrated in Figure 9.1 or a similar compression-performing product, you must obtain two synchronous modems as well as two compression devices. In comparison, you would use two lower-cost asynchronous modems when transmitting data without the use of asynchronous compression devices. Thus, from an economic perspective:

$$(\triangle M)*2 + 2*C \leqslant 660*EL$$

where:

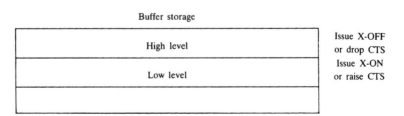

Figure 9.2 Buffer control. To prevent buffer storage overflow and the loss of data flow control occurs through the raising and lowering of the clear-to-send (CTS) control signal or the generation of XOFF and XON characters

$\triangle M$ = cost difference between synchronous and asynchronous modems

C = unit cost of each compression-performing device

EL = expected life or use of the compression-performing devices in years.

V.42 bis modems

The previously described CompressoRAD-1 can be considered as one of a series of products which were essentially rendered obsolete by the inclusion of compression technology in modems. Although several million MNP Class 5 and Class 7 compatible modems are in use worldwide, the V.42 bis modem standard based upon the BTLZ algorithm previously described in Chapter 6 has become the most popular hardware based compression performing product. Currently over a half-million modems incorporating V.42 bis compression are being manufactured monthly on a worldwide basis primarily for use with personal computers.

Due to the growth in the use of fax/data modems, a degree of elaboration concerning the compression options of this type of product are warranted. A fax/data modem that is CCITT Group 3 compatible must support modified Huffman coding for fax operations. However, if the modem is not V.42 bis compatible it may not meet your requirements for transmitting compressed data from one location to another. Thus, when selecting a fax/data modem it is important to review the method or methods of data compression supported by the modem.

Synchronous data compressors

Until 1993 compression was conspicuous by its absence in several synchronous communications applications, such as the integration of the technology into Data Service Units (DSUs) and for operating on high speed T1 and E1 transmission lines. Although numerous compression-performing hardware products reached the commercial market prior to 1993, their applicability to operate on digital synchronous circuits was typically limited to serving as an input to a multiplexing device connected to a circuit. This limitation was primarily based upon a requirement to compress only the information field of transported frames to enable frame headers to be correctly interpreted when data flowed through a network of interconnected circuits. In addition, when applied to T1 or E1 circuits, the framing and synchronization bits used by the carriers' network and user equipment could not be altered. This resulted in an additional effort required to correctly integrate compression into products designed for operation on synchronous high speed digital circuits. Two representative products introduced during 1993 which addressed the previously mentioned problems are the Codex Corporation compression-performing DSU and FastComm's Time Machine.

Compression DSU

The compression-performing DSU introduced by Codex Corporation during 1993 functions as a combined CSU/DSU for terminating a 56 or 64 kbps digital circuit as well as a data compressor. This was the first product to integrate compression into a DSU and found quite a receptive market as its use enables organizations to replace multiple 56 kbps circuits routed between locations by a single digital circuit.

Figure 9.3 illustrates the use of a pair of compression-performing DSUs to replace a transmission group of two 56 kbps digital circuits previously used to interconnect geographically separated IBM mainframe computers. Figure 9.3a illustrates the conventional method used to interconnect IBM mainframes via a transmisson group of two or more circuits linking front end processors connected to each mainframe. In this example it was assumed that each circuit operates at 56 kbps. In Figure 9.3b the use of a pair of compression-performing DSUs is illustrated. In this example the compression ratio which varies between 2:1 and 4:1 results in a variable throughput between 112 kbps and 224 kbps, more than sufficient for the replacement of two 56 kbps digital circuits by one circuit with compression-performing DSUs.

In examining Figure 9.3b note that data is transferred between each front end processor and compression-performing DSU at 256 kbps. In actuality, you can set this data rate to several multiples of 56 or 64 kbps. Setting it to 256 kbps provides you with the ability to maximize the information transfer rate over the single 56 kbps circuit when a compression ratio of 4:1 is obtained. When the compression

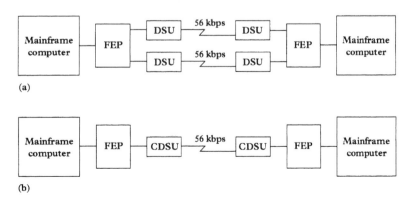

Figure 9.3 Using compression-performing DSUs

ratio falls below 4:1 the compression-performing DSU selectively initiates flow control to the attached front end processor to prevent its limited internal buffer from overflowing and the subsequent loss of data.

The Time Machine

FastComm also introduced its Time Machine stand-alone data compressor during 1993. Although this device can be used on low speed 56 and 64 kbps digital circuits, its primary utilization is to compress high speed data for transmission on T1 and E1 circuits operating at 1.544 and 2.048 Mbps, respectively.

The Time Machine is compatible with equipment using HDLC type framing, such as bridges, routers, gateways, SNA, X.25 PADs and similar products. Featuring dual input ports with an aggregate input up to 6 Mbps, the Time Machine can support a variety of high speed networking applications.

Figure 9.4 illustrates a typical Time Machine application. In this example a remote bridge connected to an Ethernet network and router connected to a Token-Ring network are configured to provide an aggregate input up to 6 Mbps into the Time Machine. The Time Machine compresses data archiving up to a 4:1 compression ratio, resulting in the use of a single T1 line operating at 1.544 Mbps to connect the two LANs at one location to LANs at another location.

Multifunctional compression devices

One of the more popular uses of compression-performing devices is to reduce the quantity of data to be multiplexed, increasing the servicing capacity of multiplexers. A natural evolution of the development of compression devices was to include both statistical multiplexing and data compression in one hardware device. A product representing this multifunctional capability is the Datagram Corporation Streamer product line which, upon that company's acquisition by Memotec Data, was relabeled as the MC series of products.

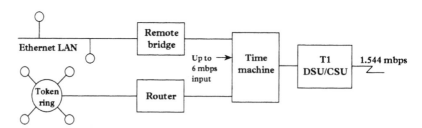

Figure 9.4 Typical Time Machine application

Streamer

Datagram Corporation's Streamer series of devices, which were renamed as the MC 504 and MC 508 when that company was acquired by Memotec Data, are statistical multiplexers that incorporate data compression, resulting in an overall statistical efficiency of up to 4:1 according to the company. In comparison to many similar products, the Streamer series of devices supports a wide variety of bit-oriented protocols to include synchronous data link control (SDLC), X.25, HDLC, Univac data link control (UDLC) and character-oriented protocols to include asynchronous and bisynchronous transmission.

Each Streamer, now marketed as the MC 504 and 508 data-compression multiplexer, analyzes incoming data on an individual port basis and employs an adaptive string compression algorithm to compress the more frequently occurring strings into shorter strings. Since the data transmitted in each direction can vary in composition, each compression-performing multiplexer maintains both inbound and outbound compression tables for each port, that are independent of each other. Once each port's data stream is compressed, the data stream of all ports is then multiplexed.

Due to the ability of the Streamer series of devices to support numerous data link protocols, this compression-performing communications device can be used in a large variety of network applications. Figure 9.5 illustrates one potential application that shows the versatility of the Streamer. The Streamer first performs typical statistical multiplexer functions to include stripping start and stop bits from asynchronous transmission and passing data only from active terminals. After the device's compressor uses an adaptive

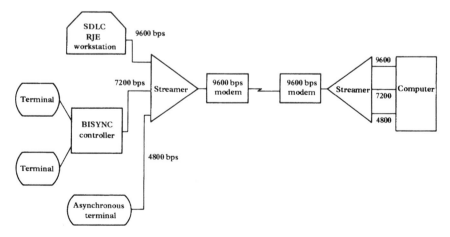

Figure 9.5 Using the Streamer. Datagram's Streamer series of compression performing multiplexers supports a large number of bit- and character-oriented protocols

compression table to encode the data, the resulting compressed data is statistically multiplexed, resulting in overall statistical and compression efficiencies of up to 4:1. One of the more interesting characteristics of the Streamer is its analysis and construction of adaptive compression tables for each direction of data flow on a port-by-port basis. This approach maximizes the potential efficiencies obtained by compressing data as it ensures different data characteristics flowing in one direction as promptly recognized.

9.2 SOFTWARE PRODUCTS

In this section we will examine three types of data-compression software products. First we will focus our attention upon teleprocessing monitor compression software designed to reduce terminal response time by reducing the quantity of data transmitted between a mainframe application and a terminal device. This will be followed by an overview of a few mainframe database compression-performing products and software designed to compress personal computer files.

Teleprocessing monitor compression products

A large number of software vendors have developed data-compression products to enhance the efficiency of a variety of IBM teleprocessing monitor systems. Some products are designed to work with specific IBM teleprocessing monitor systems, such as CICS, IMS, and TSO. Other vendors have developed products that operate at the Virtual Telecommunications Access Method (VTAM) level. When operating at the VTAM level the compression package provides compression for all applications operating on the host computer which eliminates the necessity of obtaining a separate compressionperforming product for each VTAM application. Regardless of the location where the compression program operates, each program uses a core of equivalent data-compression methods which essentially make their performance equal. Compression techniques used by teleprocessing monitor enhancement software include the elimination of all consecutive repetitive characters, outbound image-saving, and inbound mirroring.

To illustrate the effect of the operation of a teleprocessing monitor system compression program consider the network segment illustrated in Figure 9.6. In this illustration a large mainframe operating CICS, IMS and TSO is accessed by 32 terminals located at a remote site. If each terminal operator pressed a Program Function (PF) key to display a new screen image the number of bits that would be transmitted to each terminal exclusive of the protocol overhead would be $80*25*8$ or $16\,000$ per terminal. Assuming a worst case scenario

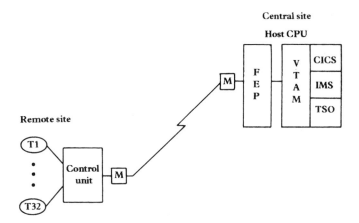

Figure 9.6 Network segment to illustrate benefits of teleprocessing monitor compression software

where 32 terminal operators simultaneously press a screen PA, PF, or Enter key, 16 000∗32 or 512 000 bits would require transmission. If the line connecting the remote location to the central computer site operates at 19.2 kbps it would require 512 000 / 19 200 or approximately 27 s to transmit all screen images. Since the preceding calculation did not consider the processing time required by the host to interpret the key request and retrieve a required screen, nor the effect of the overhead associated with the transmission protocol, the actual time would be considerably more. Even if it were not, waiting 27 s for a response would clearly be an unacceptable level of response time. Thus, any technique that can reduce response time will boost terminal operator productivity. In fact, although 32 terminal devices are illustrated in Figure 9.6 as being connected to the control unit, normally a lesser number are connected to preclude excessive response times. For large organizations with hundreds of terminals located at different offices remote from the host computer, the reduction in the number of terminals connected to each control unit will normally result in the use of additional control units, each requiring modems and a communications line. Thus, any method that reduces terminal response time can also allow more terminals to be connected to a control unit, which can result in a reduction in the cost of communications.

Compression methods

In many applications character strings that contain a letter or character are repeated several times in succession. For example, the character string '– – – . . . –Page n' is used as a page separator in many applications. This string may have a run of 60 to 70 dashes followed by a page number. In other applications a string of equal signs '======' may be used to underscore a field or separate it from other fields. Although not visible, some applications commonly use a string of blanks to delete the contents of an existing field on a screen.

Teleprocessing monitor compression products examine outbound data streams for consecutive repetitive characters. When found, repetitive characters are replaced by a four-byte sequence which uses the Repeat-to-address code recognized by terminals as the first byte in a four-byte sequence. The Repeat-to-address code is followed by a two-byte screen address which identifies where the repeated character is repeated to. This is then followed by the actual character to be repeated. Figure 9.7 illustrates the format used to compress repetitive strings. Note that each character is encoded in hexadecimal. Thus, 3C, which is the hexadecimal code for the Repeat-to-address code, represents one character.

A second method employed by teleprocessing monitor compression-performing software is commonly referred to as outbound image-saving or suppressing redundant data images. To accomplish this the compression program intercepts outbound screen images and compares the current screen image to the previously transmitted image, suppressing the transmission of common fields or portions of fields. Figures 9.8 and 9.9 illustrate the operation of image-saving. Figure 9.8 shows a typical terminal screen when an operator is using a customer maintenance mailing list program. Figure 9.9 shows how the next screen might appear. Note that only the first name, date, and amount fields changed. Thus, transmitting only those changes in place of an entirely new screen can significantly reduce the number of characters required to update the screen.

A third method used by many teleprocessing monitor compression-performing programs involves increasing the efficiency of an application program. As previously discussed in Chapter 3 in which forms-mode compression was described, when an operator enters data into a screen field, the data tag associated with that field is modified. The modified data tag (MDT) can also be turned on by an application when data is transmitted to a terminal. If this is done, the field will be trans-

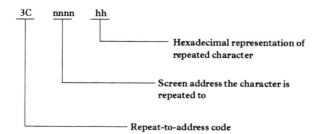

Figure 9.7   Repeated character compression format

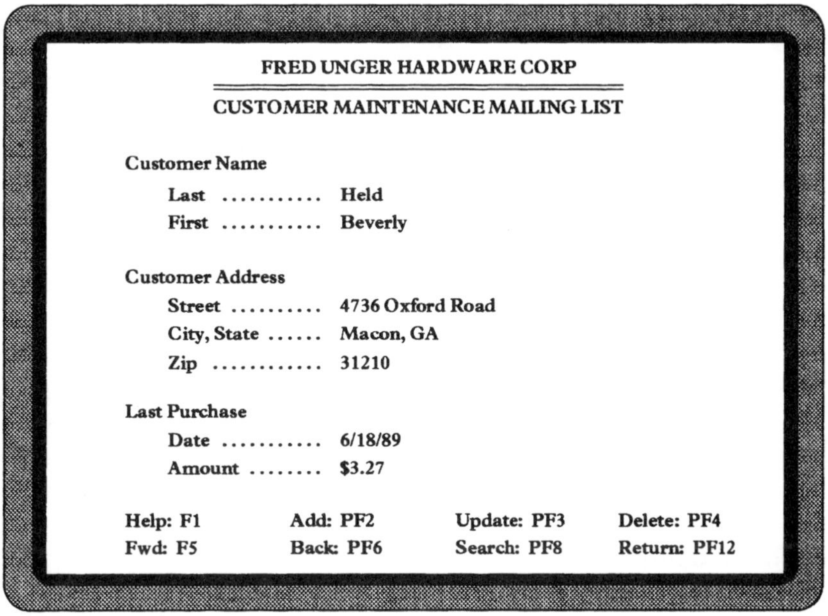

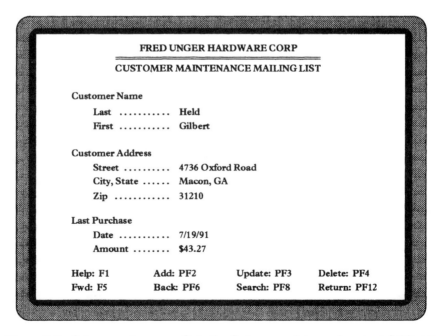

Figure 9.8 Initial screen

Figure 9.9 Subsequent screen. Only the first name, date, and amount fields are transmitted since all other fields are the same as contained in the previously transmitted screen

mitted back from the terminal even if the field is not changed by the operator.

To obtain inbound mirroring the compression program turns off all MDTs as the transmission flows to terminals, which results only in the fields modified by the operator being transmitted. By keeping an image of the screen in memory the compression program knows which fields had their associated MDTs set on. If those fields are not transmitted back from the terminal they are inserted into the downstream by the compression program prior to that program passing the data to the application program.

Two popular teleprocessing monitor enhancement compression programs are Datapacker/II marketed by H&M Systems Software of Maywood, NJ, and VTAM-EXPRESS marketed by SofTouch Systems of Oklahoma City, OK. The first program is a CICS product, while the second program supports CICS, IMS, and TSO applications.

Database compression

InfoTel Corporation of Tampa, FL, markets a series of data-compression software programs for compressing different types of databases. Programs marketed under the INFOPAK label include INFOPAK DB2, INFOPAK VSAM, INFOPAK IMS, INFOPAK IDMS, and INFOPAK SEQ.

INFOPAK DB2 reduces the amount of direct access storage device (DASD) storage required for DB2 databases. According to the company, disk space recovery rates between 50 and 75% are typically obtainable with a minimum of CPU overhead. In addition, since disk storage requirements are considerably reduced, the number of input/output (I/O) operations required for DB2 applications is also reduced.

INFOPAK DB2 uses a modified Huffman encoding technique in which a scanning program module performs a three-dimensional analysis of data and then makes compression decisions based upon the nature of the data. The scanning program uses artificial intelligence to determine which of several different compression methods to use for optimum performance. Most data is compressed using the Huffman encoding technique but other compression techniques, including repeating character substitution, bit representation of blank and zero fields, and a proprietary algorithm for certain kinds of data, are also used.

InfoTel's INFOPAK VSAM allows compression of KSDS or ESDS data sets running under IBM's MVS or MVS/XA operating systems. This program is CICS compatible and is completely transparent to the application program.

To maintain application program transparency, the compression program ensures that each key retains its exact record position. To accomplish this the uncompressed keys are saved in a separate work area during compression. After compression the keys are inserted into the compressed record in the exact positions they occupied prior to compression. Then, the new compressed record is provided to VSAM for physical storage. If the compressed data does not fill out the record after the insertion of the keys in their exact record positions the record is lengthened by the addition of bytes of binary zeros to obtain a minimum record length. This serves to maintain application program transparency.

Figure 9.10 illustrates the operation of INFOPAK VSAM on a record that has KSDS primary and alternate keys located in positions 12 and 40, respectively. Note that when the record is compressed, its compressed length depends upon the susceptibility of the data to compression.

INFOPAK IMS enables IMS databases to be compressed while INFOPAK IDMS enables compression of IDMS databases. Both compression programs result in compressed databases that are totally transparent to application programs. InfoTel's INFOPAK SEQ operates upon files that are processed sequentially. The resulting compressed sequential files are identified by a special suffix appended to the dataset name. Other than this suffix, INFOPAK SEQ is transparent to the job control language (JCL) and is completely transparent to all applications.

Due to the high cost of DASD, the INFOPAK series of products has gained wide acceptance by numerous private industry and government agencies. INFOPAK users include several US Department of Defense agencies, large universities, health care organizations, insurance companies, and many other organizations that maintain a variety of large databases. Since the reduction of just one IBM 3380 or 3390 disk system can essentially recover the cost of the program,

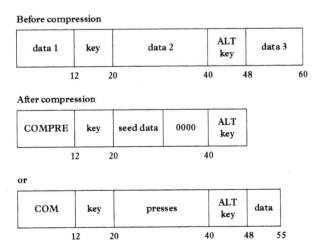

Figure 9.10 INFOPAK VSAM operation

its payback can be instantaneous. In fact, many firms can obtain savings significantly exceeding the cost of the software while obtaining additional storage capacity through the use of storage space freed from the use of the program. This can provide large data processing organizations with years of expansion capacity through the use of existing equipment.

Either due to the commercial success of third-party disk compression products or in response to customer requests, IBM recently introduced a mainframe disk compression product. Now mainframe users have an additional source for enhancing the data storage capacity of their disk systems via software.

Personal computer file compression

In tandem with the near exponential growth in the use of personal computers has been an increase in both the number of programs being marketed and their complexity. As software products proliferated many bulletin board operators quickly filled their hard disks with programs and rapidly ran out of storage space. This problem led to the development of a series of public domain compression programs that began to achieve popularity in the mid-1980s.

Among the earliest public domain compression programs were SQ and USQ, written by Dick Greenlaw. SQ (for squeeze) was based upon a Huffman encoding algorithm and compressed PC files, while USQ (for unsqueeze) decompressed previously compressed files. SQ and USQ, as well as similar programs, provided PC users with several benefits. First, program and data files could be stored using less disk storage. This, in turn, allowed bulletin boards to store more programs. In addition, it provided software vendors and end-users with the ability to distribute more software on a smaller number of diskettes. Another significant benefit was a reduction in communications time and cost, since compressed files could be downloaded quicker than uncompressed files.

Archive utility programs

An archive program takes two or more files and stores them as a single reproducible entity. The original goal behind the development of archiving programs was to reduce data storage. For example, under DOS, files are normally stored using sectors consisting of 512 bytes. Thus, a file requiring 513 bytes of storage would use two sectors, in essence wasting 511 bytes of storage. Through the addition of compression the data storage requirements of archived files are significantly reduced.

Although the archiving efficiency of archive utility programs is an important consideration, this represents only one of several factors potential users should consider when evaluating the use of this type of software product. Today, by far the most important factor to consider in addition to the compression efficiency of the program is its set of operational parameters. Those parameters define the program's capability to perform different archiving-related operations. A few examples of the effect of the operational parameters of an archiving program include its ability to operate upon hidden and system files, its ability to span multiple diskettes when used to back up a hard disk, its ability to work with archives generated by other programs and its ability to selectively update previously created archives.

Table 9.1 lists nine popular archiving and compression-performing programs developed to operate on IBM PC and compatible computers running DOS. Although it is anyone's guess as to the most popular

Table 9.1 Popular file archiving and compression-performing programs

Program	MS-DOS program names	File extension of archive	Compression algorithm	Other computer operating system platforms
ARC	ARC602.EXE	.ARC	LZW	Unix, VMS
ARJ	ARJ240.3X3	.ARJ	LZW with hashing plus static Huff-coding on a block basis	UNARJ for Unix, Macintosh and Amiga
LHA	LHA213.EXE LHA255.EXE	.LZH	LZ77 plus static Huffman coding on a block basis	Unix, Macintosh, Amiga
PAK	PAK251.EXE	.PAK	LZW	
Squeeze for MSDOS	SQZ1083e.EXE	.SQZ	LZ77	
Ultra Compressor	UC2R2.EXE	.UC2	LZ77 with static Huffman coding on a block basis and dynamic dictionaries shared among files	OS/2
PKZIP	PKZIP110.EXE	.ZIP	LZ77 plus secondary static Shannon–Fano coding on whole file	Unix, VMS, OS/2
PKZIP	PKZ204G.EXE	.ZIP	LZ77 plus secondary static coding on a block basis	OS/2, VMS, Unix, Amiga
ZOO	ZOO210.EXE	.ZOO	LZ77 with a trie data structure plus secondary static Huffman coding on a block basis	Unix, VMS, Macintosh, Amiga

program, both ARJ and PKZIP are currently very popular archiving and compressionperforming programs both in the United States and elsewhere.

ARC, developed by System Enhancement Associates (SEA) in 1985, was one of the earliest file archiving and compression performing programs to be developed. The resulting .ARC archive produced by the program's archiving process is still very popular and frequently encountered on personal computer bulletin board systems. Unfortunately, the latest shareware version of ARC, ARC6.02, dates to 1989, and although SEA introduced a commercial version of its archiving program, both the shareware and commercial versions have been superseded in popularity due to the large number of optional parameters supported by more modern archiving programs.

LHA is a popular archiving program used in Japan which due to its no-fee-license policy has been used by several software developers in the US and Europe to distribute compressed and archived programs on a lesser number of diskettes in the form of self-extracting files. As its name implies, a self-extracting file extracts the contents of the archive to restore programs to their original non-compressed form.

In Europe, Ultra Compression II, which was developed in the Netherlands, has gained popularity due to its large number of features. Although all of the archiving and compression performing programs listed in Table 9.1 obviously perform compression, in many instances the efficiency of the archiver's compression algorithm is but one of several decision criteria concerning the selection of a particular program. Other decision criteria include its license fee, its ability to generate a self-extracting program with a minimal amount of overhead, and its built-in features that govern the creation of archives and extraction of files from the archive. Concerning the latter, the program's ability to store the directory path structure of each file added to the archive and position files correctly upon extraction and the encryption of selected or all files in the archive are but two of numerous archive program features you may wish to evaluate.

In the remainder of this chapter we will focus our attention upon four archive-performing programs—ARC, PKWARE's PKZIP and PKUNZIP, ARJ and UCII. The developer of each program graciously provided permission to include the shareware version of their program on the diskette accompanying this book. This provides you with the ability to compare and contrast the operational features of each program against your specific operational requirements prior to selecting an archiving program as an organizational standard.

ARC—the archive utility that compresses data

Building upon the initial development of PC file compression and decompression programs, System Enhancement Associates originally located in Wayne, NJ, developed an archive utility for personal computers running under DOS and OS/2 that compresses data. This program, called ARC, performs an archive operation by grouping a series of specified files together into one file in such a way that the individual files may be recovered intact. In addition, ARC automatically compresses the files being archived so that the resulting archive takes up a minimum amount of space.

When ARC is used to add a file to an archive it analyzes the file to determine which of three storage methods will result in the greatest savings. These three methods are:

1. No compression; the file is stored as is.
2. Repeated-character compression; repeated sequences of the same byte value are collapsed into a three-byte code sequence.
3. Dynamic Lempel–Ziv compression; the file is stored as a series of variable-size bit codes which represent character strings and which are created 'on the fly'.

Note that since one of the three methods involves no compression at all, the resulting archive entry will never be larger than the original file.

Using ARC

ARC is invoked with a command of the following format:

ARC ⟨x⟩ ⟨arcname⟩ [⟨template⟩. . .]

where

⟨x⟩ is an ARC command letter (see below), in either upper or lower case.

⟨arcname⟩ is the name of the archive to act on, with or without an extension. If no extension is supplied, then .ARC is assumed. The archive name may include path and drive specifiers.

⟨template⟩ is one or more filename templates. The 'wildcard' characters * and ? may be used. A filename template may include a path or drive specifier, though it is not always meaningful.

If ARC is invoked with no arguments (by typing ARC and pressing Enter), then a brief command summary is displayed.

Table 9.2 contains a brief summary of available ARC commands and options. The commands and options listed in Table 9.2 are applicable to ARC Version 6.02 which is discussed later in this section.

Table 9.2 ARC Version 6.02 commands and options

ARC commands

a   = add files to archive
m = move files to archive
u   = update files in archive
f   = freshen files in archive
d   = delete files from archive
x,e = extract files from archive
r   = run files from archive
p   = copy files from archive to standard output
l   = list files in archive
v   = verbose listing of files in archive
t   = test archive integrity
c   = convert entry to new storage method

ARC options

m = move files to archive
z   = include subdirectories in archive
v   = verbose mode
b   = retain backup copy of archive
s   = suppress compression (store only)
w = suppress warning messages
n   = suppress notes and comments
o   = overwrite existing files when extracting
5   = produce only level-5 compatible archives
g   = encode or decode archive entry

The options listed in Table 9.2 are used to alter the operation of certain ARC commands.

To illustrate the operation of ARC commands, let us examine the use of the A (Add), U (Update), and F (Freshen) commands which allow files to be added to an archive:

Add always adds the file

Update differs from Add in that the file is only added if it is not already in the archive, or if it is newer than the corresponding entry in the archive.

Freshen is similar to Update, except that new files are not added to the archive; only files already in the archive are updated.

For example, if you wished to add a file named TEST.DAT to an archive named MY.ARC, you would use a command of the form:

ARC a my test.dat

If you wanted to add all files with a .C extension and all files named STUFF to an archive named JUNK.ARC you could type:

ARC a junk *.c stuff.*

If you had an archive named TEST.ARC and you wanted to add to it all of your files with an extension of .TXT which have been created or changed since they were last archived, then you would type:

ARC u text *.txt

If you have a bunch of files in your current directory with backup copies being stored in an archive named SAFE.ARC, then if you wanted to make sure that every file in the archive is the latest version of that file, you would type:

ARC f safe

A word about Update and Freshen—these are similar in that they look at the date and time of last change on the file and only add it if the file has been changed since it was last archived. They differ in that Update will add new files, while Freshen will not.

In other words, Update looks for the files on disk and adds them if they are new or have changed, while Freshen looks in the archive and tries to update the files which are already there.

To illustrate the potential efficiency of ARC, let us examine its main executable file named ARC602.EXE, which, when copied to a user's personal computer, unpacks itself. Figure 9.11 illustrates the sequence of operations performed by the author to copy ARC602.EXE to his personal computer's hard disk under the directory \ARC that he created, list his hard disk directory structure, unpack ARC602.EXE, and take a second directory listing to observe the effect of the unpacking. Note that after ARC602.EXE was copied to the author's hard disk the directory listing shows that this file contains 85 527 bytes. After entering the command ARC602 the program unpacks or uncrunches itself into four files: ARC.DOC, ARC.EXE, ARC.TXT, and ARCORDER.TXT. Note that the second directory listing indicates that these four files require 136 102 bytes of storage. Thus, ARC obtained a compression ratio of 1.59 when operated upon itself. To appreciate its compression efficiency, readers should note that ARC.EXE is an executable file while ARC.DOC, ARC.TXT, and ARCORDER.TXT are ASCII files.

Readers should also note that ARC is a copyrighted proprietary program of System Enhancement Associates, Inc., which has granted permission to the author to include it and several associated programs on the convenience disks included with this book. Purchasers

```
C>md\arc

C>cd\arc

C>copy d:arc602.exe c:
        1 File(s) copied

C>dir

 Volume in drive C is GILSGARBAGE
 Directory of  C:\ARC

 .              <DIR>       4-23-90    5:37p
 ..             <DIR>       4-23-90    5:37p
ARC602   EXE     85527      6-22-89    3:39p
        3 File(s)   1986560 bytes free

C>


C>arc602
SARC  Copyright 1986-89 by Wayne Chin and Vernon D. Buerg.  /H gives help
Archive self-extractor, Version 1.10, 03/06/89. ALL RIGHTS RESERVED.

Archive:  C:\ARC\ARC602.EXE
UnCrunching > ARC.DOC
UnCrunching > ARC.EXE
UnCrunching > ARC.TXT
UnCrunching > ARCORDER.TXT
C>dir

 Volume in drive C is GILSGARBAGE
 Directory of  C:\ARC

 .              <DIR>       4-23-90    5:37p
 ..             <DIR>       4-23-90    5:37p
ARC602   EXE     85527      6-22-89    3:39p
ARC      DOC     64321      6-22-89    3:33p
ARC      EXE     65339      3-13-89   10:30a
ARC      TXT      2299      3-14-89    9:57a
ARCORDER TXT      4143      6-22-89    3:39p
        7 File(s)   1845248 bytes free

C>
```

Figure 9.11   Installing ARC602.EXE

of this book are granted a LIMITED LICENSE to use ARC and to copy it and distribute it, provided that the following conditions are met:

1. No fee may be charged for such copying and distribution.
2. ARC may *only* be distributed in its original, unmodified state.
3. ARC may *not* be distributed, in whole or in part, as part of any commercial product or service without the express written permission of System Enhancement Associates, Inc.

Contributions for the use of this program will be appreciated and should be sent to: System Enhancement Associates, Inc., PO Box 949, Nassawadox, VA 23413, USA.

In addition to ARC602.EXE the convenience disk contains several additional programs, including ARCE.COM, ARCE.DOC, ARCP.EXE, MARC.EXE, MARCP.EXE, MKSARC.EXE, and PACK-

ING.LST. ARCP.EXE is a program which is used to extract files from an archive in a highly optimized manner and can be quicker than using ARC.EXE. ARCE.DOC contains the documentation for the program.

MARC.EXE can be used to merge archives previously created by ARC. MKSARC.EXE is used to create an archive into a self-unpacking program similar to the concept used to create ARC602.EXE. PACK-ING.LST as its name may imply is an ASCII file which describes the files contained on the ARC program disk. Figure 9.12 illustrates the list of PACKING.LST. Readers should note that all of the files listed in Figure 9.12 are contained in the directory \ARC on the convenience disk.

Readers should note that each of these programs is user-supported software. This means that you may copy it freely and give the copies away to anyone you wish, at no cost. Those recipients are, in turn, requested to send in a contribution if they decide to use it.

The user-supported software concept (often referred to as

Following is a list of what is contained on the ARC 6.02 program disk:

Filename	Description
ARC602.EXE	This is ARC version 6.02 for MS-DOS (plus documentation) in the form of a 'self-unpacking archive'. It unpacks itself when you run it.
ARCE.COM	This is a small archive extractor by Vern Buerg and Wayne Chin which is included on the ARC disk as a service to our customers.
ARCE.DOC	This is the full documentation for ARCE.
ARCP.EXE	This is the protected mode version of ARC for use with the the OS/2 operating system.
MARC.EXE	This is an archive merge/split utility. It is described in the ARC documentation.
MARCP.EXE	This is the protected mode version of MARC for use with the OS/2 operating system.
MKSARC.EXE	This is the program that creates self-extracting archives. It is described in the ARC documentation.
PACKING.LIST	This file, of course!
README	This file contains directions on how to install ARC on your system.
README.BAT	This is a batch file to print out the README file.

Figure 9.12 PACKING.LST file contents. This file lists and describes the files contained on the ARC program disk that are also contained in the \ARC directory on the convenience diskette accompanying this book

'shareware') is an attempt to provide software at low cost. The cost of offering a new product by conventional means is staggering and, hence, dissuades many independent authors and small companies from developing and promoting their ideas. User-supported software is an attempt to develop a new marketing channel where products can be introduced at low cost.

If user-supported software works, then everyone will benefit. The user will benefit by receiving quality products at low cost and by being able to 'test drive' software thoroughly before purchasing it. The author benefits by being able to enter the commercial software arena without first needing large sources of venture capital.

But it can only work with your support. We're not just talking about ARC here, but about *all* user-supported software. If you obtain a user-supported program from a friend or colleague and are still using it after a couple of weeks, then it is obviously worth something to you, and a contribution should be sent to the developer.

Using ARJ

Although the first release of ARJ only occurred in February of 1991, this archiving program has received a considerable amount of acceptance by system operations (SYSOPS) of bulletin board systems and by commercial organizations. One of the reasons for its acceptance is the large number of commands and command options the program supports which exceeds most archiving programs by a considerable margin.

ARJ was developed by Robert Jung of Norwood, MA. Similar to a large number of archiving programs, the basic design for the early versions of ARJ were based upon the work of Haruhiko Okumura incorporated into his AR002 program.

When this book was written, the current version of ARJ was 2.41A, which is included on the diskette accompanying this book under the directory ARJ. Included in that directory are the self-extracting programs ARJ241A.EXE and UNARJ241.EXE. The first program contains 25 files ranging in scope from the executable archive program to user manuals, a manual update and a conversion utility. Concerning the latter, one of the unique features of ARJ is the inclusion of an archive conversion program which facilitates the conversion of LZH, ZIP, PAK, ARC, DWC, HYP, LZS and ZOO archives to the format used by ARJ. Table 9.3 lists the contents of the ARJ241A.EXE self-extracting program.

The second self-extracting program, UNARJ242.EXE, includes technical information you may wish to read concerning file formats and data compression used by archiving programs. That archive also includes a C source code you may wish to review as a learning tool.

Similar to other archiving programs included in this book, ARJ is

Table 9.3 Contents of ARJ241A.EXE

ARJ.DOC	– ARJ user manual
ARJ.EXE	– Version 2.41a
ARJ_BBS.DOC	– List of ARJ support BBSes
ARJDISP.DOC	– Description of new ARJ display feature
ARJSORT.BAT	– Sort archive by various parameters
ARJSORT.COM	– Compiled ARJSORT.BAT file
ARJSORT.DOC	– ARJSORT user manual
CREDIT.CRD	– Form for handling credit card orders
ERRORS.DOC	– Description of ARJ error messages
INTRO.DOC	– Introduction for new users of ARJ
LICENSE.DOC	– License policy for ARJ
ORDERFRM.DOC	– Order form for ARJ
README.DOC	– File which lists contents of self-extracting program
REARJ.CFG	– Default REARJ configuration file
REARJ.DOC	– REARJ user manual
REARJ.EXE	– Archive conversion utility
REGISTER.EXE	– Program to register ARJ programs
UPDATE.DOC	– Update to ARJ user manual
WHATSNEW.DOC	– Changes to ARJ from 2.20
WHY_ARJ.DOC	– ARJ commercial message
ARJBACK.BAT	– Batch file to fully backup C: to A:
ARJREST.BAT	– Batch file to restore to C: from A:
ARJINCR.BAT	– Batch file to incrementally backup C:
REARJALL.BAT	– Batch file to REARJ ZIP, LZH to ARJ
TESTARJ.BAT	– Batch file to test system compatibility

a shareware program. After using the program during its 30-day evaluation period, you should register the program with the developer or one of his authorized distributors to continue the use of the program. Mr Jung can be reached at 2606 Village Road West, Norwood, MA 02062. Included in the documentation accompanying the program are a list of authorized distributors that provide local registration and technical support for ARJ. The format of an ARJ command is illustrated below:

$$ARJ \langle command \rangle [-\langle sw \rangle[-\langle sw \rangle. .]]\langle archive\text{-}name \rangle$$
$$[\langle file\_names \langle . . .]$$

To illustrate the use of ARJ we will use it to archive the contents of a directory on a hard disk. Figure 9.13 illustrates the operation of ARJ using its 'a' command to add files to an archive. In this example CELLO.ARJ represents the archive we wish to create, while C:\CELLO*.* tells the program we wish to archive all programs in the indicated directory. Similar to most archiving programs, the file extension is optional when creating an archive as .ARJ is the default

```
C:\ARJ2_40>arj a cello.arj c:\cello\*.*
ARJ 2.41a Copyright (c) 1990-93 Robert K Jung. Jul 10 1993
*** This SHAREWARE program is NOT REGISTERED for use in a business, commercial,
*** government, or institutional environment except for evaluation purposes.

Creating archive  : CELLO.ARJ
Adding      C:\CELLO\CELLO.SIG              40.2%
Adding      C:\CELLO\README.1ST            40.9%
Adding      C:\CELLO\CELLO.EXE             31.2%
Adding      C:\CELLO\CELLO.BMK              3.4%
Adding      C:\CELLO\CELLO.STY             43.0%
Adding      C:\CELLO\1024-768.STY          40.8%
Adding      C:\CELLO\640-480.STY           31.0%
Adding      C:\CELLO\800-600.STY           32.7%
Adding      C:\CELLO\CLOHELP.HLP           30.4%
Adding      C:\CELLO\CELLO.BAK             58.6%
Adding      C:\CELLO\DEFAULT.HTM           44.5%
Adding      C:\CELLO\CELLO.INI             59.8%
Adding      C:\CELLO\TEST.TXT              44.5%
Adding      C:\CELLO\APPLE.PIE             28.5%
Adding      C:\CELLO\GIL.TXT              50.6%
   15 file(s)
```

Figure 9.13 Using ARJ to archive the contents of the directory CELLO

extension of the program. Note that the program displays the message 'Adding' followed by the directory path of each file being added, followed by the data reduction of each file being added to the archive. A resulting directory listing indicated that the newly created archive required 293 619 bytes of storage, while the original series of 15 files required 939 153 bytes of storage. Thus, in this example ARJ achieved a compression ratio of 939 153/294 619 or approximately 3.2:1.

In addition to providing one of the highest compression ratios achieved from the use of over 15 archiving programs examined by the author of this book, ARJ provides literally hundreds of commands and switches you can use to tailor its operations to satisfy almost every imaginable application requirement you may have. Through a series of screen displays we will note a portion of its comprehensive set of commands and command switches.

Figure 9.14 illustrates the basic ARJ help screen displayed when you enter the program name without any parameters. As noted in Figure 9.14, the basic help screen indicates frequently used ARJ commands and switches as well as the ARJ command line format. Also included on this screen is a notation to enter ARJ -? for additional help. Prior to doing so, note that 'ARJ a . . .' is used to add files to an archive, while 'ARJ e . . .' is used to extract files from an archive. Thus, ARJ is similar to ARC in that one program provides both an archiving and an extraction capability.

Figure 9.15 illustrates the first in a series of screen displays generated by ARJ upon entry of the ARJ -? command. Note that this screen provides you with a series of examples of the use of ARJ commands.

Figure 9.16 illustrates the resulting display which occurs when you press the Enter key after the screen illustrated in Figure 9.15 was

```
C:\ARJ2_40>arj
ARJ 2.41a Copyright (c) 1990-93 Robert K Jung. Jul 10 1993
*** This SHAREWARE program is NOT REGISTERED for use in a business, commercial,
*** government, or institutional environment except for evaluation purposes.

List of frequently used commands and switches.  Type ARJ -? for more help.

Usage:     ARJ <command> [-<sw> [-<sw>...]] <archive_name> [<file_names>...]
Examples:  ARJ a -e archive, ARJ e archive, ARJ l archive *.doc
<Commands>
   a: Add files to archive               m: Move files to archive
   d: Delete files from archive          t: Test integrity of archive
   e: Extract files from archive         u: Update files to archive
   f: Freshen files in archive           v: Verbosely list contents of archive
   l: List contents of archive           x: eXtract files with full pathname
<Switches>
   c: skip time-stamp Check              r: Recurse subdirectories
   e: Exclude paths from names           s: set archive time-Stamp to newest
   f: Freshen existing files             u: Update files (new and newer)
   g: Garble with password               v: enable multiple Volumes
   i: with no progress Indicator         w: assign Work directory
   m: with Method 0, 1, 2, 3, 4          x: eXclude selected files
   n: only New files (not exist)         y: assume Yes on all queries

C:\ARJ2 40>
```

Figure 9.14 Basic ARJ help screen displayed by entering ARJ without any parameters

```
ARJ 2.41a Copyright (c) 1990-93 Robert K Jung. Jul 10 1993
*** This SHAREWARE program is NOT REGISTERED for use in a business, commercial,
*** government, or institutional environment except for evaluation purposes.

Example ARJ commands:
   Add files minus directory structure:       ARJ a -e archive *.*
   Add two files to archive:                  ARJ a archive name1 name2
   Add files to archive with verification:    ARJ a -jt archive *.*
   Add files with maximum compression:        ARJ a -jm archive *.*
   Comment archive header only:               ARJ c archive -zcmt.fil
   Strip archive comment only:                ARJ c archive -zNUL
   Extract files from archive:                ARJ e archive
   Extract maintaining directory structure:   ARJ x archive
   Extract new and newer files without query: ARJ e archive -u -y
   Extract subdirectory from archive:         ARJ e archive subdir\*.* -p1
   List files in archive:                     ARJ l archive
   Move files to archive:                     ARJ m archive *.doc
   Move files from archive:                   ARJ e -d archive *.doc
   Test integrity of files in archive:        ARJ t archive
   Add files to a multiple volume archive:    ARJ a -va a:archive *.*
   Create up to 999 archive volumes:          ARJ a -va a:archive.001
   Extract from a multiple volume archive:    ARJ x -v a:archive
   Convert archive to self-extractor:         ARJ y -je1 archive

Press ENTER to continue:
```

Figure 9.15 The initial help screen generated by entering the command ARJ -?

displayed. As indicated by the screen header line, this screen provides more detailed ARJ information, commencing with the program's command line format.

To provide the reader with an indication of the comprehensive series of command switches supported by the program, Figure 9.17 lists

```
More detailed ARJ information:

Usage:   ARJ <command> [{/¦-}<switch>[-¦+¦<option>]...] <archive_name>[.ARJ]
         [<base_directory_name>\] [<!list_name>¦<path_name>¦<wild_name>...]

<Commands>
   a: Add files to archive              n: reName files in archive
   b: execute Batch or dos command      o: Order files in archive
   c: Comment archive files             p: Print files to standard output
   d: Delete files from archive         r: Remove paths from filenames
   e: Extract files from archive        s: Sample files to screen with pause
   f: Freshen files in archive          t: Test integrity of archive
   g: Garble files in archive           u: Update files to archive
   i: check Integrity of ARJ.EXE        v: Verbosely list contents of archive
   j: Join archives to archive          w: Where are text strings in archive
   k: remove bacKup files in archive    x: eXtract files with full pathname
   l: List contents of archive          y: copY archive with new options
   m: Move files to archive

<Main defaults>
   Save path information in archive
   Don't save drive and root in path information
   Don't display comment ANSI sequences
   Prompt before overwriting output files
Press ENTER to continue:
```

Figure 9.16 The second screen displayed through the use of the ARJ -? command line entry

```
(Switches)
 -: disables switch char           p:  match using full Pathnames
 +: inhibits ARJ_SW usage          p1: match Pathname with subdirs
 !: sets list char (!)             q:  Query on each file
 &: set batch critical error       r:  Recurse subdirectories
    handler
 #: select files by number         s:  set archive time-Stamp to
                                        newest
 $: add/extract volume label       s1: save original time-Stamp
    $A: add/extract label to       s2: set archive time-Stamp
    drive A
 a:  allow any file Attribute      t:  set file Type (default 0)
 a1: any files and directories         t0: set binary file type
 b:  Backup changed files              t1: set C text file type
```

Figure 9.17 A small selection of the switches and shifted switches supported by ARJ

a portion of a series of screens displayed by the program concerning switches and shifted switches. Although only a handful out of approximately 100 switches supported by ARJ are listed in Figure 9.17, those in that illustration indicate the potential versatility of the program. For example, the exclamation point '!' switch provides you with the ability to set the character used to identify a list file, a file containing a list of files to be operated upon. This little feature permits you to use ARJ to operate on list files previously created for use with other archiving programs. Another interesting switch is 'r', which enables the recursion of subdirectories. When used with the odd (a) command, this switch provides you with the ability to archive the contents of subdirectories under a specified directory, a powerful labor-saving tool when you want to use an archiving program to perform a disk backup operation.

Using PKWARE

A popular set of file-compression programs developed for MS-DOS and PC-DOS compatible personal computers are frequently referred to as PKWARE, which represents the name of the firm marketing this software. Similar to ARC and ARJ, PKWARE programs can be used to both compress and store files together under one filename. However, unlike ARC and ARJ, separate programs must be used to create an archive (PKZIP) and extract files from a previously created archive (PKUNZIP).

Similar to ARJ, the latest versions of PKZIP and PKUNZIP as well as manuals and associated files are distributed on one self-extracting file. That file, PKZ204G.EXE, is contained in the directory PKZIP on the diskette included with this book. Figure 9.18 illustrates a portion of the self-extraction process after the author of this book transferred the file to the directory PKZIP on his hard drive and entered the name of the self-extracting file.

Similar to most shareware programs, if you decide to use PKWARE's programs after their trial period, you should send your registration fee to PKWARE, 9025 North Deerwood Drive, Brown Deer, WI 53223, USA.

```
C:\PKZIP>pkz204g

PKSFX (R)  FAST!  Self Extract Utility  Version 2.04g  02-01-93
Copr. 1989-1993 PKWARE Inc. All Rights Reserved. Shareware version
PKSFX Reg. U.S. Pat. and Tm. Off.

Searching EXE: C:/PKZIP/PKZ204G.EXE
  Inflating: README.DOC              -AV
  Inflating: SHAREWAR.DOC            -AV
  Inflating: WHATSNEW.204            -AV
  Inflating: V204G.NEW               -AV
  Inflating: HINTS.TXT               -AV
  Inflating: LICENSE.DOC             -AV
  Inflating: ORDER.DOC               -AV
  Inflating: ADDENDUM.DOC            -AV
  Inflating: MANUAL.DOC              -AV
  Inflating: AUTHVERI.FRM            -AV
  Inflating: PKZIP.EXE               -AV
```

Figure 9.18 A portion of the self-extraction process of PKZ204G

The file names listed in Figure 9.18 reflect the contents of each file. A few additional files not shown in Figure 9.18 are worthy of discussion.

The file PKUNZIP.EXE provides you with the ability to extract files from a previously created archive. The file PKZIPFIX.EXE can be used to fix a corrupted archive due to a disk problem or the transmission of a file using no error detection and correction or an error detection and correction method susceptible to the occurrence of undetected errors. Two additional files not shown in the portion of the self-extraction process illustrated in Figure 9.18 are PKUNZJR.COM and ZIP2-EXE.EXE. The first file permits the extraction of files using a minimum of computer memory, while the second file is used for the conversion of an archive into a self-extracting archive.

The archive creating program, PKZIP, has a basic command line format similar to several archiving programs. This format is indicated below:

PKZIP [options] zipfile [files...]

The options supported by PKZIP include commands and command switches. The 'zipfile' represents the name of the archive to be created or modified and if the optional extension .ZIP is not included in the command line, it will be automatically added by the program. Finally, the 'file list' entry represents the files to be archived.

To illustrate the operation of PKZIP as well as obtaining a general indication of its operational efficiency, let's use it to compress the directory previously compressed using ARJ. Figure 9.19 illustrates the execution of PKZIP based upon the following command line entry:

PKZIP CELLO.ZIP C:\CELLO*.*

This command causes the files in the directory CELLO to be archived onto the file CELLO.ZIP.

In examining Figure 9.19, note that the display of the archive creation process is similar to that in ARJ. PKZIP denotes files added to an archive and the percentage of data reduction. One difference from ARJ is the fact that PKZIP identifies the type of compression being performed. Here 'deflating' represents a sliding dictionary LZ compression followed by a secondary compression using Huffman and Shannon–Fano codes. The resulting archive created by PKZIP required 292 459 bytes of storage, which is slightly less than the 293 619 bytes of storage required when ARJ was used. However, due to the ability of each program to operate slightly more efficiently than the other program when compressing certain types of files, most persons evaluating the use of an archiving program may be better served by examining the operational features in the form of commands and

```
PKZIP Reg. U.S. Pat. and Tm. Off. Patent No. 5,051,745

■ 80486 CPU detected.
■ XMS version 3.00 detected.
■ Novell Netware version 3.11 detected.
■ Using Normal Compression.

Creating ZIP: CELLO.ZIP
  Adding: CELLO.SIG      Deflating (68%), done.
  Adding: README.1ST     Deflating (60%), done.
  Adding: CELLO.EXE      Deflating (70%), done.
  Adding: CELLO.BMK      Deflating (97%), done.
  Adding: CELLO.STY      Deflating (57%), done.
  Adding: 1024-768.STY   Deflating (60%), done.
  Adding: 640-480.STY    Deflating (69%), done.
  Adding: 800-600.STY    Deflating (67%), done.
  Adding: CLOHELP.HLP    Deflating (70%), done.
  Adding: CELLO.BAK      Deflating (41%), done.
  Adding: DEFAULT.HTM    Deflating (56%), done.
  Adding: CELLO.INI      Deflating (40%), done.
  Adding: TEST.TXT       Deflating (56%), done.
  Adding: APPLE.PIE      Deflating (72%), done.
  Adding: GIL.TXT        Deflating (49%), done.
```

Figure 9.19 Using PKZIP to create an archive

command switches supported by a program as well as its licensing policy.

Figure 9.20 lists the first in a series of four help screens generated by PKZIP when the command is entered by itself without any parameters. This initial help screen illustrated in Figure 9.20 indicates how a user can directly display a specific help screen, illustrates the command line format of PKZIP and enables a user to display three additional help screens.

To obtain an appreciation for the operational capability of PKZIP, let's examine two additional help screens generated by the program. Figure 9.21 illustrates the program's second help screen which describes some basic commands as well as command options which enable you to use the program to scramble an archive using a password of your choice, view a previously created archive in one of five different ways, and format diskettes during a span operation. Concerning the latter, the value of this feature becomes most appreciated if you are archiving the contents of your fixed disk or a portion of the disk to diskettes and run out of formatted disks during the backup operation. Unlike some archiving programs that force you to begin anew and use DOS to format additional diskettes, PKZIP has the ability to format disks from within the program.

Figure 9.22 illustrates the third PKZIP help screen. This screen lists two of many PKZIP options that have both a prefix and a suffix indicator. For example, from Figure 9.21, the -a option is used to add

```
PKZIP (R) FAST! Create/Update Utility  Version 2.04g  02-01-93
Copr. 1989-1993 PKWARE Inc. All Rights Reserved. Shareware Version
PKZIP Reg. U.S. Pat. and Tm. Off.  Patent No. 5,051,745

PKZIP /h[1] for basic help  PKZIP /h[2¦3¦4] for other help screens.

Usage:  PKZIP [options] zipfile [@list] [files. . .]

        Simple Usage:  PKZIP zipfile file(s). . .
                          ¦    ¦   ¦
                          ¦    ¦   ¦
Program — — — — — — — — —┘    ¦   ¦
                               ¦   ¦
                               ¦   ¦
New zipfile to create — — — — —┘   ¦
                                   ¦
File(s) you wish to compress — — — —┘

The above usage is only a very basic example of PKZIP's capability.

Press 2 for more options (including spanning & formatting), press 3 for
advanced options, 4 for trouble shooting options, any other key to
quit help.
```

Figure 9.20 Initial PKZIP help screen

files to a previously created archive. From Figure 9.22, the use of the -a+ option is used to turn off the archive attribute of files added to an archive. By clearing the archive attribute you can perform incremental backup archives, a most handy feature when you have a large number of files which are infrequently updated and you wish to maintain a backup of your files on a daily or weekly basis.

As previously discussed, PKWARE requires separate programs for creating an archive and extracting programs from a previously created archive. Concerning the latter, you would use PKUNZIP to extract files from a previously created archive.

PKUNZIP is similar to PKZIP in that it supports the same command format but uses different commands and command options. Although space does not permit an examination of the use of PKUNZIP, the convenience diskette accompanying this book includes that program in the previously mentioned self-extracting file for readers to obtain experience in its use prior to registering the program.

Using Ultra Compressor II

In concluding this chapter we will focus our attention upon the Ultra Compressor II revision 2, UC2r2, which is a copyright of Ad Infinitum Programs, the Netherlands (AIP-NL). The program UC2r2.EXE, which is a self-extracting file containing 54 archived files, is included on the diskette accompanying this book in the directory UC2. UC2r2.EXE is

```
Usage:  PKZIP [options] zipfile [@list] [files...]

-a              Add files
-b[drive]       create temp zipfile on alternative drive
-d              Delete files
-e[x,n,f,s,0]   use [eXtra¦Normal (default)¦Fast¦Super fast¦NO compression]
-f              Freshen files
-l              display software License agreement
-m[f,u]         Move files [with Freshen ¦ with Update]
-u              Update files
-p¦P            store Pathnames¦p=recursed into¦P=specified & recursed into
-r              Recurse subdirectories
-s[pwd]         Scramble with password [If no pwd is given, prompt for pwd]
-v[b][r][m][t][c] View .ZIP [Brief][Reverse][More][Technical][Comment] sort by
  [d,e,n,o,p,s] [Date¦Extension¦Name¦natural Order(default)¦Percentage¦Size]
-&[f¦l¦u        Span disks [Format¦format Low density¦Unconditional format¦
  ul¦w¦v]            Unconditional Low density¦Wipe disk¦enable dos Verify¦
  [s[drive]]           Back up entire disk w/ subdirs (-rp) [drive to back up]]

*** For more information, please consult the user manual ***
Press 1 for basic options, 3 for advanced options, 4 for trouble shooting
options, any other key to quit help.
```

Figure 9.21 Second PKZIP help screen

provided as an evaluation copy for readers for a 60-day trial period. Readers are referred to the file LICENSE.DOC to register the program if it is to be used beyond the 60-day evaluation period.

One of the more distinguishing features of UC2r2 is its use of color, which visually sets it apart from other archiving programs. Another distinguishing feature of UC2r2 is its enhanced compression performance, which results in an archive size typically a few percent less than competitive archiving programs. Another most interesting feature of this archiving program is its documentation support of 13 languages in addition to English, its ability to perform a virus scan, use of Expanded and Extended Memory and other features that many competitive programs lack.

To install Ultra Compressor II, copy the file VC2R2.EXE from the directory VC2 on the diskette accompanying this book to your hard drive. Since that file is a self-extracting file, you can enter the name of the file to initiate the self-extracting process. Unlike other archiving programs examined by this author, VC2 displays the directory location of the self-extracting program and provides you with the ability to change the location of extracted files.

Figure 9.23 illustrates the initial screen display of Ultra Compressor II revision 2 when you enter the name of the self-extracting program. Note that the install program invoked from the self-extracting archive automatically displays the directory location of the self-extracting program followed by a question mark and permits you to install the files from the archive in that directory, in a different location, or to abort the installation process.

Assuming you select option 1, the self-extracting program com-

```
Usage:   PKZIP [options] zipfile [@list] [files...]

-a+             clear archive Attribute after compression
-c              create/Edit Comments for all files
-C              add Comments for new files only
-i[-]           add files with archive Attribute set [don't turn attribute off]
-j¦J<h,r,s>     mask¦don't mask <Hidden/System/Read-only> files (default=jhrs)
-k              Keep original .ZIP file date
-o              set .ZIP file date to the latest file in .ZIP file
-q              enable ANSI codes in comments
-t[date]        take files NEWER than or EQUAL to date (default=today)
-T[date]        take files OLDER than date (default=today)
-w¦W<h,s>       include¦exclude <Hidden, System> files (default=Whs)
-x<file>        eXclude specified file
-x@list         eXclude file(s) in specified list file
-z              create or modify .ZIP comment
-!              add authenticity verification to .ZIP file (registered vers only)
-$[drive]       save Volume label in .ZIP file (default = current drive)
-@list          generate list file
-=              open file in compatibility mode (bypass share)
Press 1 for basic options, 2 for other options (including spanning and
formatting), 4 for trouble shooting options, any other key to quit help.
```

Figure 9.22 Third PKZIP help screen

mences the extraction process, using the directory where the self-extracting program is located as its destination path for extracted files. Figure 9.24 illustrates an initial portion of the extraction process. In examining Figure 9.24 note that Ultra Compressor II uses a series of squares to indicate the progress of the extraction process, first displaying a series of six dots per file and then changing the dots to squares as the extraction process for each file progresses. If you focus your attention upon the last line in Figure 9.24, note that one square covered the first dot when the screen was captured, indicating that slightly more than 16% of the extraction process for the file was completed. The backslash (\) alternates with a forward slash (/) and the broken bar (¦) character to present what appears to be a rotating line which indicates the extraction process is occurring.

Unlike most archive programs that generate a few help screens when the program name is entered without a command or with a specific command option, Ultra Compressor provides users with the ability to view the documentation online. This provides you with the ability to examine the use of UC commands, find out how to configure UC to perform a variety of functions beyond the capability of many archiving programs, display the program license and trial use license, list the most essential UC commands, and view other information. The basic command format of Ultra Compressor II is as follows:

UC command [option(s)]archive-name[files...]

UC is similar to ARJ in that one program can be used to create an archive as well as to extract files from a previously created archive.

```
AIP-NL UltraCompressor II revision 2 (tm) INSTALL PROGRAM

UltraCompressor II is a powerful datacompressor which allows you to
keep collections of files in an archive.

Some of its special features are:
 - extremely tight compression
 - fast compression, very fast updates
 - very fast decompression
 - simple user interface (integrated help)
 - reliable (e.g. archives can recover from damaged sectors)
 - ability to store multiple versions of a file in an archive
 - project oriented Version Manager NEW!
 - transparent conversion of non-UC2 archives
 - advanced filtering on contents, date/time, attributes NEW!

Please note this software can only be used, (re)distributed, etc.
according to the included license agreement (license.doc)!

Do you want to install UltraCompressor II in C:\UC2\ ?
 1 → Yes
 2 → No
 3 → Different location
CHOICE (+=Abort -=Shell) ?
```

Figure 9.23   Ultra Compressor II revision 2 initial install program display

```
Extracting files from C:\UC2\ALL.UC2 (destination path C:\UC2\)
Analyzing ■■■■■■
Decompressing C:\UC2\SWEDISH.INT ■■■■■■ OK
Decompressing C:\UC2\CHINESE.INT ■■■■■■ OK
Decompressing C:\UC2\SLOVENIA.INT ■■■■■■ OK
Decompressing C:\UC2\DUTCH.INT ■■■■■■ OK
Decompressing C:\UC2\RUSSIAN.INT ■■■■■■ OK
Decompressing C:\UC2\DANISH.INT ■■■■■■ OK
Decompressing C:\UC2\GERMAN.INT ■■■■■■ OK
Decompressing C:\UC2\ITALIAN.INT ■■■■■■ OK
Decompressing C:\UC2\HUNGARIA.INT ■■■■■■ OK
Decompressing C:\UC2\SPANISH.INT ■■■■■■ OK
Decompressing C:\UC2\FRENCH.INT ■■■■■■ OK
Decompressing C:\UC2\ICELANDI.INT ■■■■■■ OK
Decompressing C:\UC2\HEBREW.INT ■■■■■■ OK
Decompressing C:\UC2\BBS.DOC ■■■■■■ OK
Decompressing C:\UC2\BASIC.DOC ■■■■■■ OK
Decompressing C:\UC2\LICENSE.DOC ■■■■■■ OK
Decompressing C:\UC2\BACKGRND.DOC ■■■■■■ OK
Decompressing C:\UC2\CONFIG.DOC ■■■■■■ OK
Decompressing C:\UC2\EXTEND.DOC ■·······\
```

Figure 9.24   A portion of the extraction process during the installation of Ultra Compressor II revision 2

To illustrate the use of Ultra Compressor II, we will again construct an archive from the contents of the directory CELLO on the author's hard drive. Figure 9.25 illustrates the entry of the 'a' command to add files to the archive CELLO by archiving all files in the directory CELLO on drive C. Although the extension .UC2 was not specified in the archive file name, it is automatically added to the archive file by the program.

The resulting archive required 286 548 bytes of storage, representing a reduction of approximately 2% in comparison to the archives created by ARJ and PKZIP. While this archiving program in general performs slightly better than other archiving programs with respect to the size of an archive based upon archiving a series of test files, you should also consider its options, command support and memory usage.

Figure 9.26 illustrates the program's configuration menu. By examining the entries in Figure 9.26, you can obtain a good indication of many features of the program. For example, you can turn the use of color on or off, set the use of extended or expanded memory and even turn on and off the program's banner. Although Ultra Compressor II is an excellent program, its memory requirements may preclude its use by some persons with 1 Mbyte systems that are connected to a network. This is because the program requires 512 kbytes of memory.

Although space does not permit an examination of the use of Ultra Compressor II commands and options, readers are referred to the convenience diskette to examine the operation of this program. By trying each archiving program you will obtain the ability to select one

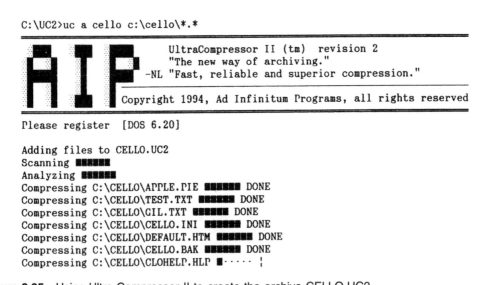

```
C:\UC2>uc a cello c:\cello\*.*
```

```
        UltraCompressor II (tm)  revision 2
        "The new way of archiving."
    -NL "Fast, reliable and superior compression."

    Copyright 1994, Ad Infinitum Programs, all rights reserved
```

```
Please register  [DOS 6.20]

Adding files to CELLO.UC2
Scanning ██████
Analyzing ██████
Compressing C:\CELLO\APPLE.PIE ██████ DONE
Compressing C:\CELLO\TEST.TXT ██████ DONE
Compressing C:\CELLO\GIL.TXT ██████ DONE
Compressing C:\CELLO\CELLO.INI ██████ DONE
Compressing C:\CELLO\DEFAULT.HTM ██████ DONE
Compressing C:\CELLO\CELLO.BAK ██████ DONE
Compressing C:\CELLO\CLOHELP.HLP █·····┆
```

Figure 9.25 Using Ultra Compressor II to create the archive CELLO.UC2

```
GENERAL OPTIONS:
    A → Default compression method   NORMAL (tight) (s-tight) (fast)
    B → Default operation            BASIC MODE (incremental mode)
    C → Reliability level            DETECT (protect) (ensure)
    D → Automatic archive conversion OFF (on)
    E → Virus scan during conversion OFF (on)
    F → Smart skipping               ON (off)
    G →Amount of output/information  NORMAL (minimal) (verbose)
    H → Show (multimedia) banners    ASK (on) (off)
    I → Store OS/2 2.x extended      OFF (on)
        attributes
    J → Store system/hidden files    ASK (on) (off)
SYSTEM OPTIONS:
    M → Video mode                   COLOR (mono) (color-bios)
                                     (mono-bios)
    N → Dynamic program swapping     ON (off)
    O → Use EMS                      4.0+ ONLY (any version) (off)
    P → Use XMS                      ON (off)
    Q → Use 386/486/Pentium/M1       ON (off)
        features
    R → Advanced networking          ON (auto-skip) (off)
    S → Location for temporary files [C:\UC2\]
    T → First loc for manuals (*.DOC) [C:\UC2\]
    U → Location for error logfile   [C:\UC2\]
    V → First loc for batch&script   [C:\UC2\]
        files
QUICK SETUP: 1→default 2→max speed 3→max compress 4→max safe
5→UNDO

CHOICE (Escape to leave configuration menu)?
```

Figure 9.26 Ultra Compressor II configuration menu

that is best suited to your operational requirements. In doing so, please remember that each program is a shareware program and the companies depend upon satisfied users to send in an appropriate fee if you intend to use the software beyond its evaluation period.

APPENDIX A

DATANALYSIS PROGRAM DESCRIPTIONS AND LISTINGS

A.1 FORTRAN PROGRAM: OPERATIONAL DESCRIPTION

The original data-compression analysis program was written in industry standard FORTRAN IV for execution on a Honeywell 66/80 computer system. It consists of one main routine which performs all file handling. Twelve independent subprograms are called by the main routine to perform specific functions such as string analysis, sorting and report formatting.

Included in this documentation are narrative descriptions of each subprogram. The reader should refer to the program listing when reading the following operational description of the main routine, subprograms and the discussion concerning program transferability and variable assignments.

A.2 MAIN ROUTINE

The routine first builds two symbolic arrays filled with those characters associated with the frequency counters. The first array contains the 127 ASCII characters followed by a summary symbol and a temporary work slot used while sorting the array. The second array contains all the ASCII character pairs normally encountered. The array, although structured in one dimension, can be viewed as a two-dimensional array directly proportional to the paired-integer array IP(28, 26). The DATA statement was chosen to minimize external I/O.

The routine then asks three questions prior to performing the actual analysis:

ENTER FILE TYPE—Up to seven characters which identifies the file to be analyzed. This is displayed on each report for future reference.

ENTER INPUT LINELENGTH—Up to 132 characters is 'strings' of data to be read. A card-type file would be 80 versus a print file of 132. This conserves run time by reducing unnecessary iterations. Longer strings can be accommodated by increasing dimensions.

TO SUPPRESS TRAILING SPACES ENTER 1—This is useful for variable line lengths, such as a BASIC program entered from a terminal. Blanks or spaces trailing the last printable character are ignored. Any other entry will cause the entire fixed line length to be used.

A record or string is then read. Each character is converted to its ordinal decimal value and the value is used to subscript into an array of tallies. Before the next record is read, three subprograms are called and the alphanumeric string is passed to each subprogram for further analysis. These subprograms are: SUB-1 to resolve purely numeric strings, SUB-7 to resolve repetitious strings, and SUB-9 to resolve pair combinations.

The remaining subprograms are called upon encountering the end-of-file to sort and print the eight reports.

Subprogram 1

This calculates sequential numeric strings which are susceptible to half-byte encoding. The following special characters normally found in financial programs are also included as part of the string:

$$044_8 = \$ = 36_{10}$$
$$052_8 = * = 42_{10}$$
$$054_8 = , = 44_{10}$$
$$055_8 = - = 45_{10}$$
$$056_8 = . = 46_{10}$$
$$057_8 = / = 47_{10}$$

Half-byte   numeric   packing   would   permit   compression

of: $***1,234.56
             12/12/80-2/11/81
             026-36-1048
             800-555-1212
             45, 675, 109, 210.86
             1941/1945/1952-1954

Subprogram 2

This prints the execution report summary.

Subprogram 3

This prints the frequency of occurrence table first in ordinal sequence, then, after sorting by SUB-6, in descending frequency of occurrence for ease of analysis. The array is folded into a table consisting of 4 columns and 32 rows to fit on a single page.

Subprogram 4

This prints the repeated character string analysis report. Only strings of *identical* characters are included. Void string lengths are omitted to conserve printing.

Subprogram 5

This prints a further analysis of that displayed by SUB-4 in that it derives the actual savings achieved by employing data-compression encoding techniques. For example, $4 \leqslant R \leqslant 256$ when compressed as:

SPECIAL CHARACTER	COUNT	COMPRESSED CHARACTER

Subprogram 6

This performs an internal parallel sort of the frequency array, with its associated ASCII symbol, in descending order of frequency. Simply stated, for every movement of an IR(N) value there is a corresponding of a B(N) symbol.

Subprogram 7

This calculates the sequential repeating characters displayed by SUB-4 and SUB-5. Three separate arrays are maintained: IY(N) for any occurrence, IN(N) for numerics $48_{10} \leqslant N \leqslant 57_{10}$, and IX(N) for spaces 32_{10}.

Subprogram 8

This prints the sequential numeric string analysis for half-byte encoding susceptibility. This subprogram uses the IM(N) array generated by SUB-1. It uses the algorithm

$$\left[\left[\frac{\text{String size} + 1}{2}\right] + 2\right] * \text{Count}$$

to derive the characters saved by half-byte encoding.

Subprogram 9

This isolates character pairs and calculates their frequency of occurrence. It begins with the first couplet in the string and continues, incrementing by two, until the string is exhausted. No attempt is made to shift boundaries of couplets.

A two-dimensional array (28×26) is maintained to count the following pair combinations, where the underline symbolizes a blank character:

_A	A_	AA	. . .	ZA
_B	B_	AB	. . .	ZB
_C	C_	AC	. . .	ZC
.	.	.		.
.	.	.		.
.	.	.		.
_Z	Z_	AZ	. . .	ZZ

An additional count is maintained for embedded pairs of blanks or spaces: e.g. bounded by none, blanks thereby not susceptible to other forms of encoding. In order to maintain a symmetric array, which is sorted by SUB-11, the Q-cell is used since this pair is never encountered in the English language.

Initial decimal values are established to represent pairs and DO loops are performed in the following manner:

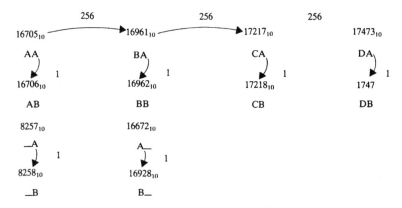

The J index is used in the program to establish initial values. Due to an unexplainable 'bug' in the system used for development, the starting values had to be inflated by 16. The values in cards 4750, 4860, 4990 and 5070 should be changed to the above to operate properly on a different system.

Subprogram 10

This prints the paired-character analysis report after SUB-11 sorts the array. One page is required to display the top 132 frequencies of occurrence. The array is folded into 22 rows of 6 columns each for ease of reference.

A final summary count of combinations is displayed at the foot of the page which may be used to derive a count of those additional combinations beyond the bounds of the table.

Subprogram 11

This performs an internal parallel sort of the paired frequency array IP (N1,N2) with character array C(N) in descending order of frequency. To facilitate the technique used in SUB-6, a linear conversion is performed to give IP(N1, N2) the appearance of a single dimension corresponding to C(N).

Subprogram 12

This computes file entropy and produces the final report:

$$\text{Entropy} = \sum_{i=1}^{M} P_i \log_2 P_i$$

where P_i is the probability of occurrence of characters in the data file: for example, if E occurs 100 times in a 1000 character file

$$P_i(E) = \frac{100}{1000} = 0.1$$

A.3 PROGRAM TRANSFERABILITY

Several instructions are hardware sensitive and, by their nature, are not directly transferable between word-oriented hardware. The FORTRAN compiler must be examined and, if need be, three or four instructions modified.

For example, the program derives the ordinal decimal value of each character inputed and uses it as an array subscript to increment counts of occurrence. The Honeywell compiler requires an integer-right-shift (IRS) function to align the left-justified ASCII character to a proper 36-bit double word boundary.

For example, card 1380 aligns datum in the following manner (datum = A = 65):

	0	8 9	17 18	26 27	35
Before:	65	0	0	0	

IRS(A(I),27)

	0	8 9	17 18	26 27	35
After:	0	0	0	65	

To operate on the DECSYSTEM 10, the following specific instructions (cards) should be changed as indicated:

CONVERSION TO DECSYSTEM 10Change the specific instructions (cards) as follows:
(1) Change card 1310 to: 2 FORMAT(132R1)
(2) Change card 1380 to: IP2 = A(I)
(3) Change card 4720 to: IV = (A(I)*256)
(4) Change card 4730 to: IV = IV + A(I + 1)
(5) Change card 4750 to: IF (IV − 8224)923, 920, 923
(6) Change card 4860 to: J = 8257
(7) Change card 4990 to: J = 16672
(8) Change card 5070 to: J = 16705
(9) Change card 5960 to: 1200 E = E + P(I)*ALOG(P(I))

Note: The generic functions ALOG and ALOG 10 are poorly discussed in DECSYSTEM 10 publications. The specific function desired is a log to the base two. If ALOG does not provide this properly, substitute as follows in card 5960:

$$\sum P_i \frac{\log_{10} P_i}{\log_{10} 2} \quad \text{in lieu of} \quad \sum P_i \log_2 P_i$$

A.4 VARIABLE ASSIGNMENTS

Name	Type	Purpose	Usage	I = Endogenous X = Exogenous
A(132)	DIM	ASCII record	Main	X
B(129)	DIM	Graphic symbols	SUB3,6	I
IA(132)	DIM	Line of ASCII character in octal notation	Main; SUB1,2	I
IN(133)	DIM	Numeric strings	SUB1,4	I
IX(133)	DIM	Blank strings	SUB2,5	I

IY(133)	DIM	All strings	SUB7,8	I
Z( )	DIM			
NI	INT	Linelength to search	Main; SUB1,2,4,5	X
X	Char.	File name	SUB3,4,5	X
JI	INT	Record count	Main; SUB4,5	I
I	INT	Index	Main; SUB1,2,3,4,5,6	I
J	INT	Index		
K	INT	Index	SUB1,2,3,6	I
M	INT	Flag	SUB3	I
P(128)	Doub. Prec.	Percentage of occurrence	SUB3,4,5	I
IR(128)	Doub. Prec.	Count of occurrence	Main; SUB3,4,5,6	I
IP1	INT	Octal equivalent of character	Main	I
L	INT	Index	Main	I
IPI	INT	Sort index	SUB6	I
IV	INT	Trailing space suppression flag	Main	X
N	INT	Adjusted linelength	Main; SUB1,2,4,5	I
IM(133)	DIM	Mixed numeric and specific characters	SUB8,1	I
NC	REAL	Trailing space counter	Main	I
IC & ICI	INT	Half-byte compression character count	SUB8	I
ITN$_x$, ITX, ITY$_x$, ITT$_x$, ITS$_x$	INT	Total cells	SUB5	I
E	INT	Entropy calculation	SUB12	I
BI	INT	Bits after compression	SUB12	I
B4	INT	Characters after compression	SUB12	I
B2	INT	Characters before compression	SUB12	I
B3	INT	Theoretical maximum compression	SUB12	I

A.5 BASIC PROGRAM: OPERATIONAL DESCRIPTION

The original DATANALYSIS program was written in 1980 for operation on large-scale computer systems. Although the use of FORTRAN IV permitted the program to be easily transported to various large-scale computers, its transferability excluded many popular minicomputers and microcomputers for a variety of reasons to include the lack of an economical FORTRAN compiler. Based upon the large number of requests received from readers of the first edition of this book, work was begun in 1985 to rewrite the program in Microsoft BASIC to accommodate the ever-expanding audience of MS-DOS and PC-DOS compatible computers.

To shorten the execution time of the BASIC program, which was several magnitudes greater than obtained from the use of the FORTRAN program on a large computer system, required several subroutines to be extensively revised. The two-dimensional pair-combination sort, for example, was totally unacceptable from an execution time perspective and was considerably abbreviated in BASIC to sequence columns only in descending order of occurrence. The interpretive version of the BASIC program, called DATANAL.BAS is listed at the end of this appendix. Unfortunately, even reducing the two-dimensional pair-combination sort still requires approximately 15–25 min to complete the analysis of a file containing approximately 20 000 characters. To substantially reduce the required execution time of the BASIC version of DATANALYSIS will require the reader to obtain a compiled version of the program. On the convenience diskette accompanying the current edition of this book you will find a program file named DATANAL.EXE. This is an executable program file that was produced by the use of the Microsoft Quick BASIC compiler and can operate on any 256K MS-DOS compatible system. Other files included on the referenced disk are: BRUNIO.LIB which is the Quick BASIC run-time module library; BRUNIO.EXE which is the Quick BASIC run-time module; DATANAL.OBJ which is the object version of the DATANALYSIS program; DATANAL.LST which is the listing of the source file produced by the compilation process; and DATANAL.MAP, which is a file that lists all external symbols in the program and their extensions.

The BASIC version of DATANALYSIS was structured to allow operations on systems with or without an attached parallel printer. Any file saved in ASCII can be analyzed by this program.

For the sake of run-time efficiency, use the program as is if the file is mostly in lower case. The program could be modified to accommodate both cases; however, this would greatly sacrifice both execution speed and memory, since it would double the size of the program's IP and C$ arrays. For those readers with 80286-based PCs and BASIC compilers that can access over 384K of memory, a program expansion to accommodate both cases might be well worthwhile. As the reader will note from reviewing the program listing at the end of this appendix, an extensive number of comments have been included to facilitate the analysis of the program's logic as well as to enable the program to be modified by the reader. Any improvements, comments or suggestions will be greatly appreciated by the author.

To execute the compiled BASIC version of DATANALYSIS requires you to first load either MS-DOS or PC-DOS and then insert the convenience disk in drive A. At the prompt A>, enter the command DATANAL. The following paragraphs describe the main routine and subroutines included in the BASIC version of DATANALYSIS.

A.6 MAIN ROUTINE

The routine first builds two symbolic arrays filled with those characters associated with the frequency counters. The first array contains the 127 ASCII characters followed by a summary symbol and a temporary work slot used while sorting the array. The second array contains all of the ASCII character pairs normally encountered. The array, although structured in one dimension, can be viewed as a two-dimensional array directly proportional to the paired-integer array IP (28,26). The DATA statement was chosen to minimize external I/O.

The routine then asks three questions prior to performing the actual analysis:

ENTER ASCII FILENAME—Any [DRIVE:]FILENAME EXTENSION which identifies the file to be analyzed, this is displayed on each report for future reference. The file must be in ASCII format.

IS FILE MOSTLY IN LOWER CASE—For pair analysis the array is initialized with lower case 97_{10} to 122_{10} in lieu of 65_{10} to 90_{10}.

DO YOU DO YOU WANT PRINTED OUTPUT—Allows the program to be run on a system without an attached parallel printer.

A record or string is then read. Each character is converted to its ordinal decimal value and the value is used to subscript into an array of tallies. Before the next record is read, three subprograms are called and the alphanumeric string is passed to each subprogram for further analysis. These subprograms are: SUB-1 to resolve purely numeric strings, SUB-7 to resolve repetitious strings, and SUB-9 to resolve pair combinations.

The remaining subprograms are called upon encountering the end-of-file to sort and print the eight reports.

Subprogram 1

This calculates sequential numeric strings which are susceptible to half-byte encoding. The following special characters normally found in financial programs are also included as part of the string:

$$
\begin{aligned}
\$ &= 36_{10}. \\
* &= 42_{10} \\
, &= 44_{10} \\
- &= 45_{10} \\
. &= 46_{10} \\
/ &= 47_{10}
\end{aligned}
$$

Half-byte numeric packing would permit compression of:

$***1,234.56
12/12/80-2/11/81
026-36-1048
800-555-1212
45, 675, 109, 210.86
1941/1945/1952–1954

Subprogram 2

This prints the execution report summary.

Subprogram 3

This prints the frequency of occurrence table first in ordinal sequence, then, after sorting by SUB-6, in descending frequency of occurrence for ease of analysis. The array is folded into a table consisting of 4 columns and 32 rows to fit on a single page.

Subprogram 4

This prints the repeated character string analysis report. Only strings of *identical* characters are included. Void string lengths are omitted to conserve printing.

Subprogram 5

This prints a further analysis of that displayed by SUB-4 in that it derives the actual savings achieved by employing data-compression encoding techniques. For example, $4 \leq R \leq 256$ when compressed as:

SPECIAL CHARACTER	COUNT	COMPRESSED CHARACTER

Subprogram 6

This performs an internal parallel sort of the frequency array, with its associated ASCII symbol, in descending order of frequency. Simply stated, for every movement of an IR(N) value there is a corresponding of a B$(N) symbol.

Subprogram 7

> This calculates the sequential repeating characters displayed by SUB-4 and SUB-5. Three separate arrays are maintained: IY(N) for any occurrence, IN(N) for numerics $48_{10} \leqslant N \leqslant 57_{10}$, and IX(N) for spaces 32_{10}.

Subprogram 8

> This prints the sequential numeric string analysis for half-byte encoding susceptibility. This subprogram uses the IM(N) array generated by SUB-1.
>
> It uses the algorithm

$$\left[\left[\frac{\text{String size} + 1}{2}\right] + 2\right] * \text{Count}$$

> to derive the characters saved by half-byte encoding.

Subprogram 9

> This isolates character pairs and calculates their frequency of occurrence. It begins with the first couplet in the string and continues, incrementing by two, until the string is exhausted. An attempt is made to shift boundaries of couplets when a non-alphabetic character is encountered.
>
> A two-dimensional array (28 $_*$ 26) is maintained to count the following pair combinations, where the underline symbolizes a blank character:

_A	A_	AA	. . .	ZA
_B	B_	AB	. . .	ZB
_C	C_	AC	. . .	ZC
.	.	.		.
.	.	.		.
.	.	.		.
_Z	Z_	AZ	. . .	ZZ

An additional count is maintained for embedded pairs of blanks or spaces: e.g. bounded by none, blanks thereby not susceptible to other forms of encoding. In order to maintain a symmetric array, which is sorted by SUB-11, the Q-cell is used since this pair is never encountered in the English language.

Subprogram 10

This prints the paired-character analysis report after SUB-11 sorts each column of the array in descending order of occurrence. One page is required to display the top 132 frequencies of occurrence. The array is folded into 22 rows of 6 columns each for ease of reference.

A final summary count of combinations is displayed at the foot of the page which may be used to derive a count of those additional combinations beyond the bounds of the table.

Subprogram 11

This performs an internal parallel sort of the paired-frequency array IP(N1,N2) with character array C$(N) in descending order of frequency. To facilitate the technique used in SUB-6, a linear conversion is performed to give IP(N1,N2) the appearance of a single dimension corresponding to C$(N).

Subprogram 12

This computes the file entropy and produces the final report:

$$\text{Entropy} = \sum_{i=1}^{M} P_i \log_2 P_i$$

where P_i is the probability of occurrence of characters in the data file: for example, if E occurs 100 times in a 1000 character file

$$P_i(E) = \frac{100}{1000} = 0.1$$

A.7 VARIABLE ASSIGNMENTS

Name	Type	Purpose	Usage	I = Endogenous X = Exogenous
A(132)	DIM	ASCII record	Main	X
B(129)	DIM	Graphic symbols	SUB3,6	I
IA(132)	DIM	Line of ASCII characters in octal notation	Main; SUB1,2	I
IN(133)	DIM	Numeric strings	SUB1,4	I
IX(133)	DIM	Blank strings	SUB2,5	I
IY(133)	DIM	All strings	SUB7,8	I
Z( )	DIM			
NI	INT	Linelength to search	Main; SUB1,2,4,5	I
F$	Char.	File name	SUB3,4,5	X
JI	INT	Record count	Main; SUB4,5	I
I	INT	Index	Main; SUB1,2,3,4,5,6	I
J	INT	Index		
K	INT	Index	SUB1,2,3,6	I
M	INT	Sort flag	SUB3	I
P(128)	DIM	Percentage of occurrence	SUB3,4,5	I
IR(128)	DIM	Count of occurrence	Main; SUB3,4,5,6	I
IP2	INT	Decimal equivalent of character	Main	I
IM(133)	DIM	Mixed numeric and specific characters	SUB8, 1	I
IC & ICI	INT	Half-byte compression character count	SUB8	I
ITN$_x$, ITX, ITY$_x$, ITT$_x$, ITS$_x$	INT	Total cells	SUB5	I
E	INT	Entropy calculation	SUB12	I
BI	INT	Bits after compression	SUB12	I
B4	INT	Characters after compression	SUB12	I
B2	INT	Characters before compression	SUB12	I
B3	INT	Theoretical maximum compression	SUB12	I
D$	CHAR	LPT1: or SCRN: option	Main	X
V$	CHAR	Upper/lower case flag	Main	X

A.8 FORTRAN DATANALYSIS PROGRAM LISTING

```
C *MAIN**MAIN FREQ ROUTINE**HELD & MARSHALL****DATANALYSIS**1980*         00000100
C *THIS PROGRAM WAS WRITTEN TO ANALYZE FILES FOR SUSCEPTABILITY *         00000110
C *TO VARIOUS COMPRESSION TECHNIQUES. IT IS WRITTEN IN INDUSTRY *         00000120
C *STANDARD FORTRAN IV FOR RUNNING IN TIME-SHARING OR BATCH MODE*         00000130
C *HARDWARE SENSITIVE INSTRUCTIONS HAVE BEEN FLAGGED BY COMMENT*          00000140
C * SEE APPENDIX A OF DOCUMENTATION FOR PROGRAM TRANSFERABILITY *          00000150
C *TO RUN H66/80 TSS: RUN # XXXX "09" (WHERE XXXX=PERMFILE NAME)          00000160
      DIMENSION A(132),B(129),IA(132),IN(133),IM(133),IX(133),IY(132)     00000170
     &,Z(132)                                                             00000180
      DIMENSION C(729),IP(28,26)                                          00000190
      DOUBLE PRECISION IR(128),P(128)                                     00000200
      CHARACTER *1 A                                                      00000210
      CHARACTER *2 B,C                                                    00000220
      CHARACTER *7 X                                                      00000230
      INTEGER OUTPUT/6/,INPUT/5/,FILIN/9/                                 00000240
C    STRUCTURE GRAPHIC SYMBOL ARRAY FOR  FREQ OF OCCURRENCE TABLE         00000250
      DATA(B(L),L=1,129)/                                                 00000260
     &2HSH,2HEX,2HSX,2HET,2HEQ,2HAK,2HBL,2HBS,2HHT,2HLF,                  00000270
     &2HVT,2HFF,2HCR,2HSO,2HSI,2HDE,2HD1,2HD2,2HD3,2HD4,                  00000280
     &2HNK,2HSY,2HEB,2HCN,2HEM,2HSB,2HEC,2HFS,2HGS,2HRS,                  00000290
     &2HUS,2HSP,2HEP,2H" ,2HW ,2HX ,2H$ ,2H' ,2H( ,                      00000300
     &2H) ,2H* ,2H+ ,2H, ,2H- ,2H. ,2H/ ,2H0 ,2H1 ,2H2 ,                 00000310
     &2H3 ,2H4 ,2H5 ,2H6 ,2H7 ,2H8 ,2H9 ,2H: ,2H; ,2H< ,                 00000320
     &2H= ,2H> ,2HQM,2H@ ,2HA ,2HB ,2HC ,2HD ,2HE ,2HF ,                 00000330
     &2HG ,2HH ,2HI ,2HJ ,2HK ,2HL ,2HM ,2HN ,2HO ,2HP ,                 00000340
     &2HQ ,2HR ,2HS ,2HT ,2HU ,2HV ,2HW ,2HX ,2HY ,2HZ ,                 00000350
     &2H[ ,2H\ ,2H] ,2H↑ ,2H← ,2H@ ,2HA ,2HB ,2HC ,2HD ,                 00000360
     &2HE ,2HF ,2HG ,2HH ,2HI ,2HJ ,2HK ,2HL ,2HM ,2HN ,                 00000370
     &2HO ,2HP ,2HQ ,2HR ,2HS ,2HT ,2HU ,2HV ,2HW ,2HX ,                 00000380
     &2HY ,2HZ ,2H[ ,2H  ,2H] ,2H   ,2HDL,2H**,2H??/                     00000390
C    STRUCTURE PAIRED SYMBOL ARRAY                                        00000400
      DATA(C(J),J=1,729)/                                                 00000410
     &2H+A,2H+B,2H+C,2H+D,2H+E,2H+F,2H+G,2H+H,2H+I,2H+J,                  00000420
     &2H+K,2H+L,2H+M,2H+N,2H+O,2H+P,2H+Q,2H+R,2H+S,2H+T,                  00000430
     &2H+U,2H+V,2H+W,2H+X,2H+Y,2H+Z,2HA+,2HB+,2HC+,2HD+,                  00000440
     &2HE+,2HF+,2HG+,2HH+,2HI+,2HJ+,2HK+,2HL+,2HM+,2HN+,                  00000450
     &2HO+,2HP+,2H++,2HR+,2HS+,2HT+,2HU+,2HV+,2HW+,2HX+,                  00000460
     &2HY+,2HZ+,2HAA,2HAB,2HAC,2HAD,2HAE,2HAF,2HAG,2HAH,                  00000470
     &2HAI,2HAJ,2HAK,2HAL,2HAM,2HAN,2HAO,2HAP,2HAQ,2HAR,                  00000480
     &2HAS,2HAT,2HAU,2HAV,2HAW,2HAX,2HAY,2HAZ,2HBA,2HBB,                  00000490
     &2HBC,2HBD,2HBE,2HBF,2HBG,2HBH,2HBI,2HBJ,2HBK,2HBL,                  00000500
     &2HBM,2HBN,2HBO,2HBP,2H3Q,2HBR,2HBS,2HBT,2HBU,2HBV,                  00000510
     &2HBW,2HBX,2HBY,2HBZ,2HCA,2HCB,2HCC,2HCD,2HCE,2HCF,2HCG,             00000520
     &2HCH,2HCI,2HCJ,2HCK,2HCL,2HCM,2HCN,2HCO,2HCP,2HCQ,                  00000530
     &2HCR,2HCS,2HCT,2HCU,2HCV,2HCW,2HCX,2HCY,2HCZ,2HDA,                  00000540
     &2HDB,2HDC,2HDD,2HDE,2HDF,2HDG,2HDH,2HDI,2HDJ,2HDK,                  00000550
     &2HDL,2HDM,2HDN,2HDO,2HDP,2HDQ,2HDR,2HDS,2HDT,2HDU,                  00000560
     &2HDV,2HDW,2HDX,2HDY,2HDZ,2HEA,2HEB,2HEC,2HED,2HEE,                  00000570
     &2HEF,2HEG,2HEH,2HEI,2HEJ,2HEK,2HEL,2HEM,2HEN,2HEO,                  00000580
     &2HEP,2HEQ,2HER,2HES,2HET,2HEU,2HEV,2HEW,2HEX,2HEY,                  00000590
     &2HEZ,2HFA,2HFB,2HFC,2HFD,2HFE,2HFF,2HFG,2HFH,2HFI,                  00000600
     &2HFJ,2HFK,2HFL,2HFM,2HFN,2HFO,2HFP,2HFQ,2HFR,2HFS,                  00000610
     &2HFT,2HFU,2HFV,2HFW,2HFX,2HFY,2HFZ,2HGA,2HGB,2HGC                   00000620
     &2HGD,2HGE,2HGF,2HGG,2HGH,2HGI,2HGJ,2HGK,2HGL,2HGM,                  00000630
     &2HGN,2HGO,2HGP,2HGQ,2HGR,2HGS,2HGT,2HGU,2HGV,2HGW,                  00000640
     &2HGX,2HGY,2HGZ,2HHA,2HHB,2HHC,2HHD,2HHE,2HHF,2HHG,                  00000650
     &2HHH,2HHI,2HHJ,2HHK,2HHL,2HHM,2HHN,2HHO,2HHP,2HHQ,                  00000660
     &2HHR,2HHS,2HHT,2HHU,2HHV,2HHW,2HHX,2HHY,2HHZ,2HIA,                  00000670
     &2HIB,2HIC,2HID,2HIE,2HIF,2HIG,2HIH,2HII,2HIJ,2HIK,                  00000680
     &2HIL,2HIM,2HIN,2HIO,2HIP,2HIQ,2HIR,2HIS,2HIT,2HIU,                  00000690
     &2HIV,2HIW,2HIX,2HIY,2HIZ,2HJA,2HJB,2HJC,2HJD,2HJE,                  00000700
     &2HJF,2HJG,2HJH,2HJI,2HJJ,2HJK,2HJL,2HJM,2HJN,2HJO,                  00000710
     &2HJP,2HJQ,2HJR,2HJS,2HJT,2HJU,2HJV,2HJW,2HJX,2HJY,                  00000720
     &2HJZ,2HKA,2HKB,2HKC,2HKD,2HKE,2HKF,2HKG,2HKH,2HKI,                  00000730
     &2HKJ,2HKK,2HKL,2HKM,2HKN,2HKO,2HKP,2HKQ,2HKR,2HKS,                  00000740
     &2HKT,2HKU,2HKV,2HKW,2HKX,2HKY,2HKZ,2HLA,2HLB,2HLC,                  00000750
     &2HLD,2HLE,2HLF,2HLG,2HLH,2HLI,2HLJ,2HLK,2HLL,2HLM,                  00000760
     &2HLN,2HLO,2HLP,2HLQ,2HLR,2HLS,2HLT,2HLU,2HLV,2HLW,                  00000770
```

```
      &2HLX,2HLY,2HLZ,2HMA,2HMB,2HMC,2HMD,2HME,2HMF,2HMG,          00000780
      &2HMH,2HMI,2HMJ,2HMK,2HML,2HMM,2HMN,2HMO,2HMP,2HMQ,          00000790
      &2HMR,2HMS,2HMT,2HMU,2HMV,2HMW,2HMX,2HMY,2HMZ,2HNA,          00000800
      &2HNB,2HNC,2HND,2HNE,2HNF,2HNG,2HNH,2HNI,2HNJ,2HNK,          00000810
      &2HNL,2HNM,2HNN,2HNO,2HNP,2HNQ,2HNR,2HNS,2HNT,2HNU,          00000820
      &2HNV,2HNW,2HNX,2HNY,2HNZ,2HOA,2HOB,2HOC,2HOD,2HOE,          00000830
      &2HOF,2HOG,2HOH,2HOI,2HOJ,2HOK,2HOL,2HOM,2HON,2HOO,          00000840
      &2HOP,2HOQ,2HOR,2HOS,2HOT,2HOU,2HOV,2HOW,2HOX,2HOY,          00000850
      &2HOZ,2HPA,2HPB,2HPC,2HPD,2HPE,2HPF,2HPG,2HPH,2HPI,          00000860
      &2HPJ,2HPK,2HPL,2HPM,2HPN,2HPO,2HPP,2HPQ,2HPR,2HPS,          00000870
      &2HPT,2HPU,2HPV,2HPW,2HPX,2HPY,2HPZ,2HQA,2HQB,2HQC,          00000880
      &2HQD,2HQE,2HQF,2HQG,2HQH,2HQI,2HQJ,2HQK,2HQL,2HQM,          00000890
      &2HQN,2HQO,2HQP,2HQQ,2HQR,2HQS,2HQT,2HQU,2HQV,2HQW,          00000900
      &2HQX,2HQY,2HQZ,2HRA,2HRB,2HRC,2HRD,2HRE,2HRF,2HRG,          00000910
      &2HRH,2HRI,2HRJ,2HRK,2HRL,2HRM,2HRN,2HRO,2HRP,2HRQ,          00000920
      &2HRR,2HRS,2HRT,2HRU,2HRV,2HRW,2HRX,2HRY,2HRZ,2HSA,          00000930
      &2HSB,2HSC,2HSD,2HSE,2HSF,2HSG,2HSH,2HSI,2HSJ,2HSK,          00000940
      &2HSL,2HSM,2HSN,2HSO,2HSP,2HSQ,2HSR,2HSS,2HST,2HSU,          00000950
      &2HSV,2HSW,2HSX,2HSY,2HSZ,2HTA,2HTB,2HTC,2HTD,2HTE,          00000960
      &2HTF,2HTG,2HTH,2HTI,2HTJ,2HTK,2HTL,2HTM,2HTN,2HTO,          00000970
      &2HTP,2HTQ,2HTR,2HTS,2HTT,2HTU,2HTV,2HTW,2HTX,2HTY,          00000980
      &2HTZ,2HUA,2HUB,2HUC,2HUD,2HUE,2HUF,2HUG,2HUH,2HUI,          00000990
      &2HUJ,2HUK,2HUL,2HUM,2HUN,2HUO,2HUP,2HUQ,2HUR,2HUS,          00001000
      &2HUT,2HUU,2HUV,2HUW,2HUX,2HUY,2HUZ,2HVA,2HVB,2HVC,          00001010
      &2HVD,2HVE,2HVF,2HVG,2HVH,2HVI,2HVJ,2HVK,2HVL,2HVM,          00001020
      &2HVN,2HVO,2HVP,2HVQ,2HVR,2HVS,2HVT,2HVU,2HVV,2HVW,          00001030
      &2HVX,2HVY,2HVZ,2HWA,2HWB,2HWC,2HWD,2HWE,2HWF,2HWG,          00001040
      &2HWH,2HWI,2HWJ,2HWK,2HWL,2HWM,2HWN,2HWO,2HWP,2HWQ,          00001050
      &2HWR,2HWS,2HWT,2HWU,2HWV,2HWW,2HWX,2HWY,2HWZ,2HXA,          00001060
      &2HXB,2HXC,2HXD,2HXE,2HXF,2HXG,2HXH,2HXI,2HXJ,2HXK,          00001070
      &2HXL,2HXM,2HXN,2HXO,2HXP,2HXQ,2HXR,2HXS,2HXT,2HXU,          00001080
      &2HXV,2HXW,2HXX,2HXY,2HXZ,2HYA,2HYB,2HYC,2HYD,2HYE,          00001090
      &2HYF,2HYG,2HYH,2HYI,2HYJ,2HYK,2HYL,2HYM,2HYN,2HYO,          00001100
      &2HYP,2HYQ,2HYR,2HYS,2HYT,2HYU,2HYV,2HYW,2HYX,2HYY,          00001110
      &2HYZ,2HZA,2HZB,2HZC,2HZD,2HZE,2HZF,2HZG,2HZH,2HZI,          00001120
      &2HZJ,2HZK,2HZL,2HZM,2HZN,2HZO,2HZP,2HZQ,2HZR,2HZS,          00001130
      &2HZT,2HZU,2HZV,2HZW,2HZX,2HZY,2HZZ,2H??/                    00001140
C        HOW MANY CHAR. TO SEARCH IN INPUT RECORD                 00001150
         WRITE(OUTPUT,23)                                         00001160
   23 FORMAT("ENTER FILE TYPE")                                   00001170
         READ(INPUT,24)X                                          00001180
   24 FORMAT(A7)                                                  00001190
C        READ IN FILE TYPE FOR REPORT                             00001200
         WRITE(OUTPUT,21)                                         00001210
   21 FORMAT("ENTER INPUT LINELENGTH")                            00001220
         READ(INPUT,22)N1                                         00001230
   22 FORMAT(I3)                                                  00001240
         WRITE(OUTPUT,25)                                         00001250
   25 FORMAT("TO SUPRESS TRAILING SPACES ENTER 1")                00001260
         READ(INPUT,26)IV                                         00001270
   26 FORMAT(I1)                                                  00001280
C     READ A RECORD FROM DESIGNATED FILE; CHECK FOR EOF           00001290
    1 READ(FILIN,2,END=4)(A(I),I=1,127)                           00001300
    2 FORMAT(132A1)                                               00001310
      J1=J1+1                                                     00001320
      NC=0                                                        00001330
C     INCREMENT COUNTERS IN THE IR ARRAY BY USING ORDINAL EQUIV OF 00001340
C     THE CHARACTER AS THE SUBSCRIPT TO THE ARRAY.                00001350
      DO 3 I=1,N1                                                 00001360
C     NOTE-FOLLOWING 2 INST HARDWARE SENSITIVE-SEE APPENDIX A.    00001370
      IP2=IRS(A(I),27)                                            00001380
      IA(I)=IP2                                                   00001390
      IR(IP2)=IR(IP2)+1                                           00001400
      IF (IV.EQ.1.AND.IA(I).EQ.32) GOTO 7                         00001410
      NC=0                                                        00001420
      GO TO 3                                                     00001430
    7 NC=NC+1                                                     00001440
    3 IR(128)=IR(128)+1                                           00001450
C     SUBTRACT TRAILING SPACE COUNT FROM SUMMARY ARRAY            00001460
      IR(32)=IR(32)-NC                                            00001470
      IR(128)=IR(128)-NC                                          00001480
```

```
C     REDUCE LINE LENGTH BY TRAILING SPACE COUNT IF OPTIONED    00001490
      N=N1-NC                                                    00001500
      CALL SUB1(IA,IM,N)                                         00001510
      CALL SUB7(IA,IX,IY,IN,N)                                   00001520
      CALL SUB9(A,IP,N)                                          00001530
      GOTO 1                                                     00001540
    4 CONTINUE                                                   00001550
      N=N1                                                       00001560
C     PAUSE "POSITION PAPER NOW"                                 00001570
      CALL SJB2(X)                                               00001580
      CALL SUB3(IR,P,B,X)                                        00001590
      CALL SUB6(IR,B)                                            00001600
      CALL SUB3(IR,P,B,X)                                        00001610
      CALL SUB4(IN,IX,IY,IR,P,X,J1,N)                            00001620
      CALL SJB5(IN,IX,IY,IR,P,X,J1,N)                            00001630
      CALL SUB8(IM,IR,P,X,N)                                     00001640
      CALL SUB11(IP,C)                                           00001650
      CALL SUB10(IP,IX,B,C)                                      00001660
      CALL SUB12(IR,P,X)                                         00001670
      STOP                                                       00001680
      END                                                        00001690
C     SUB-1**CALCULATE NUMERIC STRINGS**                         00001700
      SUBROUTINE SJB1(IA,IM,N)                                   00001710
      DIMENSION IA(132),IM(133)                                  00001720
      I=1                                                        00001730
   11 IF(IA(I).GE.48.AND.IA(I).LE.57)GO TO 20                    00001740
      IF(IA(I).GE.44.AND.IA(I).LE.47)GO TO 20                    00001750
      IF(IA(I).EQ.42)GO TO 20                                    00001760
      IF('A(I).EQ.36)GO TO 20                                    00001770
   15 I=I+1                                                      00001780
      IF(I-N)11,11,40                                            00001790
   20 K=1                                                        00001800
      I=I+1                                                      00001810
      IT=IA(I)                                                   00001820
      M=0                                                        00001830
   30 IF(IA(I).GE.48.AND.IA(I).LE.57)GO TO 25                    00001840
      IF(IA(I).GE.44.AND.IA(I).LE.47)GO TO 25                    00001850
      IF(IA(I).EQ.42)GO TO 25                                    00001860
      IF(IA(I).EQ.36)GO TO 25                                    00001870
   39 IF (M.EQ.0) GO TO 15                                       00001880
      IM(K)=IM(K)+1                                              00001890
      GO TO 15                                                   00001900
   25 I=I+1                                                      00001910
      K=K+1                                                      00001920
C     ELIMINATE FULLY REPITITIOUS STRINGS FROM CONSIDERATION     00001930
      1IF(IA(I-1).EQ.IT) GO TO 16                                00001940
      M=1                                                        00001950
   16 IF (I-N)30,30,39                                           00001960
   40 RETURN                                                     00001970
      END                                                        00001980
C     SUB-2**PRINT EXECUTION REPORT SUMMARY**                    00001990
      SUBROUTINE SUB2(X)                                         00002000
      INTEGER OUTPUT/6/                                          00002010
      CHARACTER *7 X                                             00002020
      WRITE(OUTPUT,201)                                          00002030
  201 FORMAT(1H1,30X,"DATANALYSIS")                              00002040
      WRITE(OUTPUT,204)                                          00002050
      WRITE(OUTPUT,202)                                          00002060
  202 FORMAT(1H0,24X,"EXECUTION   REPORT   SUMMARY")             00002070
      WRITE(OUTPUT,204)                                          00002080
      WRITE(OUTPUT,203)X                                         00002090
  203 FORMAT(1H0,24X,"FILE ANALYZED:",4X,A7)                     00002100
      WRITE(OUTPUT,204)                                          00002110
  204 FORMAT(1H0)                                                00002120
      WRITE(OUTPUT,205)                                          00002130
  205 FORMAT(1H0,15X,"SEGMENT",10X,"DESCRIPTION")                00002140
      WRITE(OUTPUT,206)                                          00002150
  206 FORMAT(1H0,18X,"I",13X,"SYSTEM STANDARD FREQUENCY OF OCCURRENCE  00002160
     &E")                                                        00002170
      WRITE(OUTPUT,207)                                          00002180
  207 FORMAT(1H0,18X,"II",12X,"SORTED FREQUENCY OF OCCURRENCE")   00002190
      WRITE(OUTPUT,208)                                          00002200
  208 FORMAT(1H0,18X,"III",11X,"REPEATED CHARACTER STRING ANALYSIS")  00002210
```

```
      WRITE(OUTPUT,209)                                            00002220
  209 FORMAT(1H0,18X,"IV",12X,"RUN LENGTH ENCODING SUSCEPTIBILITY") 00002230
      WRITE(OUTPUT,210)                                            00002240
  210 FORMAT(1H0,18X,"V",13X,"HALF-BYTE  ENCODING SUSCEPTIBILITY") 00002250
      WRITE(OUTPUT,211)                                            00002260
  211 FORMAT(1H0,18X,"VI",12X,"DIATOMIC COMPRESSION ANALYSIS")      00002270
      WRITE(OUTPUT,212)                                            00002280
  212 FORMAT(1H0,18X,"VII",11X,"ENTROPY ANALYSIS")                 00002290
      WRITE(OUTPUT,204)                                            00002300
      RETURN                                                       00002310
      END                                                          00002320
C     SUB-3**PRINT FREQUENCY OF OCCURRANCE TABLE**                 00002330
      SUBROUTINE SUB3(IR,P,9,X)                                    00002340
      DOUBLE PRECISION IR(128),P(128)                              00002350
      DIMENSION B(129)                                             00002360
      INTEGER OUTPUT/6/                                            00002370
      CHARACTER *2 B                                               00002380
      CHARACTER *7 X                                               00002390
C     NOW CALCULATE PERCENTAGES                                    00002400
      DO 8 K=1,128                                                 00002410
    8 P(K)=IR(K)/IR(128)                                           00002420
C     PRINT FULL TABLE - 4 COLUMNS/32 ROWS                        00002430
      IF (M.EQ.1) GO TO 9                                          00002440
      WRITE(OUTPUT,301)                                           00002450
  301 FORMAT(1H1,"SEGMENT I",11X,"SYSTEM STANDARD FREQUENCY OF    00002460
     & OCCURRANCE",6X,"DATANALYSIS")                               00002470
      GO TO 12                                                     00002480
    9 WRITE(OUTPUT,302)                                           00002490
  302 FORMAT(1H1,"SEGMENT II",11X,"SORTED FREQUENCY OF OCCUR      00002500
     &RENCE",14X,"DATANALYSIS")                                    00002510
   12 WRITE(OUTPUT,303)X                                          00002520
  303 FORMAT(18X,"TABLE OF CHARACTERS FOUND IN ",A7," FILE")       00002530
      WRITE(OUTPUT,305)                                           00002540
  305 FORMAT(1H0,"CHAR    COUNT    %    CHAR    COUNT    %         00002550
     & CHAR    COUNT    %    CHAR    COUNT    %")                  00002560
      DO 5 I=1,32                                                  00002570
    5 WRITE(OUTPUT,6)B(I),IR(I),P(I),B(I+32),IR(I+32),P(I+32),     00002580
     &B(I+64),IR(I+64),P(I+64),B(I+96),IR(I+96),P(I+96)            00002590
    6 FORMAT(4(A4,0PF8.0,2PF8.2))                                  00002600
      M=1                                                          00002610
      WRITE(OUTPUT,304)                                           00002620
  304 FORMAT(1H0,"** DENOTES TOTAL CHARACTERS IN FILE")            00002630
      RETURN                                                       00002640
      END                                                          00002650
C     SUB-4**PRINT REPITITIOUS STRING ANALYSIS **                 00002660
      SUBROUTINE SUB4(IN,IX,IY,IR,P,X,J1,N)                        00002670
      DOUBLE PRECISION IR(128),P(128)                              00002680
      DIMENSION IN(133),IX(133),IY(133)                            00002690
      INTEGER OUTPUT/6/                                            00002700
      CHARACTER *7 X                                               00002710
      WRITE(OUTPUT,401)                                           00002720
  401 FORMAT(1H1,"SEGMENT III",30X,"DATANALYSIS")                 00002730
      WRITE(OUTPUT,402)X                                          00002740
  402 FORMAT(1H0,"REPEATED CHARACTER STRING ANALYSIS FOR ",A7," FILE 00002750
     &")                                                           00002760
      WRITE(OUTPUT,403)                                           00002770
  403 FORMAT(1H0,"STRING ****NUMBER OF OCCURRENCES******")         00002780
      WRITE(OUTPUT,404)                                           00002790
  404 FORMAT("LENGTH  NUMERICS SPACES  OTHER  TOTAL")             00002800
      DO 51 I=3,N                                                  00002810
      IN(N+1)=IN(N+1)+(IN(I)*I)                                    00002820
      IX(N+1)=IX(N+1)+(IX(I)*I)                                    00002830
      IY(N+1)=IY(N+1)+(IY(I)*I)                                    00002840
      IF(IY(I).EQ.0) GO TO 51                                      00002850
   50 WRITE(OUTPUT,61)I,IN(I),IX(I),(IY(I)-IN(I)-IX(I)),IY(I)      00002860
   61 FORMAT(I4,4I8)                                               00002870
   51 CONTINUE                                                     00002880
      WRITE(OUTPUT,41)                                            00002890
   41 FORMAT(1H0,"**ANALYSIS SUMMARY**")                          00002900
      WRITE(OUTPUT,42)J1                                          00002910
   42 FORMAT("RECORDS PROCESSED ",I6)                             00002920
```

```
          WRITE(OUTPUT,43)IY(N+1)                                    00002930
       43 FORMAT("TOTAL REPETITIONS",I6)                             00002940
          WRITE(OUTPUT,71)IR(128)                                    00002950
       71 FORMAT("TOTAL CHARACTERS IN FILE",OPF8.0)                  00002960
          P(128)=IY(N+1)/IR(128)                                     00002970
          WRITE(OUTPUT,81)P(128)                                     00002980
       81 FORMAT("PERCENT REPETITIOUS",2PF8.2)                       00002990
          RETURN                                                     00003000
          END                                                        00003010
C         SUB-5**PRINT RUN-LENGTH ENCODING ANALYSIS **              00003020
          SUBROUTINE SJ95(IN,IX,IY,IR,P,X,J1,N)                      00003030
          DOUBLE PRECISION IR(128),P(128)                            00003040
          DIMENSION IN(133),IX(133),IY(133)                          00003050
          INTEGER OUTPUT/6/                                          00003060
          CHARACTER *7 X                                             00003070
          WRITE(OUTPUT,501)                                          00003080
      501 FORMAT(1H1,"SEGMENT IV",45X,"DATANALYSIS")                 00003090
          WRITE(OUTPUT,502)                                          00003100
      502 FORMAT(1H0,12X,"RUN LENGTH ENCODING COMPRESSION SUSCEPTIBILITY 00003110
         &")                                                         00003120
          WRITE(OUTPUT,503)X                                         00003130
      503 FORMAT(12X,"FOR REPEATED CHARACTER STRINGS IN ",A7," FILE") 00003140
          WRITE(OUTPUT,53)                                           00003150
       53 FORMAT(1H0,"STRING  *****REPEATED CHARACTERS*****   CHAR.   00003160
         & SAVED BY COMPRESSION")                                    00003170
          WRITE(OUTPUT,54)                                           00003180
       54 FORMAT("LENGTH  NUMERICS SPACES   OTHER   TOTAL   NUMERICS  00003190
         & SPACES OTHER TOTAL")                                      00003200
          WRITE(OUTPUT,55)                                           00003210
       55 FORMAT(1H0)                                                00003220
          DO 63 I=3,N                                                00003230
          ITN1=ITN1+(IN(I)*I)                                        00003240
          ITX1=ITX1+(IX(I)*I)                                        00003250
          ITY1=ITY1+(IY(I)*I)                                        00003260
          ITN2=(I-3)*IN(I)                                           00003270
          ITX2=(I-3)*IX(I)                                           00003280
          ITY2=(I-3)*(IY(I)-IX(I)-IN(I))                             00003290
          ITT2=(I-3)*(IY(I))                                         00003300
          ITN3=ITN3+ITN2                                             00003310
          ITX3=ITX3+ITX2                                             00003320
          ITY3=ITY3+ITY2                                             00003330
          ITT3=ITT3+ITT2                                             00003340
          IF(IY(I).EQ.0) GO TO 63                                    00003350
          IYT1=(IY(I)*I)-(IN(I)*I)-(IX(I)*I)                         00003360
       52 WRITE(OUTPUT,62)I,(IN(I)*I),(IX(I)*I),IYT1,(IY(I)*I),ITN2,ITX2 00003370
         &,ITY2,                                                     00003380
         &ITT2                                                       00003390
       62 FORMAT(I4,8I8)                                             00003400
       63 CONTINUE                                                   00003410
          ITS1=IR(128)                                               00003420
          ITS2=(ITN1/IR(128)*100)                                   00003430
          ITS3=(ITX1/IR(128)*100)                                   00003440
          ITS4=((ITY1-ITX1-ITN1)/IR(128)*100)                       00003450
          ITS5=(ITY1/IR(128)*100)                                   00003460
          ITS6=(ITN3/IR(128)*100)                                   00003470
          ITS7=(ITX3/IR(128)*100)                                   00003480
          ITS8=(ITY3/IR(128)*100)                                   00003490
          ITS9=(ITT3/IR(128)*100)                                   00003500
          I=99999                                                    00003510
          WRITE(OUTPUT,505)                                          00003520
      505 FORMAT(1H0)                                                00003530
          WRITE(OUTPUT,62)I,ITN1,ITX1,(ITY1-ITN1-ITX1),ITY1,ITN3,ITX3,IT 00003540
         &Y3,ITT3                                                    00003550
          WRITE(OUTPUT,506)ITS1                                      00003560
      506 FORMAT("SUMMARY INFORMATION: TOTAL CHAR. IN DATA FILE:",I6) 00003570
          WRITE(OUTPUT,507)                                          00003580
      507 FORMAT(" PERCENTAGE SEQUENTIAL REPEATED CHAR:")            00003590
          WRITE(OUTPUT,508)ITS2                                      00003600
      508 FORMAT("  NUMERIC:            ",I6)                        00003610
          WRITE(OUTPUT,509)ITS3                                      00003620
      509 FORMAT("  BLANK(SPACES):      ",I6)                        00003630
          WRITE(OUTPUT,510)ITS4                                      00003640
      510 FORMAT("  NONNUMERIC NONBLANK:",I6)                        00003650
```

```
          WRITE(OUTPUT,511)ITS5                                    00003660
    511 FORMAT("  TOTAL REPEATED:      ",I6)                       00003670
          WRITE(OUTPUT,512)                                        00003680
    512 FORMAT(1H0,"POTENTIAL COMPRESSION REDUCTION (%):")         00003690
          WRITE(OUTPUT,513)ITS6                                    00003700
    513 FORMAT("  NUMERIC:            ",I6)                        00003710
          WRITE(OUTPUT,514)ITS7                                    00003720
    514 FORMAT("  BLANK(SPACES):      ",I6)                        00003730
          WRITE(OUTPUT,515)ITS8                                    00003740
    515 FORMAT("  NONNUMERIC NONBLANK:",I6)                        00003750
          WRITE(OUTPUT,516)ITS9                                    00003760
    516 FORMAT("  TOTAL:              ",I6)                        00003770
          RETURN                                                   00003780
          END                                                      00003790
C     SUB-6**SORT SYMBOL ARRAY IN DESCENDING SEQUENCE**            00003800
          SUBROUTINE SUB6(IR,B)                                    00003810
          DOUBLE PRECISION IR(128)                                 00003820
          DIMENSION B(129)                                         00003830
          CHARACTER *2 B                                           00003840
C     SORT THE FREQUENCY AND SYMBOL ARRAYS IN DESCENDING ORDER;    00003850
C     WHERE IR(N) = VALUE AND B(N) = ASSOCIATED SYMBOL.            00003860
          DO 10 I = 1,126                                          00003870
          IP1 = I+1                                                00003880
          DO 10 K=IP1,126                                          00003890
          IF(IR(I).GE.IR(K))GO TO 10                               00003900
          TEMP = IR(I)                                             00003910
          IR(I) = IR(K)                                            00003920
          IR(K) = TEMP                                             00003930
          B(129) = B(I)                                            00003940
          B(I) = B(K)                                              00003950
          B(K) = B(129)                                            00003960
     10 CONTINUE                                                   00003970
          RETURN                                                   00003980
          END                                                      00003990
C     SUB-7**CALCULATE SEQUENTIAL REPEATING CHARACTERS***          00004000
          SUBROUTINE SUB7(IA,IX,IY,IN,N)                           00004010
          DIMENSION IA(132),IX(133),IY(133),IN(133)                00004020
C     CALCULATE SEQUENTIAL REPEATING CHARACTERS IN STRING          00004030
          I=1                                                      00004040
          K=1                                                      00004050
     73 IP=IA(I)                                                   00004060
          IF(IP.EQ.IA(I+1)) GO TO 75                               00004070
     74 K=1                                                        00004080
          I=I+1                                                    00004090
          IF(I-N) 73,73,79                                         00004100
     75 K=K+1                                                      00004110
          I=I+1                                                    00004120
          IF(I-N) 68,68,69                                         00004130
     68 IF(IP.EQ.IA(I+1)) GO TO 75                                 00004140
     69 IY(K)=IY(K)+1                                              00004150
          IF(IP.GE.48.AND.IP.LE.57) GO TO 76                       00004160
          IF(IP.EQ.32) GO TO 77                                    00004170
     78 GOTO 74                                                    00004180
     76 IN(K)=IN(K)+1                                              00004190
          GOTO 74                                                  00004200
     77 IX(K)=IX(K)+1                                              00004210
          GOTO 74                                                  00004220
     79 RETURN                                                     00004230
          END                                                      00004240
C       SUB-8**PRINT SEQUENTIAL NUMERIC STRING ANALYSIS**          00004250
          SUBROUTINE SUB8(IM,IR,P,X,N)                             00004260
          DOUBLE PRECISION IR(128),P(128)                          00004270
          DIMENSION IM(133)                                        00004280
          INTEGER OUTPUT/6/                                        00004290
          CHARACTER *7 X                                           00004300
          WRITE(OUTPUT,801)                                        00004310
    801 FORMAT(1H1,"SEGMENT V",50X,"DATANALYSIS")                  00004320
          WRITE(OUTPUT,802)                                        00004330
    802 FORMAT(1H0,12X,"HALF-BYTE ENCODING COMPRESSION SUSCEPTIBILITY" 00004340
         &)                                                        00004350
          WRITE(OUTPUT,803)X                                       00004360
```

```
    803 FORMAT(8X,"FOR SEQUENTIAL (NON-REPEATING) NUMERICS IN ",A7," F    00004370
       &ILE")                                                            00004380
          WRITE(OUTPUT,804)                                              00004390
    804 FORMAT(1H0,"STRING     NUMBER OF    TOTAL     COMPRESSED   CHA    00004400
       &RACTERS")                                                        00004410
          WRITE(OUTPUT,805)                                              00004420
    805 FORMAT("LENGTH     OCCURRANCES  CHARACTERS CHARACTERS     SAVED   00004430
       &")                                                               00004440
          DO 80 I=4,N                                                    00004450
          IM(N+1)=(IM(N+1)+IM(I)*I)                                      00004460
          IF(IM(I).EQ.0) GO TO 80                                        00004470
          IC=(((I+1)/2)+2)*IM(I)                                         00004480
          IC1=IC1+IC                                                     00004490
          WRITE(OUTPUT,82)I,IM(I),(IM(I)*I),IC,((IM(I)*I)-IC)            00004500
     82 FORMAT(I4,4I13)                                                  00004510
     80 CONTINUE                                                         00004520
          WRITE(OUTPUT,806)                                              00004530
    806 FORMAT(1H0,"**ANALYSIS SUMMARY**")                               00004540
          WRITE(OUTPUT,83)IR(128)                                        00004550
     83 FORMAT("TOTAL CHARACTERS IN FILE",0PF8.0)                        00004560
          WRITE(OUTPUT,807)IM(N+1)                                       00004570
    807 FORMAT("TOTAL CHAR. SUSCEPTIBLE TO COMPRESSION",I6)              00004580
          WRITE(OUTPUT,808)(IM(N+1)-IC1)                                 00004590
    808 FORMAT("TOTAL CHARACTERS SAVED  BY COMPRESSION",I6)              00004600
          P(128)=(IM(N+1)-IC1)/IR(128)                                   00004610
          WRITE(OUTPUT,84)P(128)                                         00004620
     84 FORMAT("PERCENT FILE COMPRESSION REDUCTION",2PF8.2)              00004630
          RETURN                                                         00004640
          END                                                            00004650
    C       SUB-9**ISOLATE CHARACTER PAIRS IN STRINGS (DIATOMIC)**       00004660
          SUBROUTINE SUB9(A,IP,N)                                        00004670
          DIMENSION A(132),IP(28,26)                                     00004680
          I=1                                                            00004690
    C     NOTE-"J" VALUES INFLATED BY 16 DUE TO HONEYWELL BUG.           00004700
    C     NOTE-FOLLOWING 2 INST. HARDWARE SENSITIVE-SEE APPENDIX A.      00004710
    901 IV=IRS(A(I),19)                                                  00004720
          IV=IV+(IRS(A(I+1),27))                                         00004730
    C     CHECK FOR IMBEDDED BLANK PAIRS;USE 0+ COUNTER                  00004740
          IF(IV-8240)923,920,923                                         00004750
    920 IV1=IRS(A(I+2),27)                                               00004760
          IF(IV1.EQ.32)GO TO 923                                         00004770
          IF(I-1)922,922,921                                             00004780
    921 IV1=IRS(A(I-1),27)                                               00004790
          IF(IV1.EQ.32)GO TO 923                                         00004800
    922 IP(2,17)=IP(2,17)+1                                              00004810
    923 CONTINUE                                                         00004820
          IC=1                                                           00004830
          IR=0                                                           00004840
    C     NOW CHECK FOR BLANK/LETTER                                     00004850
          J=8273                                                         00004860
    902 IR=IR+1                                                          00004870
          IF(IV-J)903,905,903                                            00004880
    903 IF(IR-26)904,907,907                                             00004890
    904 J=J+1                                                            00004900
          GO TO 902                                                      00004910
    905 IP(IC,IR)=IP(IC,IR)+1                                            00004920
    906 I=I+2                                                            00004930
          IF(I-N)901,901,918                                             00004940
    907 IC=2                                                             00004950
          IR=0                                                           00004960
    C     NOW CHECK FOR LETTER/BLANK                                     00004970
    C     USE J=24850 FOR LOWER CASE                                     00004980
          J=16688                                                        00004990
    908 IR=IR+1                                                          00005000
          IF(IV-J)909,905,909                                            00005010
    909 IF(IR-26)910,911,911                                             00005020
    910 J=J+256                                                          00005030
          GO TO 908                                                      00005040
    C     NOW CHECK FOR LETTER/LETTER                                    00005050
    C     USE J=24945 FOR LOWER CASE                                     00005060
    911 J=16721                                                          00005070
    912 IR=0                                                             00005080
```

```
          IC=IC+1                                              00005090
      913 IR=IR+1                                              00005100
          IF(IV-J)914,905,914                                  00005110
      914 IF(IR-26)915,916,916                                 00005120
      915 J=J+1                                                00005130
          GO TO 913                                            00005140
      916 IF(IC-28)917,906,906                                 00005150
      917 J=J+231                                              00005160
          GO TO 912                                            00005170
      918 RETURN                                               00005180
          END                                                  00005190
C         SUB-10**PRINT PAIRED CHARACTER ANALYSIS REPORT**     00005200
          SUBROUTINE SUB10(IP,IX,B,C)                          00005210
          DIMENSION IP(28,26),IX(133),B(129),C(729)            00005220
          INTEGER OUTPUT/6/                                    00005230
          CHARACTER *2 B,C                                     00005240
          WRITE(OUTPUT,101)                                    00005250
      101 FORMAT(1H1,"SEGMENT VI",50X,"DATANALYSIS")            00005260
          WRITE(OUTPUT,102)                                    00005270
      102 FORMAT(1H0,15X,"PAIRED CHARACTER COMPRESSION ANALYSIS") 00005280
          WRITE(OUTPUT,103)                                    00005290
      103 FORMAT(1H0," PAIR/COUNT   PAIR/COUNT   PAIR/COUNT   PAIR/COUNT 00005300
         &  PAIR/COUNT PAIR/COUNT ")                           00005310
          I=1                                                  00005320
          N=0                                                  00005330
          DO 110 J=1,28                                        00005340
          DO 110 K=1,26                                        00005350
          IF(IP(J,K).EQ.0) GO TO 110                           00005360
C         PRINT ONLY THE TOP 132 PAIR OCCURRENCES              00005370
          IF(N-22)106,106,110                                  00005380
      106 IX(I)=IP(J,K)                                        00005390
          IPS=IPS+IX(I)                                        00005400
          M=((J-1)*26)+K                                       00005410
          B(I)=C(M)                                            00005420
          I=I+1                                                00005430
          IF(I-6)110,110,109                                   00005440
      109 WRITE(OUTPUT,105)B(1),IX(1),B(2),IX(2),B(3),IX(3),B(4),IX(4), 00005450
         &B(5),IX(5),B(6),IX(6)                                00005460
      105 FORMAT(6(A6,I6))                                     00005470
          I=1                                                  00005480
          N=N+1                                                00005490
      110 CONTINUE                                             00005500
C         FINISH OFF PRINT LINE NOW IF PARTIALLY FILLED        00005510
          IF(I-1)112,112,114                                   00005520
      114 DO 113 I=I,6                                         00005530
          B(I)=B(128)                                          00005540
          IX(I)=0                                              00005550
      113 CONTINUE                                             00005560
      111 WRITE(OUTPUT,105)B(1),IX(1),B(2),IX(2),B(3),IX(3),B(4),IX(4), 00005570
         &B(5),IX(5),B(6),IX(6)                                00005580
      112 WRITE(OUTPUT,115)IPS                                 00005590
      115 FORMAT(1H0,"TOTAL COMBINATIONS FOUND:",I6)            00005600
          RETURN                                               00005610
          END                                                  00005620
C         SUB-11**SORT  PAIRED CHARACTER ARRAY ***             00005630
          SUBROUTINE SUB11(IP,C)                               00005640
          DIMENSION IP(28,26),C(729)                           00005650
          CHARACTER *2 C                                       00005660
C         SORT THE TWO-DIMENSION PAIRS ARRAY & ASSOCIATED SYMBOLS 00005670
C         WHERE IP(N1,N2) = VALUE AND C(N) = RELATED PAIR SYMBOL. 00005680
          DO 1101 J=1,28                                       00005690
          DO 1101 K=1,26                                       00005700
          KK=K                                                 00005710
          DO 1101 J1=J,28                                      00005720
          DO 1101 K1=KK,26                                     00005730
          KK=1                                                 00005740
          IF(IP(J,K).GE.IP(J1,K1)) GO TO 1101                  00005750
          TEMP=IP(J,K)                                         00005760
          IP(J,K)=IP(J1,K1)                                    00005770
          IP(J1,K1)=TEMP                                       00005780
```

```
C        PERFORM LINEAR CONVERSION OF TWO DIMENSIONS INTO ONE.      00005790
         M1=((J-1)*26)+K                                            00005800
         M2=((J1-1)*26)+K1                                          00005810
         C(729)=C(M1)                                               00005820
         C(M1)=C(M2)                                                00005830
         C(M2)=C(729)                                               00005840
 1101 CONTINUE                                                      00005850
      RETURN                                                        U0005860
      END                                                           00005870
C        SUB-12**COMPUTE ENTROPY & PRINT ANALYSIS REPORT**          00005880
      SUBROUTINE SUB12(IR,P,X)                                      00005890
      DOUBLE PRECISION IR(128),P(128),ZERO                          00005900
      INTEGER OUTPUT/6/                                             00005910
      CHARACTER *7 X                                                00005920
      ZERO=0.D0                                                     00005930
      DO 1201 I=1,127                                               00005940
      IF(P(I)-ZERO)1201,1201,1200                                   00005950
 1200 E=E+P(I)*DLOG2(P(I))                                          00005960
 1201 CONTINUE                                                      00005970
      E=-1*E                                                        00005980
      B1=E*IR(128)                                                  00005990
      B2=IR(128)*8                                                  00006000
      B3=(B2-B1)/B2*100                                             00006010
      B4=B1/8                                                       00006020
      WRITE(OUTPUT,1202)                                            00006030
 1202 FORMAT(1H1,"SEGMENT VII",50X,"DATANALYSIS")                   00006040
      WRITE(OUTPUT,1203)X                                           00006050
 1203 FORMAT(1H0,15X,"ENTROPY ANALYSIS FOR ",A7," FILE")            00006060
      WRITE(OUTPUT,1204)E                                           00006070
 1204 FORMAT(1H0,10X,"ENTROPY: ",0PF8.2," BITS/CHARACTER")          00006080
      WRITE(OUTPUT,1205)                                            00006090
 1205 FORMAT(1H0,10X,"FILE CONTAINS:")                              00006100
      WRITE(OUTPUT,1206)IR(128)                                     00006110
 1206 FORMAT(1H0,15X,0PF8.0," CHARACTERS")                          00006120
      WRITE(OUTPUT,1207)B2                                          00006130
 1207 FORMAT(1H0,15X,0PF8.0," BITS")                                00006140
      WRITE(OUTPUT,1211)B3                                          00006150
 1211 FORMAT(1H0,10X,"THEORETICAL STATISTICAL COMPRESSION: ",0PF8.2,00006160
     &"X")                                                          00006170
      WRITE(OUTPUT,1209)                                            00006180
 1209 FORMAT(1H0,10X,"FILE WOULD REQUIRE:")                         00006190
      WRITE(OUTPUT,1210)B4                                          00006200
 1210 FORMAT(1H0,15X,0PF8.0," CHARACTERS")                          00006210
      WRITE(OUTPUT,1212)B1                                          00006220
 1212 FORMAT(1H0,15X,0PF8.0," BITS")                                00006230
      RETURN                                                        00006240
      END                                                           00006250
```

A.9 BASIC LANGUAGE DATANALYSIS PROGRAM LISTING

```
10 '*******************************************************************
20 '* COPYWRITE GILBERT HELD & THOMAS MARSHALL - DATANALYSIS        *
30 '* THIS PROGRAM WAS WRITTEN TO ANALYSE/FILES FOR SUSCEPTIBILITY  *
40 '* TO VARIOUS COMPRESSION TECHNIQUES.  REFER TO DOCUMENTATION.   *
50 '*******************************************************************
60 '----- INITIALIZATION -----------
70 DIM A(132),B$(129),IA(132),IN(133),IM(133),IX(133),IY(132)
80 DIM Z(132),C$(729),IP(26,28),IR(128),P(128)
90 WIDTH 80:CLS                'CLEAR SCREEN & SET WIDTH TO 80
100 ON ERROR GOTO 4050        'ERROR HANDLER
110   '- - SETUP PROCESSING OF INPUT FILE  - -
120   LOCATE ,,1
130   PRINT "ENTER ASCII FILENAME IN THE FORMAT [DRIVE:]FILESPEC";
140   INPUT F$: OPEN F$ FOR INPUT AS #2
150 PRINT "IS FILE MOSTLY LOWER CASE(Y/N)";
160 INPUT V$
170 IF V$ = "Y" OR V$ = "y" THEN Z= 97 ELSE Z= 65
```

```
180  PRINT "DO YOU WANT PRINTED OUTPUT (Y/N)";
190  INPUT D$
200  IF D$= "y" OR D$= "Y" THEN D$="LPT1:" ELSE D$="SCRN:"
210  OPEN D$ FOR OUTPUT AS #1
220  ' STRUCTURE GRAPHIC SYMBOL ARRAY
230  PRINT "DATANALYSIS - INITIALIZATION..."
240  DATA SH,SX,EX,ET,EQ,AK,BL,BS,HT,LF,VT,FF,CR,SO,SI,DL,D1,D2,D3,D4
250  DATA NK,SY,EB,CN,EM,SB,EC,FS,GS,RS,US,SP
260  FOR I=1 TO 32:READ A$:B$(I)=A$:NEXT I
270  FOR I= 33 TO 127
280  B$(I) = CHR$(I):NEXT I:B$(127)="DL":B$(128)="**"
290  '*****************************************************
300  '*   STRUCTURE THE PAIRED SYMBOL ARRAY FIRST         *
310  '*****************************************************
320  ' 1ST SPACE/LETTER COMBINATIONS
330  K=1
340  I=32
350  FOR J= Z TO Z+25
360  C$(K)= CHR$(I)+ CHR$(J)
370  K=K+1
380  NEXT J
390  ' NOW LETTERS/SPACE  COMBINATIONS
400  J=32
410  FOR I=Z TO Z+25
420  C$(K)= CHR$(I)+CHR$(J)
430  K=K+1
440  NEXT I
450  ' NOW LETTER/LETTER COMBINATIONS
460  FOR I=Z TO Z+25
470  FOR J=Z TO Z+25
480  C$(K)= CHR$(I)+CHR$(J)
490  K=K+1
500  NEXT J
510  NEXT I
520  PRINT "PATIENCE - INPUT PROCESSING";
530  IF EOF(2) THEN 760
540  LINE INPUT #2, X$     'READ FULL LINE WITH PUNCTUATION
550  PRINT ".";            'TELL THEM WE'RE STILL ALIVE
560  K= LEN(X$)            'SAVE LENGTH OF INPUT STRING
570  FOR I= 1 TO K         'AND SET UP END OF STRING
580  A$= MID$(X$,I,1)      'PULL OUT A CHARACTER TO ANALYSE
590  '*****************************************************************
600  '*INCREMENT COUNTERS BY THE ORDINAL EQUIV OF CHAR AS SUBSCRIPT*
610  '*****************************************************************
620  IP2=ASC(A$)           'DERIVE DECIMAL REPRESENTATION
630  IR(IP2)= IR(IP2)+1    'ADD TO DISPLACEMENT COUNTER
640  IR(128)= IR(128)+1    'ADD TO THE TOTAL CHAR READ
650  IA(I)= IP2            'STRING THE RECORD OUT IN ONE DIMENSION
660  N1= N1+1              'ADD TO THE CHARACTER COUNT
670  NEXT I                'GET NEXT INPUT CHARACTER
680  GOSUB 880             'SUB 1 - CALCULATE NUMERICS
690  GOSUB 1180            'SUB 7 - CALCULATE REPEATS
700  GOSUB 1940            'SUB 9 - ISOLATE PAIRS
710  J1=J1+1               'ADD TO RECORD COUNT
720  N=N1                  'SAVE LAST RECORD LENGTH
730  N1=0                  'ZERO LENGTH FOR NEW RECORD
740  GOTO 530              'AND GO GET MORE INPUT
750  BEEP  '******** END OF MAINLINE PROGRAM ********************
760  GOSUB 4060            'SUB 2-PRINT THE EXECUTION REPORT
770  GOSUB 1420            'SUB 3-PRINT FREQ./OCCURRENCE ANALYSIS
780  GOSUB 3880            'SUB 6-SORT FREQ. ARRAY
790  GOSUB 1420            'SUB 3-REPRINT FREQ./OCCURRENCE ANALYSIS
800  GOSUB 1690            'SUB 4-PRINT STRING ANALYSIS
810  GOSUB 2770            'SUB 5-PRINT RUNLENGTH ANALYSIS
820  GOSUB 3300            'SUB 8-PRINT NUM. STRING ANALYSIS
830  GOSUB 2540            'SUB 11-SORT PAIR ARRAY
840  GOSUB 2250            'SUB 10-PRINT PAIR ARRAY
850  GOSUB 3620            'SUB 12-COMPUTE & PRINT ENTROPY
860  END                   'T-T-T-THATS ALL FOLKS
870  '*****************************************************
880  '*       SUBROUTINE 1 - CALCULATE NUMERIC STRINGS    *
890  '*****************************************************
900  I=1
910  IF IA(I)>= 48 AND IA(I)<= 57 THEN 980
920  IF IA(I)>= 44 AND IA(I)<= 47 THEN 980
930  IF IA(I) = 42 THEN 980
940  IF IA(I) = 36 THEN 980
```

```
950 I=I+1                              'NOT A NUMERIC OR SYMBOL
960 IF I> N1 THEN 1160                 'RECORD STRING PROCESS COMPLETE
970 GOTO 910
980 K=1
990 I=I+1
1000 IT=IA(I)
1010 M=0
1020 IF IA(I)>= 48 AND IA(I)<= 57 THEN 1090    'NUMERIC?
1030 IF IA(I)>= 44 AND IA(I)<= 47 THEN 1090    'SYMBOL?
1040 IF IA(I) = 42 THEN 1090                   'ASTERISK?
1050 IF IA(I) = 36 THEN 1090                   'DOLLAR?
1060 IF M=0 THEN 950
1070 IM(K)= IM(K)+1                    'ADD TO STRING SIZE COUNT
1080 GOTO 950
1090 I=I+1
1100 K=K+1
1110 ' ELIMINATE FULLY REPITITIOUS STRINGS FROM CONSIDERATION
1120 IF IA(I-1)= IT THEN 1140
1130 M=1
1140 IF I> N1 THEN 1060
1150 GOTO 1020
1160 RETURN
1170 '*****************************************************
1180 '*          SUBROUTINE 7 - CALCULATE SEQ. REPEATING CHAR. *
1190 '*****************************************************
1200 I=1
1210 K=1
1220 IC= IA(I)                         'PULL CHAR. FROM STRING
1230 IF IC= IA(I+1) THEN 1280          'SAME AS NEXT IN STRING?
1240 K=1
1250 I= I+1
1260 IF I> N1 THEN 1400                'END OF RECORD
1270 GOTO 1220
1280 K=K+1                             'BUMP REPITITION COUNT
1290 I= I+1
1300 IF I> N1 THEN 1320                'END OF STRING?
1310 IF IC = IA(I+1) THEN 1280         'LOOP ON REPITIONS
1320 IY(K) = IY(K)+1                   'ADD TO "ANY" COUNT
1330 IF IC>= 48 AND IC<= 57 THEN 1360  ' ITS NUMERIC
1340 IF IC = 32 THEN 1380              ' ITS A SPACE
1350 GOTO 1240
1360 IN(K) = IN(K)+1                   'ADD TO REPEAT COUNT
1370 GOTO 1240
1380 IX(K)= IX(K)+1                    'ADD TO SPACE COUNT
1390 GOTO 1240
1400 M=0:RETURN                        'RESET TOGGLE & RETURN
1410 '*****************************************************
1420 '*        SUBROUTINE 3 - PRINT FREQ OF OCCURRENCE TABLE    *
1430 '*****************************************************
1435 IF D$= "SCRN:" THEN WIDTH #1,255     'ELIM. DOUBLE SPACE
1440 ' 1ST CALCULATE PERCENTAGES
1450 FOR K= 1 TO 128
1460 P(K)= IR(K)/IR(128)               'IR(128) IS TOTAL CHAR STORE
1470 NEXT K
1480 ' NOW PRINT FULL TABLE - 4 COLUMNS/32 ROWS
1490 GOSUB 3580                        'SET TOP OF PAGE
1500 IF M=1 THEN 1530                  'THIS TIME IT IS SORTED
1510 PRINT #1, SPC(5)"SEGMENT I";SPC(11)"STANDARD FREQ. OF OCCURRENCE"
1520 GOTO 1540
1530 PRINT #1,SPC(5)"SEGMENT II";SPC(11)"SORTED  FREQ. OF OCCURRENCE"
1540 PRINT #1,SPC(18)"TABLE OF CHARACTERS FOUND IN ";F$;" FILE"
1550 PRINT #1,
1560 PRINT #1,"CHAR    COUNT     %   CHAR   COUNT      %   CHAR    COUNT      %";
1570 PRINT #1,"   CHAR   COUNT     %"
1580 PRINT #1,
1590 P1$= " \\ ######   .##   "
1600 FOR I = 1 TO 32
1610 PRINT #1,USING P1$;B$(I);IR(I);P(I);
1620 PRINT #1,USING P1$;B$(I+32);IR(I+32);P(I+32);
1630 PRINT #1,USING P1$;B$(I+64);IR(I+64);P(I+64);
1640 PRINT #1,USING P1$;B$(I+96);IR(I+96);P(I+96)
1650 NEXT I
1660 M=1
1670 RETURN
1680 '*****************************************************
1690 '*        SUBROUTINE 4 - PRINT REPITITIOUS STRING ANALYSIS    *
1700 '*****************************************************
1710 GOSUB 3580                        'SET TOP OF PAGE
1720 PRINT #1,"SEGMENT III";SPC(11)"REPEATED CHARACTER STRING ANALYSIS"
```

```
1730 PRINT #1,
1740 PRINT #1,"STRING *****NUMBER OF|OCCURRENCES***"
1750 PRINT #1,"LENGTH  NUMERICS SPACES  OTHER  TOTAL"
1760 PRINT #1,
1770 P2$="####    ####   ####   ####   ####"
1780 FOR I=3 TO N    . 'IGNORE STRINGS LESS THAN THREE
1790 IN(N+1)=IN(N+1)+(IN(I)*I)     'ADD TO NUMERIC TOTAL
1800 IX(N+1)=IX(N+1)+(IX(I)*I)     'ADD TO SPACE TOTAL
1810 IY(N+1)=IY(N+1)+(IY(I)*I)     'ADD TO SUM TOTAL
1820 IF IY(I)=O THEN 1840          'SKIP IF ZERO COUNT
1830 PRINT #1,USING P2$;I;IN(I);IX(I);(IY(I)-IN(I)-IX(I));IY(I)
1840 NEXT I
1850 PRINT #1,
1860 PRINT #1,"**ANALYSIS SUMMARY**"
1870 PRINT #1," RECORDS PROCESSED ";J1
1880 PRINT #1," TOTAL REPITITIONS ";IY(N+1)
1890 PRINT #1," CHARACTERS IN FILE ";IR(128)
1900 P(128)=IY(N+1)/IR(128)*100
1910 PRINT #1," PERCENT REPITITIOUS ";INT(P(128))
1920 RETURN
1930 '*******************************************************
1940 '*       SUBROUTINE 9 - ISOLATE PAIRS IN STRINGS (DIATOMIC) *
1950 '*******************************************************
1960 I=1
1970 ' CHECK FOR IMBEDED BLANK PAIRS; USE Q_ COUNTER FOR TALLY
1980 IR=17:IC=2
1990 IF IA(I) <> 32 THEN 2020
2000 IF IA(I+1) <> 32 THEN 2020
2010 GOTO 2190                     'ITS A HIT
2020 ' NOW CHECK FOR BLANK/LETTER
2030 IR=IA(I+1)-(Z-1):IC=1         'SET TO ROW N, COL 1
2040 IF IA(I) <> 32 THEN 2070
2050 IF IA(I+1) < Z OR IA(I+1) > Z+25 THEN 2070    'IS IT WITHIN RANGE?
2060 GOTO 2190                     'ITS A HIT
2070 'NOW CHECK FOR LETTER/BLANK
2080 IR=IA(I)-(Z-1):IC=2           'SET TO ROW N,COL 2
2090 IF IA(I) < Z OR IA(I) > Z+25 THEN 2120    'IS IT WITHIN RANGE?
2100 IF IA(I+1) <> 32 THEN 2120
2110 GOTO 2190                     'ITS A HIT
2120 ' NOW CHECK FOR LETTER/LETTER
2130 IC= IA(I)-(Z-3)               'SET COLUMN TO 1ST PAIR CHAR
2140 IR= IA(I+1)-(Z-1)             'SET ROW TO 2ND PAIR CHAR
2150 IF IA(I) < Z OR IA(I) > Z+25 THEN 2180     '1ST WITHIN RANGE?
2160 IF IA(I+1) < Z OR IA(I+1) > Z+25 THEN 2180 '2ND WITHIN RANGE?
2170 GOTO 2190                     'ITS A HIT
2180 I=I+1:GOTO 2210               'NO HITS, STEP JUST ONE
2190 IP(IR,IC)=IP(IR,IC)+1         'INCREMENT PAIR COUNT IN ARRAY
2200 I=I+2                         'STEP TWO FOR NEXT PAIR TEST
2210 IF I>=N1 THEN 2230
2220 GOTO 1970                     'CONTINUE TO END OF RECORD
2230 RETURN
2240 '*******************************************************
2250 '*       SUBROUTINE 10 - PRINT PAIRED CHAR. ANALYSIS    *
2260 '*******************************************************
2270 GOSUB 3580                    'SET TOP OF PAGE
2280 PRINT #1,"SEGMENT VI";SPC(11)"PAIRED CHARACTER DIATOMIC ANALYSIS"
2290 PRINT #1,
2300 PRINT #1,"PAIR/COUNT  PAIR/COUNT  PAIR/COUNT  PAIR/COUNT  PAIR/COUNT  PAIR/
COUNT"
2310 PRINT #1,
2320 I=1
2330 N=0
2340 P3$="\\ #####    \\ #####    \\ #####    \\ #####    \\ #####    \\ #####"
2350 FOR J=1 TO 26
2360 FOR K=1 TO 28
2370 IF IP(J,K)< 2 THEN 2510       'SKIP O AND 1 COUNTS
2380 ' PRINT ONLY THE TOP 132 PAIRS COUNT
2390 IF N < 23 THEN 2410
2400 J=26:K=28:GOTO 2510           'IM DONE
2410 IX(I)= IP(J,K)
2420 IPS=IPS+IX(I)
2430 M=((K-1)*26)+ J
2440 B$(I)=C$(M)
2450 IF B$(I)= "Q " OR B$(I)="q " THEN B$(I)= "  " 'IMBEDDED BLANK PAIR
2460 I=I+1
2470 IF I <=6 THEN 2510            'SETUP SIX PAIRS TO THE LINE
2480 PRINT #1,USING P3$;B$(1);IX(1);B$(2);IX(2);B$(3);IX(3);B$(4);IX(4);B$(5);IX
(5);B$(6);IX(6)
```

```
2490 I=1
2500 N=N+1
2510 NEXT K
2520 NEXT J
2530 RETURN
2540 '*********************************************************
2550 '*          SUBROUTINE 11 - SORT PAIRED CHAR. ARRAY      *
2560 '*********************************************************
2570 ' SORT THE TWO DIMENSION PAIRS ARRAY (IP) & ASSOC. SYMBOLS (C)
2580 PRINT:PRINT "(NOW SORTING PAIR ARRAY)";
2590 FOR K=1 TO 28                  ' SET COLUMN LOOP
2600  FOR J=1 TO 25                 ' SET ROW LOOP
2610   FOR JJ=J+1 TO 26             ' SET WITHIN-COLUMN LOOP
2620    IF IP(J,K) >= IP(JJ,K) THEN 2710
2630    TMP=IP(J,K)                 'TEMP SAVE THE GREATER COUNT
2640    IP(J,K)=IP(JJ,K)            'SWAP HIGHER COUNT
2650    IP(JJ,K)=TMP                'WITH LOWER COUNT
2660 M1=((K-1)*26)+J                'CALC LOWER SYMBOL ADDR
2670 M2=((K-1)*26)+JJ               'CALC HIGHER SYMBOL ADDR
2680 C$(729)=C$(M1)                 'TEMP SAVE THE SYMBOL
2690 C$(M1)=C$(M2)                  'SWAP 1
2700 C$(M2)=C$(729)                 'SWAP 2
2710    NEXT JJ                     'COMPLETE 3RD LOOP
2720    NEXT J                      'COMPLETE 2ND LOOP
2730    PRINT ".";                  'LET EM KNOW WE'RE ALIVE
2740  NEXT K                        'COMPLETE 1ST LOOP
2750 RETURN
2760 '*********************************************************
2770 '*  SUBROUTINE 5 - RUN LENGTH ENCODING ANALYSIS          *
2780 '*********************************************************
2790 GOSUB 3580              'SET TOP OF PAGE
2800 PRINT #1,"SEGMENT IV - RUN LENGTH ENCODING ANALYSIS"
2810 PRINT #1,
2820 PRINT #1,"STRING ****** REPEATED CHARACTERS *****   CHAR. SAVED"
2830 PRINT #1,"LENGTH  NUMERICS SPACES  OTHER   TOTAL   NUMERICS";
2840 PRINT #1," SPACES OTHER TOTAL"
2850 PRINT #1,
2860 P4$=" ####     ####    ####   ####  #####     ####    #### #### ####"
2870 FOR I=3 TO N
2880 ITN1=ITN1+(IN(I)*I)           'TALLY NUMERICS
2890 ITX1=ITX1+(IX(I)*I)           'TALLY SPACES
2900 ITY1=ITY1+(IY(I)*I)           'TALLY OTHER
2910 ITN2=(I-3)*IN(I)
2920 ITX2=(I-3)*IX(I)
2930 ITY2=(I-3)*(IY(I)-IX(I)-IN(I))
2940 ITT2=(I-3)*(IY(I))
2950 ITN3=ITN3+ITN2
2960 ITX3=ITX3+ITX2
2970 ITY3=ITY3+ITY2
2980 ITT3=ITT3+ITT2
2990 IF (IY(I)=0) THEN 3020        'SKIP 0 COUNTS
3000 IYT1=(IY(I)*I)-(IN(I)*I)-(IX(I)*I)
3010 PRINT#1,USING P4$;I;(IN(I)*I);(IX(I)*I);IYT1;(IY(I)*I);ITN2;ITX2;ITY2;ITT2
3020 NEXT I
3030 ITS1=IR(128)                  'PICKUP TOTAL CHAR INPUT
3040 ITS2=(ITN1/IR(128)*100)       'COMPUTE NUMERIC %
3050 ITS3=(ITX1/IR(128)*100)       'COMPUTE SPACE %
3060 ITS4=((ITY1-ITX1-ITN1)/IR(128)*100)  'COMPUTE OTHER %
3070 ITS5=(ITY1/IR(128)*100)
3080 ITS6=(ITN3/IR(128)*100)
3090 ITS7=(ITX3/IR(128)*100)
3100 ITS8=(ITY3/IR(128)*100)
3110 ITS9=(ITT3/IR(128)*100)
3120 PRINT #1,
3130 PRINT #1,USING P4$;I;ITN1;ITX1;(ITY1-ITN1-ITX1);ITY1;ITN3;ITX3;ITY3;ITT3
3140 PRINT #1,
3150 PRINT #1,"SUMMARY INFORMATION: TOTAL CHAR READ";ITS1
3160 PRINT #1,
3170 PRINT #1,"PERCENTAGE SEQUENTIAL REPEATED CHAR:"
3180 PRINT #1,"   NUMERIC:";INT(ITS2)
3190 PRINT #1,"   BLANKS(SPACES):";INT(ITS3)
3200 PRINT #1,"   NONNUMERIC/NONBLANK:";INT(ITS4)
3210 PRINT #1,"   TOTAL REPEATED:";INT(ITS5)
3220 PRINT #1,
3230 PRINT #1," POTENTIAL COMPRESSION REDUCTION(X):"
3240 PRINT #1,"   NUMERIC:";INT(ITS6)
3250 PRINT #1,"   BLANK(SPACES):";INT(ITS7)
3260 PRINT #1,"   NONNUMERIC/NONBLANK:";INT(ITS8)
3270 PRINT #1," TOTAL:";INT(ITS9);"X"
3280 RETURN
```

```
3290 '*************************************************************
3300 '*        SUBROUTINE 8 - NUMERIC STRING ANALYSIS            *
3310 '*************************************************************
3320 GOSUB 3580            'SET TOP OF PAGE
3330 PRINT #1,"SEGMENT V - HALF BYTE ENCODING ANALYSIS":PRINT
3340 PRINT #1,"FOR SEQUENTIAL (NON-REPEATING) NUMERIC DATA"
3350 PRINT #1,
3360 PRINT #1,"STRING      NUMBER OF     TOTAL      COMPRESSED   CHAR"
3370 PRINT #1,"LENGTH      OCCURANCES    CHAR.      CHAR.        SAVED"
3380 PRINT #1,
3390 P5$=" ###        ####       ####        ####       ####      ###"
3400 FOR I=4 TO N
3410 IM(N+1)=IM(N+1)+(IM(I)*I)
3420 IF (IM(I)=0) THEN 3460       'SKIP 0 COUNTS
3425 IF I> 15 THEN IC= .5 ELSE IC= 0 'EXTRA HALF-BYTE OVER 15
3430 IC= IM(I)*CINT(IC+1.5+(I/2)+.4) 'COMPUTE COMPRESSION
3440 IC1=IC1+IC
3450 PRINT #1,USING P5$;I;IM(I);(IM(I)*I);IC;((IM(I)*I)-IC)
3460 NEXT I
3470 PRINT #1,
3480 PRINT #1,"*ANALYSIS SUMMARY*"
3490 PRINT #1,"TOTAL CHAR. IN FILE=";IR(128)
3500 PRINT #1,"TOTAL CHAR. SUSCEPTIBLE| TO COMPRESSION:";IM(N+1)
3510 PRINT #1,"TOTAL CHAR. SAVED BY COMPRESSION:";INT(IM(N+1)-IC1)
3520 P(128)=(IM(N+1)-IC1)/IR(128)*100   'COMPUTE PERCENTAGE
3530 PRINT #1,"PERCENT FILE COMPRESSION REDUCTION:";INT(P(128))
3540 RETURN
3550 '*************************************************************
3560 '* SETS TOP OF FORM OR SCREEN DELAY BASED ON OPTION         *
3570 '*************************************************************
3580 IF D$= "LPT1:" THEN 3600
3590 PRINT:PRINT "STRIKE ENTER WHEN READY";:INPUT G$
3600 PRINT #1, CHR$(12)           'SET TOP OF FORM
3610 RETURN
3620 '*************************************************************
3630 '*  SUBROUTINE 12 - COMPUTE & PRINT ENTROPY                 *
3640 '*************************************************************
3650 GOSUB 3580          'SET TOP OF PAGE
3660 FOR I= 1 TO 127
3670 IF P(I) = 0 THEN 3690
3680 E= E+P(I)* LOG(P(I))          'SEE DOCUMENTATION
3690 NEXT I
3700 E= -1*E
3710 B1= E*IR(128)
3720 B2= IR(128)*8
3730 B3= (B2-B1)/B2*100
3740 B4= B1/8
3760 PRINT #1, "SEGMENT VII - ENTROPY ANALYSIS"
3770 PRINT #1,
3780 PRINT #1, "ENTROPY:  ";E;" BITS/CHARACTER"
3790 PRINT #1, "FILE CONTAINS:"
3800 PRINT #1, INT(IR(128));" CHARACTERS"
3810 PRINT #1, INT(B2);" BITS"
3820 PRINT #1, "THEORETICAL STATISTICAL COMPRESSION: ";B3
3830 PRINT #1, "FILE WOULD REQUIRE:"
3840 PRINT #1, INT(B4);"  CHARACTERS"
3850 PRINT #1, INT(B1);" BITS"
3860 GOSUB 3580           'SET TOP OF PAGE
3870 RETURN
3880 '*************************************************************
3890 '*  SUBROUTINE 6 - SORT FREQ OCCURRENCE ARRAY DESCENDING    *
3900 '*************************************************************
3910 PRINT:PRINT "(NOW SORTING FREQUENCY TABLE)";
3920 FOR I= 1 TO 125
3930 FOR K= I+1 TO 126
3940 IF (IR(I)) >= IR(K) THEN 4010       'GET GREATER COUNT
3950 TEMP= IR(I)                         'AND SWAP...
3960 IR(I)= IR(K)
3970 IR(K)= TEMP
3980 B$(129)= B$(I)
3990 B$(I)= B$(K)
4000 B$(K)= B$(129)
4010 NEXT K
4020 PRINT ".";                      'LET EM KNOW WE'RE ALIVE
4030 NEXT I
4040 RETURN
4050 PRINT "**ERROR";ERR;" ENCOUNTERED ON LINE";ERL;"***":BEEP:STOP
```

```
4060 '****************************************************
4070 '*    SUBROUTINE 2 - PRINT EXECUTION REPORT SUMMARY *
4080 '****************************************************
4090 FOR I= 1 TO 10:PRINT #1,:NEXT I     'SKIP 10 LINES
4100 PRINT #1,:PRINT #1,SPC(30)"DATANALYSIS"
4110 PRINT #1,:PRINT #1,SPC(24)"EXECUTION  REPORT  SUMMARY"
4120 PRINT #1,:PRINT #1,SPC(24)"FILE ANALYSED: ";F$
4130 PRINT #1,:PRINT #1,SPC(15)"SEGMENT";SPC(10)"DESCRIPTION"
4140 PRINT #1,:PRINT #1,SPC(18)"I";SPC(13)"SYSTEM STANDARD FREQUENCY OF OCCURRENCE
CF"
4150 PRINT #1,:PRINT #1,SPC(18)"II";SPC(12)"SORTED FREQUENCY OF OCCURRENCE"
4160 PRINT #1,:PRINT #1,SPC(18)"III";SPC(11)"REPEATED CHARACTER STRING ANALYSIS"

4170 PRINT #1,:PRINT #1,SPC(18)"IV";SPC(12)"RUN LENGTH ENCODING SUSCEPTIBILITY"
4180 PRINT #1,:PRINT #1,SPC(18)"V";SPC(13)"HALF-BYTE ENCODING SSUSCEPTIBILITY"
4190 PRINT #1,:PRINT #1,SPC(18)"VI";SPC(12)"DIATOMIC COMPRESSION ANALYSIS"
4200 PRINT #1,:PRINT #1,SPC(18)"VII";SPC(11)"ENTROPY ANALYSIS"
4210 RETURN
4220 PRINT "**ERROR";ERR;" ENCOUNTERED ON LINE";ERL;"***":BEEP:STOP
```

APPENDIX B

SHRINK PROGRAM DESCRIPTIONS AND LISTINGS

For illustrative purposes, we have combined run-length, diatomic and half-byte encoding compression routines into two programs which we collectively call SHRINK. The first program, whose listing is contained in Figure B.1, is called MERGEC.BAS. As the reader may surmise from the name of this BASIC program, it represents the merging of the three previously mentioned compression techniques. The second BASIC program presented in this appendix was developed to decompress the data compressed by MERGEC.DAT. This program is called MERGED.BAS and its listing is contained in Figure B.2.

Like all BASIC programs presented in this book, the reader should note that the understanding of the coding techniques, and not program efficiency, was of primary concern. Thus, almost every line of each BASIC program contains a comment defining the function of the line. Since interpretive BASIC checks each comment during program execution, this excessive documentation considerably slows the execution of each program. Owing to this, it is highly recommended that the reader who wishes to use these programs excludes all comments from inclusion in the program. In addition, the compilation of each program by the use of a BASIC compiler will greatly increase the operating efficiency of each program.

B.1 MERGEC.BAS AND MERGED.BAS

The MERGEC.BAS program listed in Figure B.1 has some interesting changes from the previous programs that were used to indicate a specific compression technique. First, the Jewell table has been expanded to include lower case characters as indicated in lines 1160

```
10 '**************************************************************
20 '* MERGEC.BAS PROGRAM COMBINES ALL COMPRESSION TECHNIQUES    *
30 '* INTO ONE PROGRAM.  THE COMPANION PROGRAM, MERGED.BAS IS   *
40 '* USED TO DECOMPRESS THIS PROGRAMS OUTPUT (COMPRESS.DAT)    *
50 '* INTO ITS ORIGINAL FORM.  IT ALLOWS UP TO 254 CHARACTER    *
60 '* INPUT STRINGS AND MIXED UPPER/LOWER CASE.  IF NO LOWER    *
70 '* CASE PAIRS ARE ENCOUNTERED IN THE 1ST 100 CHARACTERS THE  *
80 '* PROGRAM ADJUSTS ITS LOOKUPS TO UPPER CASE ONLY IN ORDER   *
90 '* TO SHORTEN RUN TIME.  AUTHOR: THOMAS R. MARSHALL  1986     *
100 '**************************************************************
110 DIM O$(256), C(256)
120 WIDTH 80:CLS
130 '**********MAIN ROUTINE**********************
140 '* THIS ROUTINE READS RECORDS FROM AN ASCII *
150 '* FILE INTO A STRING CALLED X$ WHICH IS    *
160 '* THEN PASSED TO SUBROUTINES FOR COMPRESSION
170 '********************************************
180 PRINT "ENTER ASCII FILENAME. EG, [DRIVE:]FILESPEC";
190 INPUT F$: OPEN F$ FOR INPUT AS #2
200 OPEN "COMPRESS.DAT" FOR OUTPUT AS #3
210 PRINT "PATIENCE - INPUT PROCESSING"
220 GOSUB 1150            'SETUP PAIR TABLE
230 IF EOF(2) THEN GOTO 1940
240 LINE INPUT #2, X$
250 N= LEN(X$)
260 GOSUB 340            'RUN LENGTH ENCODE
270 GOSUB 680:N=I-1      'SWAP I/O BUFFERS
280 GOSUB 820            'DIATOMIC ENCODE
290 GOSUB 700:N=I-1      'SWAP I/O BUFFERS
300 GOSUB 1250           'HALF BYTE ENCODE
310 GOSUB 740            'TALLY ONLY
320 GOSUB 770            'PRINT BUFFER
330 GOTO 230
340 '*****RUN LENGTH ENCODING SUBROUTINE********
350 '* THIS ROUTINE PROCESSES RECORDS FROM X$  *
360 '* AND COMPRESSES OUT REPETITIVE CHARACTERS*
370 '* USING O$ AS THE OUTPUT BUFFER.          *
380 '*******************************************
390 K=1:J=1                     'RESET INDICES
400 FOR I= 1 TO N               'STEP THRU RECORD
410 A$= MID$(X$,I,1)            'EXTRACT A CHAR
420 IF A$= MID$(X$,I+1,1) THEN 490 'SAME AS NEXT?
430 IF K>3 THEN 530             'COMPRESS
440 IF K>1 THEN 610             'DON'T COMPRESS
450 O$(J)=A$                    'STUFF IN OUTPUT BUFFER
460 J=J+1                       'BUMP BUFFER INDEX
470 NEXT I                      'GO BACK FOR MORE
480 RETURN                      'END OF STRING
490 B$=A$                       'SAVE REPEATED CHAR
500 K=K+1                       'BUMP COUNT
510 GOTO 470                    'KEEP LOOKING
520 '**********************************************************
530 'INSERT COMPRESSION NOTATION IN OUTPUT BUFFER
540 '**********************************************************
550 O$(J)=CHR$(128)            'SET FLAG FOR RUN-LENGTH
560 O$(J+1)=B$                 'INSERT REPEATED CHAR
570 O$(J+2)=CHR$(K)            'INSERT COUNT
580 IF K=13 THEN O$(J+2)=CHR$(125) 'TRANSLATE CR
590 J=J+3:K=1                  'RESET INDEX
600 GOTO 470
610 O$(J)=B$                   'STUFF 1ST REPEAT CHAR
620 O$(J+1)=B$                 'STUFF 2ND REPEAT CHAR
630 J=J+2:K=1                  'RESET INDEX
640 GOTO 470                   'DONE
650 '*****TALLY THE COMPRESSION COUNT & WRITE BUFFER******
660 '* DISPLAY BEFORE & AFTER RESULTS OF COMPRESSION       *
670 '* AND SHOW THE NET RESULTS OBTAINED BY EACH METHOD   *
680 '****************************************************
690 N1=N1+N                    'TALLY INPUT CHAR COUNT
```

Figure B.1

```
700 IF N=0 THEN X$=SPACE$(1)      'ALLOW FOR NEW PARA ONLY
710 FOR I= 1 TO J-1
720 MID$(X$,I,1)= O$(I)           'SWAP I/O FOR NEXT ROUTINE
730 NEXT I
740 T=N-J+1                       'NET DIFFERENCE IN BUFFERS
750 T1=T1+T                       'SAVE COUNT FOR SUMMARY
760 RETURN
770 FOR I=1 TO J-1
780 PRINT #3, O$(I);              'WRITE OUT BUFFER
790 NEXT I
800 PRINT #3, ""                  'FORCE END OF WRITE
810 RETURN
820 '*****DIATOMIC COMPRESSION SUBROUTINE*******
830 '* THIS ROUTINE PROCESSES RECORDS FROM X$   *
840 '* AND COMPRESSES OUT COMMON PAIRS          *
850 '* USING O$ AS THE OUTPUT BUFFER.           *
860 '*******************************************
870 I=1                           'RESET INDICES
880 FOR J= 1 TO N-1               'STEP THRU RECORD
890 A$= MID$(X$,J,2)              'EXTRACT A PAIR
900 IF N1>100 AND LCC=0 THEN LC=25 'NO LOWER CASE
910 FOR K = 1 TO LC               'SETUP PAIR TABLE LOOP
920 IF A$=P$(K) THEN GOSUB 1030   'IS INPUT PAIR IN TABLE?
930 NEXT K                        'NO - TRY NEXT
940 IF M = 1 THEN 960             'IF MATCH FLAG SET?
950 O$(I) = MID$(A$,1,1)          'NO-STUFF 1ST CHAR IN BUFFER
960 I=I+1                         'BUMP INPUT STRING INDEX
970 M=0                           'RESET MATCH FLAG
980 NEXT J                        'GO BACK FOR MORE
990 IF J=N+1 THEN J=I:GOTO 1020   'ONE TOO MANY
1000 U$(I)= MID$(X$,J,1)          'GET LAST CHAR
1010 J=I+1                        'RESET INDEX
1020 RETURN                       'DONE
1030 M=1                          'SET PAIR MATCH FLAG
1040 '**************************************************
1050 'INSERT COMPRESSION NOTATION IN OUTPUT BUFFER
1060 V = K + 199                  'INDEX OUT TO SUBSTITUTE CHAR
1070 O$(I)=CHR$(V)                'INSERT PAIR SUBSTITUTION
1080 J=J+1                        'FORCE INPUT SHIFT 2 OVER PAIR
1090 IF K>25 THEN LCC=1           'FOUND LOWER CASE PAIR
1100 K = LC                       'FORCE END OF PAIR SEARCH
1110 RETURN                       'GO BACK FOR MORE
1120 '**************************************************
1130 '*   CONSTRUCT PAIR COMBINATION TABLE            *
1140 '**************************************************
1150 DIM P$(50)                   'JEWELL CHAR. COMBINATION PAIRS
1160 DATA "E "," T",TH," A","S ",RE,IN,HE,ER," I"," O","N ",ES,
1170 DATA " B",ON,"T ",TI,AN,"D ",AT,TE," C"," S",OR,"R "
1180 DATA "e "," t",th," a","s ",re,in,he,er," i"," o","n ",es,
1190 DATA " b",on,"t ",ti,an,"d ",at,te," c"," s",or,"r "
1200 FOR I = 1 TO 50              'SETUP PAIR TABLE
1210 READ Z$                      'GET COMMON PAIR
1220 P$(I) = Z$: NEXT I           'AND STUFF INTO PAIR TABLE
1230 LC=50:LCC=0                  'SETUP LOOP COUNT & LC FLAG
1240 RETURN                       'DONE - TABLE COMPLETE
1250 '****HALF-BYTE ENCODING SUBROUTINE*********
1260 '* THIS ROUTINE PROCESSES RECORDS FROM X$  *
1270 '* AND ENCODES  MIXED  STRINGS OF DATA INTO*
1280 '* HALF-BYTE OR 4 BIT REPRESENTATION USING *
1290 '* DOUBLE BUFFERING WITH O$ AS OUTPUT BUFF.*
1300 '*******************************************
1310 K=1:J=1                      'RESET INDICES
1320 FOR I=1 TO N STEP 2          'STEP THRU RECORD
1330 IF I=N-1 THEN  1530          'EVEN STRING SIZE
1340 C(I)=0:C(I+1)=0              'RESET ENCODE FLAGS
1350 A$= MID$(X$,I,1)             'GET 1ST BYTE
1360 B$= MID$(X$,I+1,1)           'GET 2ND BYTE
1370 IF A$= "$" THEN C(I)= 1      'SET 1ST ENCODE FLAG
1380 IF A$= "," THEN C(I)= 2
```

Figure B.1 *Continued*

```
1390 IF A$= "." THEN C(I)= 3
1400 IF A$= "*" THEN C(I)= 4
1410 IF A$< "0" OR A$> "9" THEN 1430    'SKIP OTHERS
1420 C(I)= 5
1430 IF B$= "$" THEN C(I+1)= 1    'SET 2ND ENCODE FLAG
1440 IF B$= "," THEN C(I+1)= 2
1450 IF B$= "." THEN C(I+1)= 3
1460 IF B$= "*" THEN C(I+1)= 4
1470 IF B$< "0" OR B$> "9" THEN 1490    'SKIP OTHERS
1480 C(I+1)= 5
1490 IF C(I)= 0 OR C(I+1)= 0 THEN 1530    'NOT CANDIDATE
1500 K=K+2                        'BOTH NUMERIC-BUMP COUNT
1510 NEXT I                       'GO BACK FOR MORE
1520 RETURN                       'END OF STRING
1530 IF K > 4 THEN GOSUB 1620     'ENOUGH TO ENCODE
1540 IF K > 1 THEN GOSUB 1880     'DON'T ENCODE
1550 O$(J) = MID$(X$,I,1)         'OUTPUT 1ST CHAR.
1560 O$(J+1) = MID$(X$,I+1,1)     'OUTPUT 2ND CHAR.
1570 J=J+2:K=1                    'BUMP OUTPUT-RESET COUNT
1580 IF I=N THEN J=J-1            'ONE TOO MANY
1590 IF J=N+2 THEN J=N+1          'EVEN STRING, SUB 1
1600 GOTO 1510                    'AND GO FOR MORE
1610 '***** SUBROUTINE TO PERFORM HALF-BYTE ENCODING *****
1620 O$(J)=CHR$(129)              'FLAG FOR BYTE PACKING
1630 MASK1= &HF0                  '11110000
1640 MASK2= &HF                   '00001111
1650 O$(J+1)=CHR$(K-1)            'INSERT LENGTH OF STRING
1660 J=J+2                        'BUMP OUTPUT INDEX
1670 FOR L=(I-K)+1 TO I-2 STEP 2  'SETUP ENCODE LOOP
1680 ON C(L) GOTO 1690,1700,1710,1720,1730 'USE FLAG TO ENCODE
1690 X=&HA0:GOTO 1750             '10100000
1700 X=&HB0:GOTO 1750             '10110000
1710 X=&HC0:GOTO 1750             '11000000
1720 X=&HD0:GOTO 1750             '11010000
1730 X=VAL(MID$(X$,L,1))          'GET NUM VALUE OF BYTE 1
1740 X=X*16                       'SHIFT 4 BITS LEFT
1750 X=X AND MASK1                'MASK LOWER HALF-BYTE
1760 ON C(L+1) GOTO 1770,1780,1790,1800,1810 'USE ENCODE FLAG
1770 Y=&HA:GOTO 1820              '00001010
1780 Y=&HB:GOTO 1820              '00001011
1790 Y=&HC:GOTO 1820              '00001100
1800 Y=&HD:GOTO 1820              '00001101
1810 Y=VAL(MID$(X$,L+1,1))        'GET NUM VALUE OF BYTE 2
1820 Y=Y AND MASK2                'MASK UPPER HALF-BYTE
1830 Z= X OR Y                    'OR THE TWO TOGETHER
1840 O$(J)= CHR$(Z)               'OUTPUT BYTE TO BUFFER
1850 J=J+1                        'BUMP OUTPUT INDEX
1860 NEXT L                       'GO BACK FOR MORE
1870 K=1:RETURN                   'RESET COUNT AND RETURN
1880 '***** SUBROUTINE FOR STRING NOT WORTH ENCODING *****
1890 FOR L=(I-K)+1 TO I-2         'PICKUP SHORT STRING
1900 O$(J)=MID$(X$,L,1)           'STUFF IN OUTPUT BUFFER
1910 J=J+1                        'BUMP OUTPUT INDEX
1920 NEXT L                       'GO BACK FOR MORE
1930 K=1:RETURN                   'RESET COUNT AND RETURN
1940 CLOSE: OPEN F$ FOR INPUT AS #2
1950 PRINT "FILE ";F$;" BEFORE COMPRESSION:"
1960 LINE INPUT #2,X$
1970 IF EOF(2) THEN 2000
1980 PRINT X$
1990 GOTO 1960
2000 PRINT X$:OPEN "COMPRESS.DAT" FOR INPUT AS #3
2010 PRINT "FILE ";F$;" AFTER COMPRESSION:"
2020 LINE INPUT #3,O$
2030 IF EOF(3) THEN 2060
2040 PRINT O$
2050 GOTO 2020
2060 PRINT O$:PRINT T1;" TOTAL CHARACTERS ELIMINATED FROM ";
2070 PRINT N1;"OR ";INT((T1/N1)*100);"%":CLOSE:END
```

Figure B.1 *Continued*

```
10  '*****************************************************
20  '* MERGED.BAS PROGRAM WAS WRITTEN TO DECOMPRESS      *
30  '* FILES ENCODED BY ITS COMPANION PROG MERGEC.BAS.   *
40  '* THE INPUT FILE IS NORMALLY COMPRESS.DAT BUT CAN   *
50  '* BE CHANGED TO ACCOMMODATE ANY ENCODED FILE.       *
60  '* THE DECOMPRESSED OUTPUT FILE IS DECOMP.DAT AND    *
70  '* SHOULD BE SAVED UNDER ANOTHER NAME AFTER EACH     *
80  '* RUN OF MERGED.BAS. AUTHOR: THOMAS R. MARSHALL     *
90  '*****************************************************
100 DIM O$(256)
110 WIDTH 80:CLS
120 '**********MAIN ROUTINE***********************
130 '* THIS ROUTINE READS RECORDS FROM AN ASCII *
140 '* FILE INTO A STRING CALLED X$ WHICH IS    *
150 '* THEN PASSED TO DECOMPRESSION SUBROUTINE  *
160 '********************************************
170 PRINT "ENTER ASCII FILENAME. EG, COMPRESS.DAT";
180 INPUT F$: OPEN F$ FOR INPUT AS #2
190 OPEN "DECOMP.DAT" FOR OUTPUT AS #3
200 GOSUB 1030            'CONSTRUCT PAIR TABLE
210 PRINT "PATIENCE - INPUT PROCESSING"
220 IF EOF(2) THEN GOTO 1530
230 LINE INPUT #2, X$
240 N= LEN(X$)
250 GOSUB 330                    'RUN LENGTH DECODE
260 GOSUB 620:N=I-1              'SWAP I/O BUFFERS
270 GOSUB 760                    'DIATOMIC DECODE
280 GOSUB 640:N=I-1              'SWAP I/O BUFFERS
290 GOSUB 1120                   'HALF BYTE DECODE
300 GOSUB 680                    'PICKUP LAST COUNT
310 GOSUB 710                    'PRINT BUFFERS
320 GOTO 220
330 '*****RUN LENGTH DECODING SUBROUTINE********
340 '* THIS ROUTINE PROCESSES RECORDS FROM X$   *
350 '* AND DECOMPRESSES RUN-ENCODED CHARACTERS  *
360 '* USING O$ AS THE OUTPUT BUFFER.           *
370 '********************************************
380 K=1:J=1                      'RESET INDICES
390 FOR I= 1 TO N                'STEP THRU RECORD
400 A$= MID$(X$,I,1)             'EXTRACT A CHAR
410 IF A$= CHR$(128) THEN 470    'COMPRESSION FLAG?
420 O$(J)=A$                     'STUFF IN OUTPUT BUFFER
430 J=J+1                        'BUMP BUFFER INDEX
440 NEXT I                       'GO BACK FOR MORE
450 RETURN                       'END OF STRING
460 '**********************************************************
470 'DECODE COMPRESSION NOTATION TO OUTPUT BUFFER
480 '**********************************************************
490 K$= MID$(X$,I+2,1)           'GET REPEAT COUNT
500 A$= MID$(X$,I+1,1)           'GET REPEAT CHAR
510 K= ASC(K$)                   'SET UP INDEX
520 IF K=125 THEN K=13           'TRANSLATE CR
530 FOR L= J TO J+K-1            'SET OUTPUT LOOP
540 O$(L)= A$                    'STUFF REPEAT CHAR
550 NEXT L                       'KEEP GOING
560 J= L                         'BUMP OUTPUT INDEX
570 I= I+3                       'BUMP INPUT INDEX
580 GOTO 400                     'DONE
590 '*****TALLY THE DECOMPRESSION COUNT & WRITE BUFFER****
600 '* DISPLAY BEFORE & AFTER RESULTS OF DECOMPRESSION   *
610 '* AND SHOW THE NET RESULTS OBTAINED BY EACH METHOD  *
620 '*****************************************************
630 N1=N1+N                      'TALLY INPUT CHAR COUNT
640 X$=SPACE$(J-1)               'EXPAND INPUT BUFFER
650 FOR I= 1 TO J-1
660 MID$(X$,I,1)= O$(I)          'SWAP I/O FOR NEXT ROUTINE
670 NEXT I
680 T=J-1-N                      'NET DIFFERENCE IN BUFFERS
690 T1=T1+T                      'SAVE COUNT FOR SUMMARY
700 RETURN
710 FOR I=1 TO J-1
720 PRINT #3, O$(I);             'WRITE OUT BUFFER
```

Figure B.2

```
730 NEXT I
740 PRINT #3, ""                    'END WRITE
750 RETURN
760 '*****DIATOMIC   DECODING SUBROUTINE********
770 '* THIS ROUTINE PROCESSES RECORDS FROM X$ *
780 '* AND DECOMPRESSES PAIR-ENCODED CHARACTERS*
790 '* USING O$ AS THE OUTPUT BUFFER.          *
800 '**********************************************
810 K=1:J=1:V=0                     'RESET INDICES
820 FOR I= 1 TO N                   'STEP THRU RECORD
830 A$= MID$(X$,I,1)                'EXTRACT A CHAR
840 IF A$> CHR$(199) THEN 900       'COMPRESSED PAIR?
850 O$(J)=A$                        'STUFF IN OUTPUT BUFFER
860 J=J+1                           'BUMP BUFFER INDEX
870 NEXT I                          'GO BACK FOR MORE
880 RETURN                          'END OF STRING
890 '**********************************************
900 'DECODE COMPRESSION NOTATION TO OUTPUT BUFFER
910 '**********************************************
920 K= ASC(A$)                      'GET ORDINAL EQUIV.
930 K= K-199                        'SUBTRACT FOR INDEX
940 T$= P$(K)                       'STUFF PAIR IN BUFFER
950 O$(J)=MID$(T$,1,1)
960 O$(J+1)=MID$(T$,2,1)
970 J= J+2                          'BUMP OUTPUT INDEX
980 V= V+1                          'SUM VARIABLE COUNT
990 GOTO 870                        'DONE
1000 '**********************************************
1010 "* CONSTRUCT PAIR COMBINATION TABLE        *
1020 '**********************************************
1030 DIM P$(50)                     'JEWELL CHAR. COMBINATION PAIRS
1040 DATA "E "," T",TH," A","S ",RE,IN,HE,ER," I"," O","N ",ES,
1050 DATA " B",ON,"T ",TI,AN,"D ",AT,TE," C","S ",QR,"R "
1060 DATA "e "," t",th," a","s ",re,in,he,er," i"," o","n ",es,
1070 DATA " b",on,"t ",ti,an,"d ",at,te," c"," s",or,"r "
1080 FOR I = 1 TO 50                'SET UP PAIR TABLE
1090 READ Z$                        'GET COMMON PAIR
1100 P$(I) = Z$: NEXT I             'AND STUFF INTO PAIR TABLE
1110 RETURN                         'DONE - TABLE COMPLETE
1120 '*****HALF BYTE  DECODING SUBROUTINE********
1130 '* THIS ROUTINE PROCESSES RECORDS FROM X$ *
1140 '* AND DECOMPRESSES BYTE-ENCODED CHARACTERS*
1150 '* USING O$ AS THE OUTPUT BUFFER.          *
1160 '**********************************************
1170 J=1                            'RESET INDEX
1180 FOR I= 1 TO N                  'STEP THRU RECORD
1190 A$= MID$(X$,I,1)               'EXTRACT A CHAR
1200 IF A$= CHR$(129) THEN 1280     'COMPRESSION FLAG?
1210 O$(J)=A$                       'STUFF IN OUTPUT BUFFER
1220 J=J+1                          'BUMP BUFFER INDEX
1230 NEXT I                         'GO BACK FOR MORE
1240 RETURN                         'END OF STRING
1250 '**********************************************
1260 'DECODE COMPRESSION NOTATION TO OUTPUT BUFFER   *
1270 '**********************************************
1280 MASK1= &HF0                    '11110000
1290 MASK2= &HF                     '00001111
1300 K= ASC(MID$(X$,I+1,1))         'GET STRING LENGTH
1310 M= I+2+(K/2)-1                 'SET END OF STRING
1320 FOR L=I+2 TO M                 'SETUP LOOP TO DECODE
1330 Z= ASC(MID$(X$,L,1))           'GET BYTE
1340 X= (Z AND MASK1)/16            'MASK LOWER HALF-BYTE
1350 IF X< 10 THEN 1410             'ITS NUMERIC
1360 IF X= 10 THEN O$(J)= "$"       'SPECIAL
1370 IF X= 11 THEN O$(J)= ","       'SPECIAL
1380 IF X= 12 THEN O$(J)= "."       'SPECIAL
1390 IF X= 13 THEN O$(J)= "*"       'SPECIAL
1400 GOTO 1420                      'SKIP IF SPECIAL
1410 O$(J)= CHR$(X+48)              'OUTPUT 1ST NUMERIC
1420 Y= Z AND MASK2                 'MASK UPPER HALF-BYTE
1430 IF Y< 10 THEN 1490             'ITS NUMERIC
1440 IF Y= 10 THEN O$(J+1)= "$"     'SPECIAL
```

Figure B.2 *Continued*

```
1450 IF Y= 11 THEN O$(J+1)= ","     'SPECIAL
1460 IF Y= 12 THEN O$(J+1)= "."     'SPECIAL
1470 IF Y= 13 THEN O$(J+1)= "*"     'SPECIAL
1480 GOTO 1500                      'SKIP IF SPECIAL
1490 O$(J+1)= CHR$(Y+48)            'OUTPUT 2ND NUMERIC
1500 J= J+2                         'BUMP OUTPUT BY TWO
1510 NEXT L:I= M                    'CONTINUE, BUMP INPUT INDEX
1520 GOTO 1230                      'GO BACK FOR MORE
1530 CLOSE: OPEN F$ FOR INPUT AS #2
1540 PRINT K$
1550 PRINT "FILE ";F$;" BEFORE DECOMPRESSION:"
1560 LINE INPUT #2,X$
1570 IF EOF(2) THEN 1600
1580 PRINT X$
1590 GOTO 1560
1600 PRINT X$:OPEN "DECOMP.DAT" FOR INPUT AS #3
1610 PRINT "FILE ";F$;" AFTER DECOMPRESSION:"
1620 LINE INPUT #3,O$
1630 IF EOF(3) THEN 1660
1640 PRINT O$
1650 GOTO 1620
1660 PRINT O$:PRINT T1;" TOTAL CHARACTERS INSERTED"
1670 CLOSE:END
```

Figure B.2 *Continued*

to 1190. To reduce the processing time of the program, the first 100 characters in a file to be compressed are examined in the diatomic compression subroutine. If a lower case pair is encountered, Jewell character combination pairs to include lower case pairs are matched against the data file. Otherwise, it is assumed that the file consists of upper case text and data is only matched against upper case pairs during processing.

Because of the expanded Jewell character-pair table, the base of the characters used to replace pairs of characters has been changed from ASCII 224 to ASCII 199. Thus, ASCII codes from 199 to 249 are now used to represent paired-character substitution. Another change from previous examples was the change in the run-length flag, from ASCII 125 to ASCII 128. This was done to ensure all compression-indicating characters were above ASCII 127.

It should be noted that files with graphic characters above ASCII 127 can cause unexpected results to occur. This is due to the possibility of those characters being misinterpreted as a flag character, which will provide a false indication of a particular type of compression when the decompression program operates upon previously compressed data. This problem can be alleviated by modifying the MERGEC.BAS program to insert a second compression-indicating character after one occurs naturally in the input data. Then, the MERGED.BAS program can be modified to check the character following every compression-indicating character. If the second character is the same as the first, the program would then strip one character and ignore the second as it represents naturally occurring data and does not signify that compression occurred. For readers desiring to expand the scope of these programs, the previously described modifications may represent an interesting challenge as well as result in a program that can compress tokenized BASIC programs.

B.2 PROGRAM OPERATIONS

Unless altered by the reader, MERGEC.BAS will create a file named COMPRESS.DAT which represents a compressed form of the input file specified when the program is executed.

Figure B.3 represents a typical electronic mail message one might transmit. This message contains 776 characters. Since many electronic mail services charge for both connect time and the number of characters transmitted, any reduction in data may result in a commensurate reduction in transmission cost.

Figure B.4 illustrates the resulting compressed file after MERGEC.BAS operated upon the sample data file listed in Figure B.3. The reader should note that some of the compression flag characters are equivalent to very interesting printer control characters, which is the reason for the listing of the compressed file only vaguely resembling the original data file. As indicated, 245 characters representing 31% of the original data were eliminated, due to the mixture of the three compression routines used in the program.

Several important items concerning the compressed file warrant further discussion, especially if one desires to transmit such data over an electronic mail service. First, when you compress the data, be sure to remove any destination information from the file. This is because the electronic mail system cannot understand your routing requests if they are compressed. Secondly, prior to transmitting compressed data, the reader should ascertain if the electronic mail service supports the extended ASCII character set, which is sometimes called 8-bit ASCII. If the electronic mail service does not support extended

```
        J.J. ASTOR
        ASTORIA, OREGON

        MY GOOD SIR;

        I AM RESPONDING TO YOUR CORRESPONDENCE LAST REGARDING THE DISPOSITION
        OF PROCEEDS FROM SALE OF SAID PROPERTY.   THE GROSS PROCEEDS REALIZED
        WERE $832,746,381.99.

        FROM THIS, WE HAVE TAKEN THE LIBERTY OF DEDUCTING MINOR EXPENSES IN THE
        FOLLOWING MANNER:

        A. TRANSFER FEES        $136,941
        B. REALITY FEES         $8,327,436
        C. LEGAL FEES           $9,938,862
        D. ADVERTISING          $422,977
        E. TAXES                $1,646,311
        F. SUNDRY               $7,139,774
        G. MISCELLANEOUS        $462,114,283

        THE NET PROCEEDS RESIDE IN A NON-INTEREST BEARING ACCOUNT UNTIL
        INSTRUCTIONS TO THE CONTRARY ARE RECEIVED.   TO OUR VALUED CLIENT WE
        REMAIN-

                                VERY TRULY YOURS,

                                DEWEY, CHEATHAM & HOWE
```

Figure B.3

```
FILE astor AFTER COMPRESSION:
J.J. STα
ASTαYA, π GH

MY GOOD SIR;

I M SP DING O YOUR x SP DENC LAS GARD G DISPOSI
OF PROC ED ROM AL OF AI PROP TY.   GRO PROCEED ALIZED
WI LS + + K  .

FROM HIS, WI HAV TAKE U IB TY F DEDUC NG M α EXPENS U E
FO LOW G M NL:

A. R SF L FE $k1
B. AI ITY FE
C. LEGAL FE      ¿|'|6
                    $
                    2
D. DV LS G
        $B+7
E. AX  $
F. UNDRY )$4
G. MISCELL  EOUS




Rb
 LL NE PROC ED SID L  N I S BEAR G CCOUN UN L
 STRU J TO TRARY CEIVED. O R VALUE CLIEN WE
 MA

 'V LY RULY YOURS,

 'DEWEY, L HAM & HOWE
 245  TOTAL CHARACTERS ELIMINATED FROM  776 OR  31 %
Ok
```

Figure B.4

ASCII you will not be able to transmit compressed data since compression-indicating characters are all above ASCII 127. Lastly, let us pose a question to the reader. From examining Figure B.4, could you determine what the original message stated?

While compression is no substitute for the utilization of encryption devices, it may make it much more difficult for an unauthorized reader to decipher a message.

Prior to leaving the reader to SHRINK IT themselves, let us review the complete compression and decompression of a small file. This is illustrated in Figure B.5. At the top of Figure B.5 our four-line file to which we assigned the name tst is listed. Next, upon loading and executing the program MERGEC.BAS we are prompted to enter the ASCII file we wish to compress. As indicated, the program lists the original file as well as its compressed form.

The second half of Figure B.5 illustrates the execution of MERGED.BAS, which will convert a compressed file back into its original form. Since MERGEC.BAS automatically used the filename com-

```
Lets take a look at the benefits of compression when we
have different types of data to include large numbers
such as $123,456,789.98 and repeating strings in a file.
such as --------------.

ENTER ASCII FILENAME. EG, [DRIVE:]FILESPEC? tst
PATIENCE - INPUT PROCESSING
FILE tst BEFORE COMPRESSION:
Lets take a look at the benefits of compression when we
have different types of data to include large numbers
such as $123,456,789.98 and repeating strings in a file,
such as --------------.
FILE tst AFTER COMPRESSION:
Letptakra looko⫞Σrbenefitμof°omprssi± wθθwe
havrdiffΩen⫞typ€⬦f d‡aⲙoδncludⲅlargⲅnumbℛs
suchⲣí #┤∨┐             ┌8σnⳑ⊤pe‡☒g·t⊤☒gμ☒σ  file,

suchⲣᵖ-⎕

   53  TOTAL CHARACTERS ELIMINATED FROM  187 OR  28 %
Ok
```

Running MERGEC.BAS

```
ENTER ASCII FILENAME. EG, COMPRESS.DAT? compress.dat
PATIENCE - INPUT PROCESSING

FILE compress.dat BEFORE DECOMPRESSION:
Letptakra looko⫞Σrbenefitμof°omprssi± wθθwe
havrdiffΩen⫞typ€⬦f d‡aⲙoδncludⲅlargⲅnumbℛs
suchⲣí #┤∨┐             ┌8σnⳑ⊤pe‡☒g·t⊤☒gμ☒σ  file,

suchⲣᵖ-⎕

FILE compress.dat AFTER DECOMPRESSION:
Lets take a look at the benefits of compression when we
have different types of data to include large numbers
such as $123,456,78920T8 and repeating strings in a file,
such as --------------.
   54  TOTAL CHARACTERS INSERTED
Ok
```

Running MERGED.BAS

Figure B.5

press.dat to store compressed data, we used that file as input to MER-GED.BAS. As indicated, the file is faithfully reconstructed into its original form; however, from careful examination, 53 characters were eliminated and 54 were inserted according to the printout. In case the reader is puzzled, it should be noted that in the case of half-byte encoding, the saving of an odd number of bytes was rounded down while the insertion of actual characters is counted, resulting upon occasion in a one-character discrepancy in comparing character removals and character insertions. This, however, has no effect upon the decoding of previously compressed data.

REFERENCES

Aronson, J. (1977). Data compression—a comparison of methods. *National Bureau of Standards*, PB-269 296, June.

Bell, T. C., Cleary, J. G., and Witten, I. H. (1990). *Text Compression*, Prentice Hall, Englewood Cliffs, NJ.

Dishon, Y. (1977). Data compaction in computer systems. *Computer Design*, April, 85–90.

Gage, P. (1994). A new algorithm for data compression. *The C Users Journal*, **12**, No. 2, February.

Jakobsson, M. (1985). Compression of character strings by an adaptive dictionary. *BIT*, **25**, No. 4, 593–603.

Jewell, G. C. (1976). Text compaction for information retrieval systems. *IEEE SMC Newsletter*, **5**, No. 1.

Lempel, A. and Ziv, J. (1977). A universal algorithm for sequential data compression. *IEEE Transactions on Information Theory*, **IT-23**, No. 5.

Langdon, G. G. (1984). An introduction to arithmetic coding. *IBM Journal of Research and Development*, March.

Mandelbrot, B. (1982). *The Fractal Geometry of Nature*, W. H. Freeman, San Francisco.

McCullough, T. (1977). Data compression in high-speed digital facsimile. *Telecommunications*, July, 40–43.

Mitchell, J. L., and Pennebaken, W. B. (1988). Software implementations of the Q-coder. *IBM Journal of Research and Development*, **32**, November, 753–774.

Moilanen, U. (1978). Information preserving codes compress binary pictorial data. *Computer Design*, November, 134–136.

Peterson, J. L., Bitner, J. R., and Howard, J. H. (1978). The selection of optimal tab settings. *Communications of the ACM*, **21**, No. 12, 1004–1007.

Preuss, D. (1976). *Comparison of two-dimensional facsimile coding schemes*, IEEE Press,

Rissanen, J., and Langdon, G. G. (1979). Arithmetic coding. *IBM Journal of Research and Development*, March.

Rubin, F. (1976). Experiments in text file compression. *Communications of the ACM*, **19**, No. 11, 617–623.

Ruth, S., and Kreutzer, P. (1972). Data compression for large business files. *Datamation*, **18**, No. 9, 62–66.

Snyderman, M., and Hunt, B. (1970). The myriad virtues of text compaction. *Datamation*, **1**, December, 36–40.

Storer, J. A., and Szymanski, T. G. (1982). Data compression via textual substitution. *Journal of the Association for Computing Machinery*, **29**, No. 4, 928–951.

Tischer, P. (1987). A modified LZW data compression scheme. *Australian Computer Science Communications*, **9**, No. 1, 262–272.

Welch, T. (1984). A technique for high-performance data compression. *IEEE Computer*, June.

Witten, I. H., Neal, R., and Cleary, J. G. (1987). Arithmetic coding for data compression. *Communications of the ACM*, **30**, June, 520–540.

Ziv, J., and Lempel, A. (1977). A universal algorithm for sequential data compression. *IEEE Transactions on Information Theory*, **IT-23**, No. 3.

Ziv, J., and Lempel, A. (1978). Compression of individual sequences via variable-rate coding. *IEEE Transactions on Information Theory*, **IT-24**, No. 5, 530–536.

FURTHER READING

Andrews, C. A., Davies, J. M., and Schwarz, E. (1967). Adaptive data compression. *Proceedings of the IEEE*, **55**, No. 3.

Ash, R. (1965). *Information Theory*, Interscience, New York.

Barton, I. J., Creasey, S. E., Lynch, M. F., and Snell, M. J. (1974). An information-theoretic approach to text searching in direct access systems. *Communications of the ACM*, **17**, No. 6, 345–350.

Bemer, R. W. (1960). Do it by the numbers—digital shorthand. *Communications of the ACM*, **3**, N810, 530–536.

Bentley, J. L., Skator, D. D., Tarjan, R. E., and Wei, V. K. (1986). A locally adaptive data compression scheme. *Communications of the ACM*, **29**, No. 4, 320–330.

Blasbalg, H., and Van Blerkom, R. (1972). Message compression. *IRE Transactions on Space Electronics and Telemetry*, September, 228–238.

Bookstein, A., and Fouty, G. (1976). A mathematical model for estimating the effectiveness of bigram coding. *Information Proceedings and Management*, **12**, 111–116.

Bray, J. M., Nelson, V. P., deMaine, P. A. D., and Irwin, J. D. (1985). Data-compression techniques ease storage problems. *Computer Design*, October, 102–105.

Brent, R. P. (1987). A linear algorithm for data compression. *The Australian Computer Journal*, **19**, No. 2, 64–68.

Clare, A. G., Cook, G. M., and Lynch, M. F. (1972). The identification of variable-length, equifrequency character strings in a natural language data base. *Computer Journal*, **15**.

Corbin, H. (1981). An introduction to data compression. *BYTE*, April, 218–250.

Cortesi, D. (1982). An effective text-compression algorithm. *BYTE*, January, 397–403.

Costlow, T. (1989). Compression doubles QIC capacity. *Electronic Engineering Times*, January, 4.

Costlow, T. (1989). What's new in data compression. *Electronic Engineering Times*, January, 53–57.

Cullum, R. D. (1972). A method for the removal of redundancy in printed text. *NTIS*, AD751 407, September.

Davisson, L. D. (1966). Theory of adaptive data compression. In A. V. Balakrishinan (Ed.), *Advances in Communications Systems*, Academic Press, New York, 173–192.

Davisson, L. D. (1967). An approximate theory of prediction for data compression. *IEEE Transactions on Information Theory*, **IT-13**, No. 2, 274–278.

Davisson, L. D. (1968). Data compression using straight line interpolation. *IEEE Transactions on Information Theory*, **IT-14**, No. 3, 300–304.

Davisson, L. D. (1968). The theoretical analysis of data compression systems. *Proceedings of the IEEE*, **56**, No. 2, 176–186.

Davisson, L. D., and Gray, R. M. (1976). *Data Compression*, Dowden, Hutchinson and Ross, Dowden.

De Main, P. A. D., Kloss, K., and Marron, B. A. (1967). The SOLID System III. alphanumeric compression. Washington: US Government Printing Office, NBS Technical Note 413, August.

Doherty, R. (1989). System puts real-time squeeze on color videos. *Electronic Engineering Times*, February.

Ehrman, L. (1967). Analysis of some redundancy removal bandwidth compression techniques. *Proceedings of the IEEE*, **55**, No. 3, 278–287.

Ellias, P. (1955). Predictive coding. *IRE Transactions*, **IT-1**, 16–44.

Fano, R. M. (1949). The transmission of information. Research Laboratory for Electronics, Massachusetts Institute of Technology, Technical Report, No. 65.

Fano, R. M. (1961). *The Transmission of Information*, John Wiley & Sons,

Fiala, E. R., and Greene, D. H. (1989). Data compression with finite windows. *Communications of the ACM*, **32**, No. 4, 490–505.

Forney, G. D., and Tao, W. Y. (1976). Data compression increases throughout. *Data Communications*, May/June.

Gilbert, E. N., and Moore, E. F. (1959). Variable-length binary encodings. *Bell System Technical Journal*, April, 933–967.

Gottlieb, B., Hagereth, S. E., Denot, P. G. H., and Rabinowitz, H. S. (1975). A classification of compression methods and their usefulness for a large data processing center. *Proceedings of the National Computer Conference*, 453–458.

Hann, B. (1974). A new technique for compression and storage of data. *Communications of the ACM*, **17**, No. 8.

Harker, J. (1982). Byte oriented data compression techniques. *Computer Design*, October, 95–100.

Heaps, H. S. (1972). Storage analysis of a compression coding for document data bases. *Information*, **10**, No. 1.

Held, G. (1979). Eliminating those blanks and zeros in data transmission. *Data Communications*, **8**, No. 9, 75–77.

Honien, Liu. (1968). A file management system for a large corporate information system data bank. *Fall Joint Computer Conference*, Vol. 33, Part I, 145–156.

Hu, T. C., and Tucker, A. C. (1971). Optimal computer search trees and variable-length alphabetical codes. *SIAM Journal of Applied Mathematics*, **21**, No. 4, 514–532.

Huffman, D. A. (1952). A method for the construction of minimum redundancy codes. *Proceedings of the IRE*, **40**, 1098–1101.

Karp, R. M. (1961). Minimum-redundancy coding for the discrete noiseless channel. *IEEE Transactions on Information Theory*, **IT-7**, 27–38.

Knuth, D. E. (1985). Dynamic Huffman Coding. *Journal of Algorithms*, **6**, 163–180.

Kortman, C. M. (1965). Data compression and adaptive telemetry. *IEEE Western Electronic Show and Convention (WESCON)*, Vol. 9, Paper 14.4.

Kurmiss, J. M. (1974). An experiment in adaptive encoding. IBM Technical Report Troo. 2524, Poughkeepsie, NY.

Larmore, L. L., and Hirschberg, D. S. (1990). A fast algorithm for optimum length-limited Huffman codes. *Journal of the ACM*, **37**, No. 3, 464–473.

Lesk, M. E. (1970). Compressed text storage. *Computing Science Technical Report*, No. 3, Bell Telephone Laboratories.

Ling, H., and Palermo, F. P. (1975). Block-oriented information compression. *IBM Journal of Research and Development*, March.

Lynch, F. L., Petrie, H. J., and Snell, M. J. (1973). Analysis of the microstructure of titles in the INSPEC data-base. *Information Storage and Retrieval*, **9**, 331–337.

Lynch, M. F. (1973). Compression of bibliographic files using an adaption of run-length coding. *Information Storage and Retrieval*, **9**, 207–214.

Marron, B. A., and De Maine, P. A. D. (1967). Automatic data compression. *Communications of the ACM*, **10**, No. 3, 711–715.

Mayne, A., and James, E. B. (1975). Information compression by factorising common strings. *Computer Journal*, **18**, No. 2, 157–160.

McCarthy, J. P. (1973). Automatic file compression. *Proceedings of the International Computer Symposium*, North-Holland, Amsterdam, 511–516.

Moilanen, U. (1978). Information preserving codes compress binary pictorial data. *Computer Design*, November, 134–136.

Mommens, J. H., and Ravir, J. (1974). Coding for data compaction. *IBM Research Report*, RC 5150, T. J. Watson Research Center, Yorktown Heights, NY.

Mulford, J. B., and Ridall, R. K. (1971). Data compression techniques for economic processing of large commercial files. *ACM Symposium on Information Storage and Retrieval*, 207–215.

Nelson, M. (1991). *The Data Compression Book*, M&T Books,

Nordling, K. (1982). A data compression modem. *Telecommunications*, September, 67–70.

Oliver, B. M. (1952). Efficient coding. *Bell System Technical Journal*, **21**, No. 4, 724–750.

Ott, G. (1967). Compact encoding of stationary Markov sources. *IEEE Transactions on Information Theory*, **IT-13**, 82–86.

Peterson, J. L. (1979). Text compression. *BYTE*, December, 106–118.

Pountain, D. (1987). Run-length encoding. *BYTE*, June, 317–320.

Powell, D. (1989). The hidden benefits of data compression. *Networking Management*, October, 46–54.

Ruth, S. R., and Villers, J. M. (1972). *Data Compression and Data Compaction*, Government Clearing House, Study Number AD 723525.

Schieber, W. D., and Thomas, G. (1971). An algorithm for the compaction of alphanumeric data. *Journal of Library Automation*, **4**, 198–206.

Schuegraf, E. F., and Heaps, H. S. (1973). Selection of equifrequent word fragments for information retrieval. *Information Storage and Retrieval*, **9**, 697–711.

Schuegraf, E. F., and Heaps, H. S. (1974). A comparison of algorithms for data base compression by the use of fragments as language elements. *Information Storage and Retrieval*, **10**, 309–319.

Schwartz, E. S., and Kalleck, B. (1964). Generating a canonical prefix encoding. *Communications of the ACM*, **7**, 166–169.

Seither, M. (1989). Data compression doubles capacity of ¼-inch tape drives. *System Integration*, May.

Shannon, C. E. (1948). A mathematical theory of communications. *Bell System Technical Journal*, **27**, 379–423 and 623–656.

Storer, J. A. (1988). *Data Compression: Methods and Theory*, Computer Science Press,

Storer, J. A. (1992). *Image and Text Compression*, Kluwer Academic Publishers,

Thiel, L. H., and Heaps, H. S. (1972). Program design for retrospective searches on large data bases. *Information Storage and Retrieval*, **8**, 1–20.

Tropper, R. (1982). Binary-coded text: a text-compression method. *BYTE*, April, 398–412.

Vitter, J. S. (1987). Design and analysis of dynamic Huffman codes. *Journal of the ACM*, **34**, No. 4, 825–845.

Wagner, Robert A. (1973). Common phrases and minimum-space text storage. *Communications of the ACM*, **16**, No. 3, 148–152.

Wells, M. (1972). File compression using variable length encodings. *Computer Journal*, April, 308–313.

Williams, R. C. (1990). *Adaptive Data Compression*, Kluwer Books,

INDEX

ABOUT THE AUTHOR

Having gained a B.S.E.E. from Pennsylvania Military College, Gilbert Held majored in computer science for his M.S.E.E. from New York University. He also holds an M.B.A. and the M.S.T.M. degree from the American University.

Gibert Held is Chief of Data Communications for the United States Office of Personnel Management. He serves as a consultant to a number of companies and teaches several college courses in Computers and Decision Theory, Management Information Systems and Data Communications. He is the author of a large number of books and articles.

The only person to win the Interface Karp award for excellence in technical writing twice, Gilbert Held is also the winner of the Association of American Publishers Professional and Scholarly Publishing Division award. He has been selected as one of the Federal Computer Week top 100 professionals in Government, Industry and Academia who have made an outstanding contribution in the field of computer science. He has also received several Government awards for exceptional performance.

LOCAL AREA NETWORKING

PROTECTING LAN RESOURCES
A Comprehensive Guide to Securing, Protecting and Rebuilding a Network

With the evolution of distributed computing, security is now a key issue for network users. This comprehensive guide will provide network managers and users with a detailed knowledge of the techniques and tools they can use to secure their data against unauthorised users. Gil Held also provides guidance on how to prevent disasters such as self-corruption of data and computer viruses.
1995 0 471 95407 1

LAN PERFORMANCE
Issues and Answers

The performance of LANs depends upon a large number of variables, including the access method, the media and cable length, the bridging and the gateway methods. This text covers all these variables to enable the reader to select and design equipment for reliability and high performance.
1994 0 471 94223 5

TOKEN- RING NETWORKS
Characteristics, Operation, Construction and Management

This timely book provides the reader with a comprehensive understanding of how Token-Ring networks operate, the constraints and performance issues that affect their implementation, and how their growth and use can be managed both locally and as part of an Enterprise network.
1993 0 471 94041 0

ETHERNET NETWORKS
Design, Implementation, Operation, and Management
1994 0 471 59717 1

REFERENCE

DICTIONARY OF COMMUNICATIONS TECHNOLOGY
Terms, Definitions and Abbreviations
1995 0 471 95126 9 (Paper)
       0 471 95542 6 (Cloth)

THE COMPLETE MODEM REFERENCE
2nd Edition
1994 0 471 00852 4

THE COMPLETE PC AT AND COMPATIBLES REFERENCE MANUAL
1991 0 471 53315 7